An Empire of Magnetism

An Empire of Magnetism

Global Science and the British Magnetic Enterprise in the Age of Imperialism

EDWARD J. GILLIN

OXFORD
UNIVERSITY PRESS

Great Clarendon Street, Oxford, OX2 6DP,
United Kingdom

Oxford University Press is a department of the University of Oxford.
It furthers the University's objective of excellence in research, scholarship,
and education by publishing worldwide. Oxford is a registered trade mark of
Oxford University Press in the UK and in certain other countries

© Edward J. Gillin 2024

The moral rights of the author have been asserted

All rights reserved. No part of this publication may be reproduced, stored in
a retrieval system, or transmitted, in any form or by any means, without the
prior permission in writing of Oxford University Press, or as expressly permitted
by law, by licence or under terms agreed with the appropriate reprographics
rights organization. Enquiries concerning reproduction outside the scope of the
above should be sent to the Rights Department, Oxford University Press, at the
address above

You must not circulate this work in any other form
and you must impose this same condition on any acquirer

Published in the United States of America by Oxford University Press
198 Madison Avenue, New York, NY 10016, United States of America

British Library Cataloguing in Publication Data
Data available

Library of Congress Control Number: 2023938961

ISBN 978–0–19–889095–9

DOI: 10.1093/oso/9780198890959.001.0001

Printed and bound by
CPI Group (UK) Ltd, Croydon, CR0 4YY

Links to third party websites are provided by Oxford in good faith and
for information only. Oxford disclaims any responsibility for the materials
contained in any third party website referenced in this work.

For my mum, Louise Spencer

Acknowledgements

In 2020, I was fortunate enough to take a Fox-type dipping needle, built in the 1840s at Falmouth by William George, to sea for three months. Tracing the route of James Clark Ross's 1839–43 expedition as far as Cape Town before circumnavigating Africa with detours to the Seychelles, Sri Lanka, and India, we followed the celebrated voyages of HMS *Erebus*, HMS *Terror*, HMS *Beagle*, HMS *Rattlesnake*, HMS *Sulphur*, and HMS *Samarang*. With the able assistance of my hardy subaltern, Crosbie Smith, I re-enacted magnetic experiments with the Fox type, measuring the Earth's magnetic qualities over vast expanses of time and space. We came to appreciate the instrument's qualities and failings; its value for scrutinizing terrestrial magnetism and its fragility. There were natural challenges, including the accumulation of salt, an ants' nest on the docks of East London, and the needle's loss of magnetic strength during rough weather off Land's End. And there were human challenges too, from my own lack of experimental skill to a near arrest by Lisbon's harbour authorities. These trials echoed those that nineteenth-century users recorded as they circumnavigated the world. By the time we reached St Helena, we knew our Fox type's mysterious character well, going as far as to name her 'Caroline', after Fox's daughter, the celebrated Cornish diarist. Wherever she went, from Cape Verde to Cape Town, to Colombo and Mumbai, the Fox dipping needle always attracted a crowd.

For the trust of this instrument, I must thank all those at the Royal Cornwall Polytechnic Society. Mike Jenks, Sue Radmore, Louis Turner, Henrietta Boex, and Lizzie Cook all provided invaluable support, as did Lucie Nottingham, who financed the expedition through the Tanner Trust. No one has brought this project to life in quite the way Michael Carver has, and I cannot thank him enough for all the inspiration he has provided: I hope this book will meet his high standards. A particular debt is owed to Hilary Watson for saving, and then lending me, a cache of Fox's letters. Meeting members of Fox's family has been an especially wonderful part of this project. I have enjoyed many brilliant discussions with Charles Fox at Glendurgan and Rachel Morin in Penjerrick garden: I thank them both for being so helpful and constructive. Thanks go also to Carolyn Kennett, Sam Robson, Coreen Bellis, Micah Mackay, Caroline Carver, Simon Schaffer, Jim Secord, William Whyte, Oliver Carpenter, David Trippett, Melissa Van Drie, Alexander Ibell, Justin Brogan, Chris Reece, Jack Matthews, Anna Marie Roos, Laura Treloar, Daniel Belteki, Ginny Button, Tim Holt, Fanny Gribenski, Silke Muylaert, Caroline Fox, Amie Varney, Chlöe Gamlin, Tilda, Bertram, Tim Marshall, and David Lewis. They have together offered rich intellectual discussion and invaluable friendship. I am grateful to Jenny Bulstrode for kindly sharing her superb 2013 MPhil thesis on Fox and to Matthew Goodman for discussing his brilliant work with me at an early stage. This work is part of the Leverhulme Early Career Research Project, 'The State of Science: Governing Knowledge of Nature in Victorian Britain', hosted at the University of Leeds and University College London. I would like to thank my colleagues there for all their support over the past couple of years, through the trials of global pandemic; in particular, Graeme Gooday, Ellen Clarke, Julian Dodd, Steven French, Katherine Rawling, Gregory Radick, Juha Saatsi, Pekka Vayrynen, Adrian Wilson, Josh Hillman, and Jonathan Topham. I have been lucky enough to present elements of this work to supportive audiences at ShanghaiTech University, Hong Kong University of Science and Technology, and the School of Philosophy, Religion, and History of Science at Leeds. It has also been an honour to deliver some of my research to the Royal Cornwall Polytechnic Society as the 2021 Smales Lecture, as part of a day of celebrations for Fox's work, culminating in the unveiling of a blue plaque at his Falmouth home, Rosehill. This work was completed at the Bartlett School of Sustainable Construction and I would like to thank all my students and colleagues for their brilliant support, especially D'Maris Coffman, John Kelsey, Priti Parikh, Helen Pascoe, Anita Treso, and Judy Stephenson. At OUP, I am very grateful to all the help and support of Thomas Stottor, Jo Spillane, Jothi Aloysia Stephenson, and Rowena Anketell.

Apologies to anyone I have forgotten, but one group that I cannot omit thanking is that of the Twitter community. Over the past few years, there has been a rapid escalation in online interest in the nineteenth-century natural sciences, especially concerning polar exploration and the voyages of Franklin and James Clark Ross. It is hard to overstate how much enthusiasm and knowledge this network possesses and it has made this book seem far more relevant than it would otherwise have been. A special mention goes to Alexa Price, who is *the* authority on James Clark Ross, Jillian Price, Alice Oates, Kathryn Stutz, Daniella McCahey, Allegra Rosenberg, Rebekah Marks, and the many anonymous specialists and fans of all things related to HMS *Terror* and HMS *Erebus*. Thanks also to all the crew and passengers who circumnavigated Africa aboard the MS *Marco Polo* between January and March 2020. They showed wonderful interest, inviting me to lecture on the dipping needle while off Africa's southern coast, and some even participated in experiments. Notably Kaya Hutchinson, Michael Drew, Rosie Howlett, Jim Ralstin, Barry Dahms, Kim Jeffs, and Geoff and Jane were all tremendous. The crew were especially supportive, in particular, Stella Wais, Agnelo Rodrigues, Essa Bhd, and Septya.

Thanks to the staff of London's Royal Society Library, the National Portrait Gallery, Cambridge University Library, Christ College Cambridge Library, the Scott Polar Research Institute Library, the Royal Polytechnic Society of Cornwall Library, the National Archives, the Library and Archives of Canada, the Bodleian

Library in Oxford, Oxford University Museum Library, Kresen Kernow in Redruth, Trinity College Cambridge Library, the British Library, the Seeley Library, Edinburgh University Centre for Research Collections, Cambridge's Whipple Library, University of Leeds Library, and the National Maritime Museum at Greenwich.

I am grateful to my family and friends in Devon for everything they have done for me and all the patience they have shown over the past few years with my work and eccentric habits. Thanks go to my family, Steve Spencer, Alexander Teague, Robert White, Steve and Pauline Teague, and Alfred Spencer, as well as Victoria, Warren, Bassi, and Reuben Putt-Gillin. And to my friends Tommy Grimshaw, Crosbie Smith, Harry Mace, Peter Roberts, Ben Allen, Don Alessandro Fraquelli, and Charlie Brown. Contrary to the rumours, I do not have a ghostwriter, but I do owe the completion of this book to three very special individuals: the first, Christopher Yabsley, an 'intellectual powerhouse'; the second, Sid Sharma, for whom 'this is it'; and finally, Louise Spencer, my incredible mother, to whom I dedicate this work.

Contents

List of Figures xiii
List of Abbreviations xvii
Preface xix

    Introduction: Empires of Magnetism   1

1. Steam-Engine Economy and the Heat of the Mine in Early Nineteenth-Century Cornwall   26

2. The Earth's Laboratory: Underground Experiments, Philosophical Miners, and Knowledge from the Mine   56

3. Survey and Science: Polar Expeditions, Terrestrial Magnetism, and the Instruments of Empire, 1815–1839   95

4. The Antarctic Foxes: Dipping Needles on James Clark Ross's South Pole Expedition, 1838–1843   133

5. Expedition and Experiment: The British Magnetic Scheme, 1841–1843   167

6. Discovery, Disaster, and the Dipping Needle: Britain's Global Magnetic System, 1843–1850   189

7. The Twilight of Cornish Science and the Systematization of Oceanic Navigation, 1850–1907   227

    Epilogue: Global Science in an Age of Empire   257

Select Bibliography 269
Index 285

List of Figures

0.1. Sabine's magnetic chart for intensity in the northern hemisphere for 1840–5, published in 1872 (Author's image, 2021). The chart folds out of Edward Sabine's 'Contributions to terrestrial magnetism. No. XI', *Philosophical Transactions of the Royal Society*, Vol. 158 (1868), pp. 371–416.   5

1.1. Botallack Mine during the 1850s, the most iconic of West Cornwall's mines, stretching out for miles beneath the Atlantic Ocean (Author's image, 2023).   27

1.2. Watt's steam engine, with 'H' marking the separate condenser, receiving a spray of cold water (Author's image, 2021). Fig. 1.2, is from Amédée Guillemin, *La Vapeur*, (Paris: Hachette, 1873).   31

1.3. Robert Were Fox (Author's image, 2021).   42

1.4. Robert Were Fox's subtropical garden at Rosehill, located centrally in Falmouth (Author's image, 2020).   44

1.5. Located just outside Falmouth, Fox's Penjerrick garden became a place of dense subtropical growth and rich botanical specimens (Author's image, 2020).   45

2.1. Fox's tables, published in *Philosophical Transactions*, detailing the movements of two magnetic needles during the aurora borealis at Falmouth between August 1830 and February 1831 (Author's image, 2023).   64

2.2. A copy of the questionnaire that Fox sent to Cornish mine owners and captains in 1836, including some forty questions and a wealth of local mining terms (Author's image, 2022).   76

2.3. Fox's experimental arrangement for measuring the strength and direction of subterranean electrical currents, as published in the report of the BAAS's 1834 meeting (Author's image, 2020).   82

2.4. The results of Fox's BAAS-funded report on the measure of heat in mines (Author's image, 2020).   85

2.5. George Wightwick's architectural drawing for the RCPS's neoclassical building in the centre of Falmouth (Image by Michael Carver, 2020).   87

2.6. The RCPS's completed building, located in Falmouth's main street (Author's image, 2020).   88

2.7. The front cover of De la Beche's *Mining Chronicle*, with Fox's 'Experiment Extraordinary' appearing in the top-right corner (Reproduced by permission of the Royal Society, 2023).   90

2.8. Inside the *Mining Chronicle*, a depiction of Fox's heat experiments, with miners discarding clothing as they descend deeper into a mine (Reproduced by permission of the Royal Society, 2023).   92

xiv LIST OF FIGURES

3.1. A chart (*c.*1822) of the Arctic expeditions of Ross, Parry, and Franklin between 1818 and 1822, by John Thomson (Author's image). 104

3.2. The romantic image of polar exploration. James Clark Ross, which John Robert Wildman (1788–1843) originally painted in 1834, with the Pole Star shining in the top-right corner and his magnetic dipping needle in the bottom-right corner (Reproduced by permission of the National Portrait Gallery, 2023). 109

3.3. The dipping needle as it appeared in Fox's 1835 *Description of R. W. Fox's Dipping Needle Deflector* (Author's Image, 2020). 116

3.4. Fox's Falmouth-built dipping needle as portrayed in Jordan's article in *Annals of Electricity* in 1839 (Author's image, 2020). 117

3.5. Fox's magnetic chart for dip across the British Isles in 1837 (Image by Michael Carver, 2019). 127

4.1. Rossbank Observatory as depicted in Ross's published account of the Antarctic Expedition (Image in author's possession, 2021). 155

4.2. Stephen Pearce's 1871 portrait of James Clark Ross with his Fox-type dipping needle on the desk (Reproduced by permission of the National Portrait Gallery, 2023). 159

5.1. The Niger Expedition before its departure in 1841 (Author's image, 2021). 174

6.1. The jewelled pivots holding in place the needle of a George-built Fox type, dating from around 1845. Note the groove around the edge of the needle's central wheel (Author's picture, 2020). 194

6.2. The back of a Fox type showing the insertion of one of the cylindrically encased deflector needles. At the top of the circle is a second hole for the second deflector, which can be screwed in. Note that both sit on a disc that can be rotated to an angle in reference to dip (Author's image, 2020). 195

6.3. Sabine's published chart for magnetic declination in the southern hemisphere. Each measure is mapped out, showing Belcher's Indian Ocean crossings of 1842 and 1843, his *Sulphur* voyage across the Pacific, as well as those of Ross between 1839 and 1843 (Author's image, 2021). 198

6.4. HMS *Samarang*, capsized in the Sarawack, Borneo, on 17 July 1843, before the crew were able to refloat her (Image in author's possession, 2021). 199

6.5. Sabine's chart for magnetic intensity between 1840 and 1845, published in 1876, including Elliot's measures from Java and the Far East (Author's image, 2021). 207

6.6. Owen Stanley's depiction of Dayman performing magnetic measurements with his Fox type at the summit of the Pico dos Bodes on Madeira (Public Domain, 2022). 209

6.7. Owen Stanley's depiction of inhabitants from Brumi Island, New Guinea, performing a dance on HMS *Rattlesnake* in 1849 (Public Domain, 2022). 211

6.8. Edwin Landseer's romanticized portrayal of the fate of Franklin's expedition, *Man Proposes, God Disposes*, completed in 1864 (Public Domain, 2023). 218

6.9. Stephen Pearce's *The Arctic Council Planning a Search for Sir John Franklin*, 1851. From left to right, Franklin, Fitzjames, and John Barrow are included as portraits on the wall, with George Back, Edward Parry, Edward Bird, James Clark Ross, Francis Beaufort (sitting), John Barrow junior, Sabine, William Hamilton, John Richardson, and Frederick Beechey (sitting) all present (Reproduced by permission of the National Portrait Gallery, 2023). 219

6.10. The Franklin Expedition's 6-inch Robinson dipping needle that McClintock retrieved, now in the National Maritime Museum at Greenwich (Author's Image, 2019). 221

6.11. The *Illustrated London News*'s depiction of the relics recovered from the Franklin Expedition, including the Robinson dipping needle to the right (Author's image, 2023). 223

6.12. An intensity experiment with a George-built Fox type, showing the suspension of weights by silk thread from the grooved wheel of the dipping needle. Weights could be added to the hanging hooks to coerce the needle to its original dip (Author's image, 2020). 225

7.1. Fox's Wilton-built dipping needle that went on display at the 1851 Great Exhibition (Author's image, 2021). 233

7.2. Sabine's magnetic chart for declination in the northern hemisphere for 1840–5, published in 1872 (Author's image, 2021). 234

7.3. Evans's study of the *Great Eastern*, mapping the magnetic character of the vast iron ship (Author's image, 2021). 250

7.4. Sabine's global magnetic charts of 1860 for (*a*) dip, (*b*) horizontal force, and (*c*) variation, as represented in Smith and Evans's *Admiralty Manual* and revealing the magnetic system of the Earth (Author's images, 2021). 252

8.1. By 1882, Anna Maria was Fox's sole surviving child but remained a central figure at the RCPS long after her father's death (Author's image, 2021). 258

8.2. Performing a magnetic measurement with a Fox-type dipping needle on a calm Indian Ocean, off the east coast of Madagascar in February 2020 (Author's image, 2020). 266

8.3. The capricious nature of magnetic survey work. Having travelled about 5,000 miles from Bristol, rough sea prevented a disembarkation at St Helena, confining our measurements to our ship (Author's image, 2020). 266

List of Abbreviations

ACS	African Civilization Society
BAAS	British Association for the Advancement of Science
BGS	British Geological Survey
BMS	British Magnetic Scheme
BOD	Bodleian Library, Oxford
CAN	Library and Archives Canada, Ottawa
CUL	Cambridge University Library
HMS	Her or His Majesty's Ship
KKW	Kresen Kernow (Cornwall Centre), Redruth
OUM	Oxford University Museum Library
P&O	Peninsula and Oriental Steam Navigation Company
psi	per square inch
RCPS	Royal Cornwall Polytechnic Society
RGSC	Royal Geological Society of Cornwall
RIC	Royal Institution of Cornwall
RMS	Royal Mail Ship
RMSP	Royal Mail Steam Packet Company
RS	Royal Society Library, London
Scott	Scott Polar Research Institute Library, Cambridge
SS	Steamship
TNA	The National Archives, Kew

Preface

0° 24′ North, 330° 19′ East: 3 December 1839

The compass's magnetic needle oscillates gently and gradually comes to rest indicating north. The experimentalist aligns a second magnetized needle, suspended vertically in a circular metal case, so that it sits along a meridian running from north to south. This instrument, known as a 'dipping needle', vibrates mysteriously in response to the Earth's magnetic field before settling. The experimentalist reads off the angle displayed, 37° 41′.[1] This measure of the Earth's magnetic pull, or 'dip', would be 0° were the instrument at the north or south magnetic pole, where the planet's magnetic influence is so strong as to pull the needle straight down. The experimentalist repeats this dip measurement twice before turning the instrument 180° and replicates his experiment three more times, giving six readings in total from which he calculates an average. The next matter to determine is the strength, or 'intensity', of this magnetic influence. The experimentalist takes out two magnets, of known magnetic strength, and screws these into the back of the circular case, recording their impact on the dipping needle. By observing their influence and comparing this to past readings at different locations, the Earth's relative magnetic intensity can be deduced. He affixes a silk thread to the dipping needle and adds tiny weights, coercing it back to its original dip and weighing the Earth's magnetic strength at this location. With an earlier measurement taken in London used as an index of 1.000, the intensity recorded here is 0.712, revealing a relatively weaker magnetic effect and greater distance from the magnetic north pole. Day after day, week after week, month after month, the experimentalist re-enacts this laborious regime as his ship voyages over thousands of nautical miles, navigating the furthest parts of the globe. The data collected provides a representation of how the Earth's magnetic qualities vary over time and space. The experimentalist is one of a network of practitioners, armed with similar dipping needles, taking uniformed magnetic readings on ships and at observatories around the world: together they are performing a vast synchronized experimental inquiry to determine the magnetic character of the entire planet. In the mid-nineteenth century, this is a scientific enterprise of unprecedented scale.

---

[1] Dip is measured in degrees (°), minutes (′), and seconds (″). For instance, 41° 30′ 15″ would be forty-one degrees, thirty minutes, and fifteen seconds as denoted on a circular vernier of 360°.

Yet despite the systematic regime, this is far from a simple process. As the dipping needle oscillates, the sway of the ship on the rolling sea makes reading the vernier dial difficult and the needle is loath to settle, requiring steadying by the rubbing of a pin, in contact with the needle, with an ivory disc. Making all this even more challenging is that the iron on-board the ship is constantly interfering with the dipping needle: the experimentalist must keep constant check on this and distinguish it from the Earth's magnetic influence. Adding to the experimentalist's difficulties is that the dipping needle's own magnetic character is liable to change, being sensitive to vibrations or disturbances, which are frequent at sea. Throughout the voyage, the needle has lost all magnetic sensitivity on several occasions, forcing the experimentalist to remove it from the circle and remagnetize the instrument by stroking it with a permanent magnet. The experimentalist hates this process because it risks damaging the inconceivably fine pins attached to the needle which allow it to oscillate so smoothly: should these break, the entire apparatus will be rendered useless. Above all though, this experimental programme is hard to carry out because the Earth's magnetic field is itself capricious, with the magnetic poles and equator constantly changing location. In endeavouring to produce precise measurements, the experimentalist must contend with a transient phenomenon and a catalogue of ever-shifting variables, all while working on a moving warship with some of the most temperamental and delicate scientific apparatus in existence. This is an age of empire and industry but it is no exaggeration to say that it is also an age of ambition.

However, for all these threats to the accuracy of his measurements, our experimentalist's magnetic investigations are at the forefront of the great scientific question of the age. At stake here is more than the behaviour of a natural phenomenon: it is the navigation of the world's premier imperial power, for the science of terrestrial magnetism is, above all, the science of empire. It represents a crisis in imperial hubris. The British Empire was an oceanic empire, comprised of a network of overseas territories connected by an immense fleet of commercial shipping and protected by the vast Royal Navy. All of this depended on accurate navigation with reliable magnetic compasses. But as the Earth's magnetic field fluctuates over time and space, its influence on a magnetized needle varies. In particular, the tendency of magnetic poles to wander meant that the difference between magnetic and geographical north, as shown on charts, was constantly changing. Such navigational challenges had been well known to European mariners since the late fifteenth century but grew increasingly urgent as Britain's colonial territories expanded during the nineteenth century: building empire was as much about ordering and controlling nature as it was about disciplining and managing diverse populations. Projectors of imperial and scientific ventures promoted the exploration of uncharted regions, seeking new trade routes, entrepôts, and understandings of natural phenomena. This all required precise navigation and reliable knowledge of terrestrial magnetism, especially in the polar regions where the Earth's magnetic field was so strong. It was, after all, the hope of a North-West Passage, between Canada and the Arctic, to the Pacific, that represented the ultimate prize for nineteenth-century adventurers. Intensifying this concern for a global survey of the Earth's magnetism was the rising use of iron in shipbuilding. As European shipbuilders incorporated larger amounts of the ferrous metal into their vessels, with full-iron hulls introduced for commercial shipping from the 1840s and into the Royal Navy from 1860, so the disturbing influence of terrestrial magnetism on a ship's compass became more erratic and exaggerated.

We return to our experimentalist, far out on the ocean. Having completed his experimental measurements, he removes the magnetic dipping needle from its circle and mounts a second needle in its place. He repeats his dip and intensity observations and logs the day's results. It is through hundreds of thousands of these magnetic experiments, over almost two decades of exploration, that an international network of natural philosophers, naval officers, and amateur specialists will chart the Earth's magnetism in the greatest scientific venture of the age or, indeed, any preceding age. It will take place on-board famous naval vessels, including HMS *Terror*, HMS *Erebus*, HMS *Sulphur*, HMS *Beagle*, HMS *Samarang*, and HMS *Rattlesnake*, with the expeditions publically celebrated back home in Britain; it will transform the naval officers responsible for this survey—such as James Clark Ross, Robert FitzRoy, Francis Crozier, Edward Belcher, Owen Stanley, and John Franklin—into household names. Some voyages became celebrated national endeavours, such as James Clark Ross's expedition to the Antarctic in 1841, or Charles Darwin's five years aboard the *Beagle*. Others were remembered only for their failure, like John Ross's Arctic venture of 1818, where he mistook low-lying cloud for an impassable mountain range. In the worst cases, these expeditions ended in catastrophe and revealed the limits of imperial ambition, most famously Franklin's bid to discover the North-West Passage aboard the *Terror* and the *Erebus*, which sailed into the Arctic in 1845, never to be seen again. This was a profound moment of scientific discovery and romanticized adventure. Our experimentalist could be any one of these heroic celebrities, serving on-board any one of these ships, or he might be one of the many obscure contributors to this scientific undertaking, working on any one of the many uncelebrated surveying voyages. Content with the day's measurements, he returns his dipping needle to its wooden box and carries the heavy encased apparatus to his cabin, ready for tomorrow's experiments.

xxii PREFACE

Introduction

Empires of Magnetism

> It had known and served all the men of whom the nation is proud, from Sir Francis Drake to Sir John Franklin...It had borne all the ships whose names are like jewels flashing in the night of time, from the Golden Hind...to the Erebus and Terror, bound on other conquests...What greatness had not floated on the ebb of that river into the mystery of an unknown earth!...The dreams of men, the seed of commonwealth, the germs of empires.
>
> Joseph Conrad, on the Thames, in *Heart of Darkness* (1899)

The Earth's magnetic influence on a magnetized needle changes over time and space. Not only do the magnetic poles wander, but the strength and direction of the planet's magnetism is in constant flux. In mid-nineteenth-century Britain, understanding how this natural phenomenon operated constituted the great scientific problem of the age, especially given the importance of magnetic compass needles for navigation and the nation's immense maritime commerce and naval might. In 1839, responding to a campaign from Britain's leading scientific authorities, the British government initiated a global survey of the Earth's magnetic properties, sending out naval expeditions to record terrestrial magnetism around the world. To measure this phenomenon, they carried with them delicate instruments, known as 'dipping needles', capable of recording the angle and strength of the planet's magnetic influence. In 2019, in a museum cellar in Falmouth, a group of local historians unearthed one of these devices, dating to the early 1840s, and built to the designs of Cornish natural philosopher Robert Were Fox (1789–1877). Fox's instruments played a central role in the nineteenth-century investigation into the Earth's magnetic properties: through delicate measurements with these devices, a global account of the planet's magnetism would be produced. During the first three months of 2020, together with fellow historian Crosbie Smith, I was given permission to take this Fox-type dipping needle on a voyage around Africa and across the Indian Ocean, to re-work the experiments of Britain's magnetic enterprise and assess the worth of this celebrated instrument of Victorian science. From Bristol to St Helena, Cape Town, Zanzibar, Colombo, and Mumbai, we followed the routes of historic magnetic survey expeditions, performing experiments along the way. All the while, the world's economies

An Empire of Magnetism: Global Science and the British Magnetic Enterprise in the Age of Imperialism. Edward J. Gillin, Oxford University Press. © Edward J. Gillin 2024. DOI: 10.1093/oso/9780198890959.003.0001

slowly, then rapidly, collapsed amid the panic of global pandemic. As historians, there was a startling poignancy to this. We were examining the experimental practices of one of the earliest endeavours to treat the Earth as an object of philosophical scrutiny: this was a moment in which the world appeared to be more connected in trade, communication, and scientific investigation than ever before. Yet just as we were coming to understand the global ambition of this magnetic enterprise, the world around us descended into isolation: ports closed, commercial centres became quiet, and the shipping of the world's great trade routes—the arteries of modern capitalism—grew sparse in number. Throughout our voyage, the fragility of the modern global economy under the pressure of COVID-19 accompanied our increasing appreciation of the challenges of conducting a worldwide scientific investigation in the nineteenth century. *An Empire of Magnetism* is the story of this enterprise, told through its instruments and experiments, such as those with which we became acquainted during our long months at sea. It argues that this constituted one of the earliest state-orchestrated scientific enterprises that was actively celebrated for its global extent: this was a defining moment in the history of science, as well as in the mobilization of scientific knowledge for statecraft and governance.

In the peace that followed the final defeat of French emperor Napoleon Bonaparte in 1815, the British Royal Navy undertook a series of voyages to the furthest parts of the world. Tasked with mapping out new trade routes and collecting geological, zoological, and botanical specimens, these expeditions were also equipped to perform metrological, astronomical, and tidal measurements, as well as acquire new knowledge of geodesy. Increasingly, however, it was the recording of magnetic data that was to dominate this culture of scientific exploration. Throughout the 1830s, Britain's science elites lobbied Parliament to finance a survey of the world's magnetic properties, commencing with an expedition to the Antarctic and establishment of a series of overseas observatories. Questions surrounding terrestrial magnetism were not just philosophical, but also engendered urgent concerns over the accurate navigation on which Britain's commercial and colonial power depended. In 1839, this campaign succeeded in securing state support, marking the official beginning of what, in 1979, historian John Cawood termed the 'Magnetic Crusade', or what Matthew Goodman has more accurately described as the British Magnetic Scheme (BMS).[1] Over the next decade and a

---

[1] The BMS has also been used to describe the British Magnetic Survey of the 1830s, which endeavored to determine the magnetic qualities of the British Isles. For the sake of economy, the term 'BMS' is used throughout this book to denote the British Magnetic Scheme in terms of the global surveying of terrestrial magnetism, commencing in 1839. The term 'Magnetic Crusade' originated in North America in the early 1840s and was not generally used to denominate the British magnetic campaign until John Cawood's influential 1979 article. No evidence has been found that it was used in Britain until April 1842 and, even then, it did not become a commonly used category. Examined in Matthew Goodman, 'From "Magnetic Fever" to "Magnetical Insanity": Historical Geographies of British Terrestrial Magnetic Research, 1833–1857', PhD thesis, University of Glasgow, 2018, 16–17;

half, a vast network of expeditionary ships and observatories, organized with state financing and administrative bureaucracy, performed thousands of experimental measurements of the Earth's magnetism. This was not the only nineteenth-century survey of this nature, but with Britain's rapidly expanding empire and unrivalled fiscal-military resources, it was the most ambitiously global in its remit. Largely forgotten today, the challenge that the Earth's varying magnetic properties presented, both to natural philosophers keen to demystify magnetic phenomena and to Britain's maritime interests, united the nation's scientific and political elites. With the rising use of iron in shipbuilding, the Earth's magnetic influence on ships and their compasses became even more erratic: with vessels going missing and reports of unreliable compass readings abounding, public and political demands for a solution escalated. The hope was that, through the BMS's surveying expeditions and fixed observatories, magnetic phenomena could be measured, mapped, and reduced to the order of charts. Yet to achieve this ambition required immensely sensitive instruments and the formation of a disciplined body of experimental practices, so as to systematically record magnetic data around the world. For all that understanding magnetic phenomena engendered the measurement of the surface of the entire planet, *An Empire of Magnetism* demonstrates that the experimental techniques and apparatus with which knowledge of terrestrial magnetism would be advanced were unmistakably provincial in origins.

Specifically, it was in Cornwall that the experimental resources to realize Britain's magnetic ambitions were to be found. It is not 40 miles from the port of Falmouth to Land's End, England's most westerly point, and yet this tiny area was the absolute centre of steam-engine technology and coal-powered industry for the first fifty years of the nineteenth century. West Cornwall's deep mines were rich in valuable copper and tin but lacked coal. As Cornish miners dug ever deeper into the depths of the Earth, so they became more reliant on steam engines to keep these subterranean excavations pumped free of water. Around this industry, engineers conceived of machines of increasing economy: nowhere was the challenge of extracting work from coal so critical as in Cornwall and, by the 1820s, the county's engines were renowned for their efficiency. Cornwall was not only at the forefront of European industrialization, but was also part of an increasingly globalized economy, marked by a growing network of colonial territories. For British naval and merchant vessels departing for the Atlantic, bound for the Americas, Pacific, Africa, and Asia, Falmouth represented the final port of call and first place of arrival on return. Around this commercial and industrial activity, Cornish audiences grew ever more interested in Nature, with local entrepreneurs and natural philosophers fashioning a rich scientific culture. Questions over the Earth's subterranean structure, of meteorology, botany, and the operation of heat

---

see also John Cawood, 'The Magnetic Crusade: Science and Politics in Early Victorian Britain', *Isis*, 70/4 (Dec. 1979), 493–518.

and pressure intersected philosophical and industrial interests. And among all this philosophical inquiry, the examination of the Earth's magnetic properties featured prominently. Cornwall's deep mines provided unique places to observe subterranean electromagnetic activity and develop robust instruments of magnetic measurement. It was here that Fox fashioned the leading magnetic survey apparatus of the age, the Fox-type dipping needle, or dip circle. This book follows these Cornish instruments from their subterranean, industrial origins to their voyages around the world, as Britain's scientific and political elites mobilized Fox's dipping needles within the global investigation of terrestrial magnetism.

Fox's instruments might seem an obscure focal point for a study of nineteenth-century magnetic science. He was not the only builder of dipping needles. Charles Robinson of London, Henri-Prudence Gambey of Paris, and Christopher Hansteen of Oslo all contributed apparatus to the worldwide examination of terrestrial magnetism. While a horizontally positioned magnetic needle, such as a compass, could determine the direction of north or south, it required a vertically suspended needle to measure the angle and strength of the Earth's magnetic pull. These dipping needles, or 'dip circles', on account of the circular container in which they oscillated, were notoriously delicate, liable to break, and frequently became demagnetized. Fox's rivals often produced instruments of superior precision and magnetic sensitivity, but although these worked well in controlled observatories and laboratories, they travelled poorly on oceanic and overland expeditions. In contrast, Fox's instruments became celebrated for their robustness. It was their mobile quality that ensured they were the premier choice for magnetic survey work conducted over geographical space. They were not the only dipping needles used in this manner, but they rapidly became the standard equipment for measuring magnetic phenomena during maritime voyages. And this is why they are the ideal objects through which to examine the BMS as a global endeavour. Indeed, it was their potential for expeditionary work that made Fox's instruments so attractive to the BMS's director, the domineering artillery officer and magnetic fanatic Edward Sabine (1788–1883). From Woolwich Arsenal's Magnetic Department, established in 1841, Sabine orchestrated Britain's magnetic enterprise, instructing officers in their magnetic duties, overseeing the preparation of overseas observatories and expeditions, and converting incoming magnetic data into maps, tables, and publications (Figure 0.1). He conscripted Fox's instruments and experience within this industrious programme, having experienced the Cornishman's dipping needles during his own magnetic measurements in the 1830s. *An Empire of Magnetism* analyses how a group of magnetic enthusiasts took a very local, distinctly Cornish, array of techniques and instruments and integrated them within a worldwide system of knowledge production. It explains how Fox's provincial natural philosophy became a celebrated science of empire. And, crucially, it argues that this transformation marked a dramatic escalation in state-orchestrated science. Never before had a government invested so extensively in a scientific enterprise; never before had such a geographically vast investigation into a natural phenomenon been envisaged. The

EMPIRES OF MAGNETISM 5

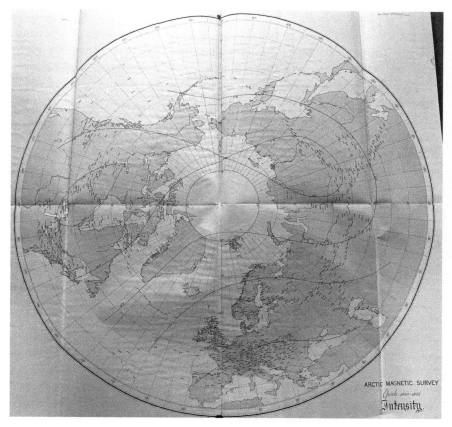

Figure 0.1 Sabine's magnetic chart for intensity in the northern hemisphere for 1840–5, published in 1872 (Author's image, 2021).
Source: Edward Sabine's, 'Contributions to terrestrial magnetism. No. XI', *Philosophical Transactions of the Royal Society*, Vol. 158 (1868), pp. 371–416.

apparatus of the nineteenth-century British state would prove essential to elevating Fox's techniques and apparatus from a provincial to a global inquiry. As this book emphasizes, this venture had broader implications over the formation of the 'modern state' and nineteenth-century imperialism.

0.1. A History of a Global Science

Cornwall might seem an inauspicious place to focus the history of a global science. The county boasted no great institutions, universities, or eminent research laboratories, nor was it a centre of government. It could not rival the rich mathematical traditions of Paris and Cambridge, Edinburgh's natural philosophy or the experimental culture of Berlin and Göttingen. As an Atlantic province without any large cities, it was far removed from the great European capitals of science which were home to flourishing networks of prestigious philosophical societies, such as the Royal Institution and Royal Society in London or Paris's Académie des sciences.[2] Nevertheless, Cornwall was a hub of philosophical activity. Since the 1990s, historians of science have stressed the spatial contingency of past knowledge production: where scientific activity has happened matters to its form, as much as when it happened and who was doing it. There is now, rightly, a wealth of detailed regionally focused histories of science, including on Cornwall, emphasizing the importance of place and localized cultural contexts.[3] However, it has become difficult for historians to draw together such diverse specialist case studies and deliver effective global histories of science, without losing the contexts of time and place. The question, then, for a historian of nineteenth-century magnetic science, is how do we get from Cornwall to the world: from the regional to the global?

One solution is to take a geopolitically pluralistic vision of the history of science, or what James Delbourgo has termed '*the knowing world*', this being 'a planet made up of a multiple scientific cultures and long memories of past judgements about who possess the requisite faculties and resources to count as scientific and therefore modern, and who does not'.[4] The analytical emphasis here is on the connections and exchanges between culturally diverse scientific regions and shifts our understanding of the development of 'modern science' away from its traditional Euro-American focus. This 'global turn' endeavours to see the world as a connected network of scientific interactions: the movement of ideas, people, and objects between different geographical regions has been crucial to the historic formation of new knowledge of Nature.[5] Such analysis promises fresh insights into the development of past scientific disciplines. When it came to geology, for instance, the European conception of 'deep time' through the discovery of plant and animal fossils, and human flint tools, resulted in increasingly disciplined

---

[2] For metropolitan centres of science, see Maurice Crosland, *Science Under Control: The French Academy of Science, 1795–1914* (Cambridge: Cambridge University Press, 1992); Martin J. S. Rudwick, *The Great Devonian Controversy: The Shaping of Scientific Knowledge among Gentlemanly Specialities* (Chicago: University of Chicago Press, 1985), 21–3.

[3] Simon Naylor, *Regionalizing Science: Placing Knowledge in Victorian England* (London: Pickering & Chatto, 2010).

[4] James Delbourgo, 'The Knowing World: A New Global History of Science', *History of Science*, 57/3 (2019), 373–99, at 377.

[5] Fa-Ti Fan, 'The Global Turn in the History of Science', *East Asian Science, Technology and Society: An International Journal*, 6 (2012), 249–58, at 251–2; Sivasundaram raised the question of how to write global histories of science which avoid perpetuating this notion of Western knowledge as something separate and epistemologically superior to that of the rest of the world. His proposed solution was to examine non-textual sources, engaging with material culture beyond the purely written. See Sujit Sivasundaram, 'Science and the Global: On Methods, Questions, and Theory', *Isis*, 101/1 (Mar. 2010), 146–58, at 146–9. For responses to this challenge, see Suman Seth, 'Colonial History and Postcolonial Science Studies', *Radical History Review*, 127 (Jan. 2017), 63–85; Lissa Roberts, 'Situating Science in Global History: Local Exchanges and Networks of Circulation', *Itinerario*, 33/1 (2009), 9–30; James Poskett, *Horizons: A Global History of Science* (London: Viking, 2022).

specialist study. However, in nineteenth-century India, geology emerged as what Pratik Chakrabarti has described as 'a hybrid enterprize', tied to antiquarian and archaeological research, and with implications for the identity of colonial India.[6] Likewise, in her study of the rise of climatology as a global science, Deborah Coen has stressed the varying role of empire in the production of climatic knowledge. While in British India the idea of a singular 'Indian climate' was an attractive proposition, suggesting a region capable of mastery through science, the recognition of the Austro-Hungarian Empire's diverse climate was connected to broader political concerns over unifying the Habsburgs' eclectic array of duchies, kingdoms, and principalities.[7] As with nineteenth-century climatology and geology, the science of terrestrial magnetism engendered similar challenges of reconciling regional variations within a global system of natural phenomena.

Growing up in Devon, next door to Cornwall, I originally approached this subject as a regional history of Cornish science. I first came across 'Mr. Fox of Falmouth' and his 'highly important results respecting the electricity of metalliferous veins in the mines of Cornwall' in Michael Faraday's 1832 Bakerian Lecture on 'Terrestrial Magneto-Electric Induction', delivered to the Royal Society.[8] I was, predictably, instantly drawn to the local nature of this reference. However, on finding a working Fox-type dipping needle at the Royal Cornwall Polytechnic Society and securing permission to take it on an oceanic voyage from Falmouth, down the Atlantic to Lisbon, St Helena, and Cape Town, across the Indian Ocean to Colombo, and then back to the British Isles via Mumbai, the Persian Gulf, Suez, the Mediterranean, and Gibraltar, I have become increasingly aware of the extent to which this book must necessarily be a global history of science.[9] The months at sea that magnetic survey work entailed, where life is not measured so much in time as in space and distance travelled, gave me a curious appreciation of the daunting prospect that a global magnetic survey presented to its promoters and practitioners. Lacking modern communications, expeditionary naval officers had no way of checking their results or obtaining guidance on how to manage their instruments. I encountered a similar detachment from outside aid on

---

[6] Pratik Chakrabarti, *Inscriptions of Nature: Geology and the Naturalization of Antiquity* (Baltimore: Johns Hopkins University Press, 2020), 5–6 and 19; contrast this to twentieth-century variations in notions of being 'nuclear', dependent on location. As Hecht argues, despite being sites of uranium extraction, 'African' mines were not defined as 'nuclear workplaces', despite being technological and industrially crucial to sustaining nuclear weapons and energy systems. See Gabrielle Hecht, *Being Nuclear: Africans and the Global Uranium Trade* (Cambridge, MA: MIT Press, 2012), 14.

[7] Deborah R. Coen, *Climate in Motion: Science, Empire, and the Problem of Scale* (Chicago: University of Chicago Press, 2018), 12–13.

[8] Michael Faraday, *Experimental Researches in Electricity* (London: Richard and John Edward Taylor, 1839), 54.

[9] For a full account of this voyage, see Edward J. Gillin, 'The Instruments of Expeditionary Science and the Reworking of Nineteenth-Century Magnetic Experiment', *Notes and Records of the Royal Society*, 76/3 (Sep., 2022), 565–592; compare with Tom Griffiths, *Slicing the Silence: Voyaging to Antarctica* (Cambridge, MA: Harvard University Press, 2007).

the Atlantic and Indian oceans, without internet or basic communications. Here, I was constantly anxious should my instrument be damaged beyond any immediate source of repair. Along with the challenges that salt, sea spray, wind, heat, and rain presented to the instrument, as well as the magnetically devastating impact stormy weather and rough swell had on the needle itself, this experimental isolation amplified my perception of what a fragile, if ambitious, enterprise the BMS constituted.

The great challenge for the BMS was, as Sabine remarked, of how to perform 'simultaneous observations at all parts of the globe' and coordinate these with those of the ongoing Göttingen Magnetischer Verein.[10] Carl Gauss (1777–1855) and Wilhelm Weber (1804–91) had established this European-led venture in 1834 at Göttingen as an international programme of magnetic investigation, but its eventual collaboration with the BMS would be limited. British magnetic specialists were far more concerned with oceanic research than their Continental counterparts and focused their efforts on maritime expeditions and colonial observations. This is hardly surprising, given Britain's considerable maritime interests, as well as the immense naval resources at its disposal. Nevertheless, on Gauss and Weber's fixed Term Days, Britain's observatories and expeditions took measurements on a five-minute cycle, theoretically in synchronization with this continental European scheme, which included about fifty observation sites around the world. Such work was immensely problematic. Organizing synchronized experimental observations of magnetic phenomena was a key problem for promoters of magnetic science. In terms of these difficulties over coordinating experiments around the world, this book unites archival and textual sources with my own experiences of magnetic survey work.[11]

The experience of this voyage has, specifically, raised the question over how far the BMS fits as an 'imperial', as opposed to 'global', endeavour. At present, the foremost publication on nineteenth-century magnetic surveys is Christopher Carter's excellent *Magnetic Fever* (2009). Carter argued that the British Empire 'provided the necessary resources for the creation of a universal inductive geoscience' and defines both the BMS and the United States' counterpart scheme as

---

[10] Edward Sabine, 'Instructions for Magnetic Observations in Africa', *Friend of Africa*, 1/4 (25 Feb. 1841), 55–7, at 56.

[11] To put this challenge of synchronization into context, consider Vanessa Ogle's claims that, while the nineteenth century is often celebrated as an age of increasingly integrated international time regulation, there was no uniformed global system of timekeeping until the twentieth century. While many colonial governments established mean times during the 1880s and 1890s, these were national, as opposed to universal, times, and were rarely calculated in reference to Greenwich as the prime meridian. See Vanessa Ogle, *The Global Transformation of Time, 1870–1950* (Cambridge, MA: Harvard University Press, 2015), 75–7.

examples of 'imperial science'.[12] This is not wrong, but Britain's magnetic enterprise cannot be understood within the confines of empire. While Carter's study focused on American and British efforts within colonial observatories and a few celebrated polar expeditions, the BMS envisioned the surface of the entire planet as a subject of inquiry. If we follow the Fox-type dipping needles' voyages in their entirety, we see that the work Carter has charted was accompanied by a vast network of expeditions around the world, across the Atlantic, Indian, and Pacific oceans, deep into Africa, around South, Central, and North America, the Caribbean, the coasts of subcontinental Asia and the Far East, Australia, Antarctica, the Arctic, New Zealand, Korea, Japan, the Philippines, and Borneo, as well as measurements in observatories beyond the sphere of Britain's direct influence. The BMS collaborated with observatories in Milan, Prague, Philadelphia, Cambridge in Massachusetts, Algiers, Breslau, Munich, Cadiz, and Brussels, as well as with the Göttingen Magnetischer Verein. The Raja of Trivandrum took the initiative to establish his own magnetic observatory at Travancore, as did the king of the princely state of Oudh, at Lucknow, and the Pasha of Egypt in Cairo, with each of these contributing to Britain's global enterprise.[13] Despite its name, the BMS was not limited to the imperial confines of the British Empire: its history is a story with a diverse cast of natural philosophers, politicians, mathematicians, naval officers, soldiers, rajas, pashas, pirates, colonial governors, mechanics, and miners.[14]

There is another, personal, reason for thinking globally, rather than imperially, about this magnetic enterprise. Terms like 'empire' and 'imperial', rightly or wrongly, hold connotations of order and control. Reflecting on the establishment of observation posts for the transit of Venus in 1761, 1769, 1874, and 1882, Jessica Ratcliff explained that 'astronomical expeditions...stayed comfortably within the confines of imperial boundaries'.[15] Magnetic survey work transcended these 'boundaries' and was anything but comfortable. Imperial order provided little guarantee of a successful magnetic experiment. This is not only borne out from

---

[12] Christopher Carter, *Magnetic Fever: Global Imperialism and Empiricism in the Nineteenth Century* (Philadelphia: American Philosophical Society, 2009), pp. xxiv–xxv; for comparative studies on the complex relationship between scientific knowledge and empire, see Zaheer Baber, *The Science of Empire: Scientific Knowledge, Civilization, and Colonial Rule in India* (Albany: State University of New York, 1996); Sarah Irving, *Natural Science and the Origins of the British Empire* (London: Routledge, 2016); Brett M. Bennett and Joseph M. Hodge (eds.), *Science and Empire: Knowledge and Networks of Science across the British Empire, 1800–1970* (Basingstoke: Palgrave Macmillan, 2011); Pratik Chakrabarti, *Medicine and Empire, 1600–1960* (Basingstoke: Palgrave Macmillan, 2014).
[13] Ratcliff, *The Transit of Venus Enterprise in Victorian Britain* (London: Pickering & Chatto, 2008), 24.
[14] Thanks to Sarah Qidwai for discussions on the diverse networks of nineteenth-century science. See Sarah Ahmed Qidwai, 'Sir Syed (1817–1898) and Science: Popularization in Nineteenth-Century India', PhD thesis, University of Toronto, 2021.
[15] Ratcliff, *The Transit of Venus Enterprise in Victorian Britain*, 14.

the historical documentation, but was obvious in my own experimental magnetic experiences. At St Helena, rough swell prevented us landing our dipping needle to perform land-based measurements comparable to those taken at sea. In equatorial waters the needle became inert, resting at 0° with little inclination to move. Wind and spray prevented more delicate measures of magnetic strength using Fox's system of tiny weights. Security concerns at Cape Town and failing light limited our experimental series at this location. And the integrity of the needle was at constant risk from the sea's roll in bad weather. As Crosbie Smith later reflected in our discussion of our voyage, oceanic travel made magnetic survey work 'capricious' and ensured that any notion of a scientific programme secured within the comforts of imperial order was unrealized. The thing with predetermined Term Days was that naval crews, if near a coast, would endeavour to stop and take on-land magnetic measurements, regardless of location. Imperial entrepôts like Cape Town, Singapore, Sydney, and Hong Kong made attractive observation posts, but the whole point of magnetic survey work was to record magnetic phenomena in previously unmeasured locations, be that on the banks of the River Niger, the coast of Borneo, the middle of the Pacific Ocean, or on an iceberg in the Arctic. Often, Term Days arrived at awkward moments, with expeditions compelled to establish temporary observatories in swamps, on coral reefs, or in lands with hostile inhabitants. Efforts to synchronize experiments around the world ensured that this enterprise was very far from an orderly imperial project.

My own sense of the global nature of the BMS is not ahistorical: it was something that the venture's original promoters themselves emphasized. In a similar fashion, James Poskett has warned that 'empire' and 'imperial' were not units that confined or limited scientific activity: historic actors frequently travelled between these imperial spaces and engaged with extra-imperial intellectuals.[16] It is important, at this stage, to distinguish between the global as a tool of analysis and the global as an actor category—a point that Poskett emphasized in his study of nineteenth-century phrenology, the 'science of the mind'. While phrenologists envisioned their science as a connected discipline, uncovering worldwide phenomena and universal principles, the actual work of building this knowledge was facilitated through global networks of correspondence, including letter writing and publishing.[17] When it came to magnetic science, this was equally true: the notion that this was a global science was something contemporary actors promoted, and they did so to emphasize that this expansive quality was something new. In 1841, Edward Sabine defined the BMS's remit as determining 'the magnetic state of the whole globe...by systematic

---

[16] James Poskett, *Materials of the Mind: Phrenology, Race, and the Global History of Science, 1815–1920* (Chicago: University of Chicago Press, 2019), 10 and 13.

[17] James Poskett, 'Phrenology, Correspondence, and the Global Politics of Reform, 1815–1848', *Historical Journal*, 60/2 (2017), 409–42, at 411–12 and 414.

observations made nearly contemporaneously at almost every accessible part of its surface'.[18] His collaborators agreed, with Fox writing of 'the importance of continuing the system of simultaneous observations in different parts of the world'.[19] In his presidential speech to the 1845 meeting of the British Association for the Advancement of Science, John Herschel praised the BMS's 'system of observatories all over the world', declaring this to be 'by far the greatest and most prolonged effort of scientific co-operation which the world has ever witnessed'.[20] In the audience was the great polymath of the age and coiner of the word 'scientist', William Whewell. In his revised edition of *History of the Inductive Sciences*, published the same year as Herschel's address, Whewell echoed this rhetoric, boasting that the 'manner in which the business of magnetic observation has been taken up by the government of our time makes this by far the greatest scientific undertaking which the world has ever seen'.[21] Such language conflated the global extent of the project with idealized notions of this being a science of value to the world. Captain Edward Belcher, for instance, recalled meeting with the Spanish governor of the Philippines in Manila, during magnetic experiments in 1844. According to Belcher, the governor understood 'the nature of our intended operations within his government, and the advantages which would result to the civilized world'.[22] Though couched in contemporary imperialistic notions of what parts of the world would benefit from the BMS's endeavour, there was a solid consensus that this was an enterprise that transcended the limits of Britain's empire. Undoubtedly, much of this was hyperbole, but such language was inseparable from, and indeed shaped, global ambitions and the treatment of terrestrial magnetism as a worldwide phenomenon. In 1849, ten years after the BMS's commencement, Sabine reflected on the importance of the magnetic charts that resulted from the data accrued since 1839, giving 'the basis of a systematic view of terrestrial magnetism as it manifests itself to us on the surface of the globe'.[23] Clearly then, nineteenth-century British magnetic specialists were thinking globally, in terms of the value of their work to the world, as well as in the extent of their investigation and commitment to treating the planet as a connected, singular magnetic

---

[18] Sabine, 'Instructions for Magnetic Observations in Africa', 55.
[19] TNA, Sabine Papers, BJ3/19/153–4: Robert Were Fox to Edward Sabine (3 Mar. 1845), 153.
[20] Anon., 'Fifteenth Meeting of the British Association for the Advancement of Science: Cambridge, 18th June 1845', *Calcutta Journal of Natural History: And Miscellany of the Arts and Sciences in India*, 7 (1847), 81–142, at 93.
[21] William Whewell, *History of the Inductive Sciences, from the Earliest to the Present Time*, 3 vols. (3rd edn., London: John W. Parker and Son, 1857), iii. 55.
[22] Edward Belcher, *Narrative of the Voyage of H.M.S. Samarang, During the Years 1843–1846; Employed Surveying the Islands of the Eastern Archipelago*, 2 vols. (London: Reeve, Benham, and Reeve, 1848), i. 101.
[23] Edward Sabine, 'Terrestrial Magnetism', in John F. W. Herschel (ed.), *Manual of Scientific Enquiry; Prepared for the Use of Her Majesty's Navy: And Adapted for Travellers in General* (London: John Murray, 1849), 14–53, at 27.

system. While the use of the term 'global' in the sense of an expansive worldwide perspective did not appear in English until much later in the century, contemporaries employed terms like 'globe' and 'world' to stress the extent of the BMS's ambition. Such grand enterprise was not unprecedented in the nineteenth century, but the BMS's promoters were eager to emphasize that this project was something new and unrivalled.

0.2. The State of Science

This book is a history of a global science, rich in regional contexts, in which historic actors actively conceived of the world as a scientific object in itself. But it is also the history of the development of state science. Orchestrated through government organs, notably the Admiralty, army, and Treasury, as well as the Royal Society—which, while not a government organ, included a host of eminent MPs and Lords, and operated under the patronage of the head of state—the BMS benefited from unparalleled state investment and organization, in terms of manpower, infrastructure in the form of ships and observatories, and finance. By 1843, the Antarctic Expedition alone had cost £109,768 since its inception in 1839.[24] However, the BMS was by no means the first large-scale government-financed scientific investigation. The Board of Longitude's establishment in 1714 represented a sustained investment in the natural sciences; it was concerned with questions of navigation until its termination in 1828. Throughout the nineteenth century, the Admiralty remained, as Jessica Ratcliff put it, 'the most important patron of science', spending in excess of £100,000 annually on 'scientific' activity.[25] Likewise, the BMS was not the first scientific venture of global vision. Famously, the French state established astronomical observation stations in Siberia, Vienna, Pondicherry, and Rodriguez, just off Madagascar, for the 1761 transit of Venus. The British government responded for the 1769 transit with expeditions to Hudson's Bay and Tahiti, the latter of these as part of Captain James Cook's celebrated voyage aboard HMS *Endeavour*, with these collaborating with six stations under Russian administration and three in British America. Similar efforts were witnessed in Denmark, Spain, the German lands, the Netherlands, and Sweden.[26] At the same time, the BMS was not alone in treating the Earth itself as an object of scientific investigation. As Simon Naylor and Simon Schaffer have astutely observed, the study of terrestrial magnetism was just one of several nineteenth-century 'survey sciences', defined as the conception of the globe as an object of measurement and scrutiny over space, capable of governance and

---

[24] Ratcliff, *The Transit of Venus Enterprise in Victorian Britain*, 23.
[25] Ibid. 26. [26] Ibid. 14–17.

reduction to manageable systems and charting.[27] Other schemes included the Ordnance Survey from 1790-1, followed by the Great Trigonometric Survey of India from 1802, the British Geological Survey in 1835, and a growing network of colonial observatories around the world for recording astronomical and metrological phenomena. Increasingly, British natural philosophers engaged in worldwide inquiries into astronomical, geological, physical, and metrological phenomena. Much of this drew inspiration from Humboldt's model of science produced through the extensive collection of data over large geographical spaces.[28] Within this framework, it was important to study connected natural phenomena across the world to determine governing laws and causes. Survey science relied on uniformed networks of mobile observers and instruments to collect scientific data over huge expanses of territory. These 'regimes of observation' were attempts to control apparatus and human agents at great distance.

So what was the significance of the BMS? In many respects, this enterprise feels familiar and, as Ratcliff observes, has many of the characteristics associated with twentieth-century 'Big Science'. Particle accelerators, space shuttles, the Manhattan Project, and nuclear energy all come to mind when we think of large-scale research, with its origins in the United States during, and after, the Second World War. Yet, as Ratcliff reminds us, this is all historically relative, 'with every period' having had 'its own "big science"'. The challenge for historians is to identify the 'bigger science', rather than the big science, of an age.[29] The BMS was, unquestionably, the bigger science of the mid-nineteenth century but, more than this, represented an important escalation in state–science relations and, of all the survey sciences, most obviously exhibited many of the qualities wrongly thought new to twentieth-century big science. To substantiate this claim, we should return to Alvin Weinberg's classic 1961 essay on the subject. Written in response to exiting president Dwight D. Eisenhower's warning that the United States' 'military-industrial complex' and Federal government's dominance of scientific research was having a corrupting influence on the nation's political system, Weinberg expressed his own concerns that such scientific culture encouraged inefficiency through overfunding, while threatening intellectual freedoms. By 'big science', however, he meant more than just big research, but something more symbolic, embodying the aspirations of societies on a monumental scale. Historians of the twentieth century would, he predicted, 'find in the monuments of Big Science ... symbols of our time just as surely as one finds in Notre Dame a symbol of the Middle Ages'.[30] Along with its monumentalism, military demands shaped this scientific culture, together with state investment and accountability,

---

[27] Simon Naylor and Simon Schaffer, 'Nineteenth-Century Survey Sciences: Enterprises, Expeditions and Exhibitions: Introduction', *Notes and Records*, 73/2 (2019), 135–47, at 141–2.
[28] Ibid. 135–7.   [29] Ratcliff, *The Transit of Venus Enterprise in Victorian Britain*, 21.
[30] Alvin M. Weinberg, 'Impact of Large-Scale Science in the United States', *Science*, NS 134/3473 (21 July 1961), 161–4, at 161.

bureaucratic administration and hierarchy, and increasing teamwork, constituting a new way of organizing scientific activity around central government.[31]

As the BMS shows, none of this was new to post-war America. It was global in vision and involved synchronized experiments over great time and physical space, performed in reference to those of the Göttingen Magnetischer Verein. When Sabine proposed departing from this continental European practice of five-minute intervals between Term Day observations, Herschel stressed the critical importance of international collaboration, warning that 'there ought not *in science* and *in the same nation* to be two reckonings of time'.[32] Here, magnetic science was not just big, but bigger than any parochial nationalism. It was also significant as a model of what, I would argue, is best thought of as 'industrial-capitalistic science'. By this, I do not mean scientific activity in terms of industrial research, but more in the sense of its organization and systematization. The BMS represented the industrialization of knowledge production, applied on a global scale.[33] To systematize experiments required the production of standardized instruments and experimental techniques, much according to the disciplined rigour of factory production, as did the reduction of the enormous amount of data, collected over the course of over fifteen years, to intelligible magnetic charts. And behind all of this, sustaining Britain's economy and the manufacturing of instrument parts, was the industrial consumption of coal. The BMS constituted an industrial venture in the organization of its instruments, practices, and labour. What makes it historically significant is not simply its global extent, its state support, its immense research scale, or its industrialized organization, but also the way in which it incorporated each of these features in what was, as contemporaries were quick to observe, the great scientific undertaking of the age.

What becomes clear throughout an analysis of the BMS, however, is that when dealing with nineteenth-century government–science relations, we need to develop a robust understanding of 'the state'. In any historical treatment of government and political power there is always a danger of 'the state' taking on a disembodied and nebulous character. This is especially true when discussing the nineteenth-century British government because of the dominance of its self-avowedly liberal political framework in which low-state interference and laissez-faire governance were celebrated as fundamental to the nation's socio-economic fortunes. The true extent of this commitment to freedoms and liberalism has, of course, been subject to debate, notably through the careful work of Peter Mandler and Jonathan Parry, but what is undeniable is that Britain's political elites boasted of the liberal nature of their government and that there was a strong consensus

---

[31] Peter Galison, 'The Many Faces of Big Science', in Peter Galison and Bruce Hevly (eds.), *Big Science: The Growth of Large-Scale Research* (Stanford, CA: Stanford University Press, 1992), 1–17, at 1–2; Steven Shapin, *The Scientific Life: A Moral History of a Late Modern Vocation* (Chicago: University of Chicago Press, 2008), 80–4 and 165–73.
[32] John Herschel, quoted in Carter, *Magnetic Fever*, 117.
[33] On scaling, see Coen, *Climate in Motion*, 16.

throughout society that Britain's parliamentary system was built on values of toleration and individual freedoms.[34] This laissez-faire attitude extended to notions of scientific progress, with leading philosophical authorities and popularizers of science keen to emphasize the heroic, individualistic, and entrepreneurial nature of scientific investigation. The cultivation of the natural sciences was, ideally, to be entrusted to private enterprise, developing most effectively within a liberal economy, through the voluntary direction of capitalistic businesses and independent societies and institutions.[35]

All of this makes it very difficult to qualify the state's actual role in the BMS and yet, without question, the government was crucial to this global undertaking. Throughout this book, it becomes apparent that the nineteenth-century British state was comprised of an often informal network of social actors and material resources. Although operating within a framework of organized bureaucracy, this was governance characterized by familiarity. This analysis therefore adopts the broad conceptualization of the state that historian Patrick Joyce has developed, building on Michel Foucault's conception of 'governmentality'. Focusing on the techniques and material systems that sustain political power, Joyce's account of 'the state' encourages a historical examination of the mundane: of the shared cultural values of governing classes, of their methods for exerting political agency, and of the physical networks through which power operates. For nineteenth-century Britain, he identified a shared ruling-class mentality, cultivated through the educational traditions of public schools and Oxbridge, combined with technocratic systems such as the postal services and office filing, and a liberal political ideology that stressed self-regulation and individual freedoms as crucial to the state's operation. Through these mechanisms, Joyce argued that power acted via impersonal and personal means. Any study of the state's role in nineteenth-century science must, therefore, consider the relationship between both: in other words, between the individuals in positions of governance and the material means at their disposal.[36] For the government's orchestration of the global magnetic survey, there was a broad commitment to philosophical inquiry

[34] Peter Mandler, 'Introduction: State and Society in Victorian Britain', in Peter Mandler (ed.), *Liberty and Authority in Victorian Britain* (Oxford: Oxford University Press, 2006), 1–21; J. Parry, 'Liberalism and Liberty', in Mandler (ed.), *Liberty and Authority in Victorian Britain*, 71–100; for studies of British liberalism, see James Vernon, *Politics and the People: A Study in English Political Vulture, c.1815–1867* (Cambridge: Cambridge University Press, 1993); Simon Gunn and James Vernon, 'Introduction: What Was Liberal Modernity and Why Was It Peculiar in Imperial Britain?', in Simon Gunn and James Vernon (eds.), *The Peculiarities of Liberal Modernity in Imperial Britain* (Berkeley and Los Angeles: University of California Press, 2011), 1–18.
[35] Martin Daunton, *State and Market in Victorian Britain* (Woodbridge: Boydell Press, 2008), 4–5 and 18–21; on science and the state, see also Peter Alter, *The Reluctant Patron: Science and the State in Britain, 1850–1920* (Oxford: Berg, 1987).
[36] Patrick Joyce, *The State of Freedom: A Social History of the British State since 1800* (Cambridge: Cambridge University Press, 2013), 3–4 and 27–32; see also Michel Foucault, 'Governmentality', in Graham Burchell, Colin Gordon, and Peter Miller (eds.), *The Foucault Effect: Studies in Governmentality with Two Lectures by and an Interview with Michel Foucault* (Chicago: University of Chicago Press: Chicago, 1991), 87–104.

shared among several individuals either close to, or actually in, positions of political power. These personnel had access to the tools of what John Brewer has termed the 'fiscal-military state', comprised of Britain's armed forces, especially its enormous navy, and its fiscal strength, built on a reliable system of taxation and solid financial credit.[37]

These national resources were organized through government institutions, with Parliament and the Cabinet responsible for legislation; the Admiralty for the Royal Navy; and various offices in Whitehall, including the Colonial Office and Exchequer, for administration. The administrators, strategists, and statesmen of these organs delivered the financial, military, and colonial means required to execute the BMS. Yet these were not abstract centres of bureaucratic power, but networks of individuals, often with scientific interests or ambitions for new exploratory triumphs. Politicians, army and naval officers, and government officials were all open to persuasion. In particular, between 1838 and 1839 John Herschel was able to promote the BMS with the prime minister, the Chancellor of the Exchequer, Queen Victoria, and a host of socially eminent individuals, obtaining state funding for an expedition to the Antarctic. Likewise, experienced naval and army officers were able to win Admiralty patronage for the adoption of new scientific instruments and equipment to facilitate the collection of magnetic data from around the world. The contribution of the nineteenth-century British state to scientific activity was very much managed through these social channels. Especially prominent was a group of officers who became increasingly concerned with the examination of natural phenomena throughout a series of polar expeditions. Many of these became household names, such as James Clark Ross, John Ross, Francis Crozier, Francis Beaufort, John Franklin, Robert FitzRoy, and Edward Sabine. They not only emphasized to Parliament and the Admiralty the importance of recording natural phenomena through naval expeditions, but themselves acted as patrons of the natural sciences as their careers advanced and they progressed to the upper echelons of the armed forces.

Significantly, it was not just military officers and scientific practitioners who sought philosophical acclaim: the politicians to whom they appealed for state support often boasted scientific credentials of their own. Indeed, many of Britain's leading nineteenth-century politicians possessed considerable philosophical understanding. While Lord Melbourne (1779–1848), prime minister from 1836 until 1841, lacked any philosophical training, he was a fellow of the Royal Society. His successor, Robert Peel (1788–1850), was also a fellow of the Royal Society, but could also claim great scientific learning, having taken the first double-first in classics and mathematics to be awarded from the University of Oxford following the reform of its undergraduate system. When Peel fell from power in 1846, the

---

[37] For this concept of the 'fiscal-military state', see John Brewer, *The Sinews of Power: War, Money and the English State, 1688–1783* (Cambridge, MA: Harvard University Press, 1989).

new Whig premier, John Russell (1792–1878), could not boast such an outstanding university education, but had studied the natural sciences under Lyon Playfair at the University of Edinburgh and secured election to Royal Society. Thomas Spring Rice (1790–1866), Chancellor of the Exchequer with responsibility for government expenditure between 1835 and 1839, was a fellow of the Royal Society, as was his Tory replacement, Francis Baring (1796–1866), who served in this position from 1841 until 1846. Collectively, these occupants of the two leading positions of political authority ensured that there was a significant degree of scientific understanding at the helm of government throughout the 1830s and 1840s, regardless of party divides between Tories and Whigs. Similarly, individuals like Davies Gilbert (1767–1839) and Charles Lemon (1784–1868) transcended political and scientific networks, being both MPs and members of the Royal Society, with Gilbert serving as its president from 1827 to 1830. Beyond Parliament too, the nation's political elites were directly engaged in scientific enterprise. Gilbert's successor as president of the Royal Society was the Duke of Sussex (1773–1843), younger brother to George IV and William IV, followed in 1838 by the Marquess of Northampton (1790–1851), who possessed impeccable aristocratic credentials. Little wonder then, that Britain's scientific community found the government such a powerful sponsor of magnetic investigation.

This informality that surrounded the cultivation of state science helps to explain how it was that such ostensibly governmental projects as the disastrous Franklin Expedition of 1845 or the locating of the south magnetic pole in 1841 fitted within Britain's politically liberal culture: for all the state support, it was the individual instrument maker or the daring naval officer who was celebrated as the great promoter of philosophical knowledge or discoverer of some new natural wonder. As much as it depended on government investment, the BMS was publicized in terms of heroic feats and romanticized voyages of discovery, rather than bureaucratic data collection and state interference. By recognizing the overlapping interests between individuals and the government, between personal ambition and national achievement, and between heroic exploits and state triumphs, this book delivers a nuanced analysis of the nineteenth-century British state's relationship with the natural sciences. This increasing governmental engagement with the cultivation of science was characteristic of a broader association of modern governance with the use of specialist, supposedly objective, knowledge. As much as centralized governance is often seen in Foucauldian terms of discipline and regulation, *An Empire of Magnetism* demonstrates how the exertion of state power could also prove constructive, particularly for the development of scientific culture and the formation of new knowledge of Nature.[38] By cultivating productive relationships with the nation's political and scientific audiences, Fox was

[38] Joyce, *The State of Freedom*, 27.

able to take advantage of the British state's material resources and growing network of overseas territories to advance his own philosophical investigation.

0.3. From Regional Experiment to Global System

How then can we transform what is, at first glance, a book about a distinctly Cornish contribution to a state-sponsored scientific project, into a 'global' history? One solution is to adopt historian Stuart McCook's proposed analysis by 'following', this being the 'deceptively simple method of following something, as it moves around the world'.[39] This book, therefore, charts the journeys of Fox-type dipping needles, as well as the ideas, experiments, social networks, publications, and users that sustained their construction and performance. It begins by exploring, in Chapter 1, the Cornish origins of these instruments on which the BMS would later develop. Cornwall's mining industry and engineering culture sustained investigations into the heat of the mine, with natural philosophers and miners measuring how temperature escalated with subterranean depth. This fits with what historians of science have recently identified as a 'vertical turn' in the ways in which past scientific practitioners have thought about the world around them. Rather than conceiving of spatiality in historic knowledge production in terms of horizontal geographic space, this has involved an emphasis in how notions of height and depth have shaped the natural sciences. Be it the ascent of a mountain or the descent into a mine, thinking vertically about the structure of Nature has been important to the formation of new knowledge of the Earth's structure.[40] Revealingly, Alexander von Humboldt (1769–1859), a contemporary whom Fox venerated and eventually met, owed much to the philosophical value of the mine, having trained at Freiberg's Mining Academy during the 1790s. Along with invaluable opportunities to study subterranean fossilized plants and develop drawing skills for presenting Nature's verticality, in terms of its layering with depth, the mine shaped his conception of flora's change over time.[41] In Cornwall, Fox too took advantage of the philosophical opportunity the mine presented.

Nevertheless, the mine remained of curious epistemological value. Chapter 2 puts this Cornish conception of the globe's interior as a place of heat in the context of wider investigations, across Europe, into electricity and magnetism. In 1820, the Danish experimentalist Christian Ørsted demonstrated that magnetic and electrical phenomena were connected. By causing a magnetized needle to

---

[39] Stuart McCook, 'Introduction', *Isis*, 104/4 (Dec. 2013), 773–6, at 776.
[40] Wilko Graf von Hardenberg and Martin Mahony, 'Introduction—Up, Down, Round and Round: Verticalities in the History of Science', *Centaurus*, 62 (2020), 595–611.
[41] Patrick Anthony, 'Mining as the Working World of Alexander von Humboldt's Plant Geography and Vertical Cartography', *Isis*, 109/1 (Mar. 2018), 28–55, at 32 and 48.

twitch by the passing of a current through a nearby copper wire, Ørsted's sensational experiment initiated over a decade of philosophical excitement, promising a new cosmological vision of the universe in which all natural forces were united. In Britain, Michael Faraday built on this discovery, experimentally determining the principle of electromagnetic rotation in 1821 and electromagnetic induction in 1831, confirming Ørsted's claim that electricity and magnetism could sustain motion. Both of Faraday's iconic experimental arrangements took place in the Royal Institution's laboratory on Albemarle Street, amid London's political and philosophical elites.[42] Faraday's experimental success was due, in part, to his ability to manage apparatus so as to replicate natural phenomena without the contrivance of such operations being evident: in other words, to bring Nature itself into the laboratory and lecture room and let it speak for itself.[43] Faraday's scientific practice, however, was just one approach to the study of the natural world. Far from the Royal Institution, Robert Were Fox also dealt in electromagnetic phenomena amid the industry of West Cornwall. Despite having no laboratory, Fox's social and business connections secured him access to local mines. It was here that he conducted electromagnetic experiments: not in a disciplined laboratory with refined copper wires, but surrounded by labour and dirt, with naturally occurring veins of copper ore.[44] Here, Fox fashioned new understandings over how electricity and magnetism might operate in the Earth's subterranean regions.

Importantly, however, Chapter 2 reveals how the epistemological value of knowledge produced in mines was curious and protean.[45] On the one hand, Fox appeared to be examining Nature unadulterated by human design. Though his experiments were arranged, the results obtained carried unique authority in being taken from the world beyond the laboratory. At the same time, in this place of industry and disruption, Fox's data appeared subject to capricious variables

[42] On Faraday at the Royal Institution, see Sophie Forgan, 'Faraday—From Servant to Savant: The Institutional Context', in David Gooding and Frank A. J. L. James (eds.), *Faraday Rediscovered: Essays on the Life and Work of Michael Faraday, 1791–1867* (Basingstoke: Macmillan Press Ltd, 1985), 51–67.

[43] On Faraday's experimental techniques, see David Gooding, '"In Nature's Schools": Faraday as an experimentalist', in Gooding and James (eds.), *Faraday Rediscovered*, 105–35.

[44] Laboratory facilities were rare in early nineteenth-century Europe. It was not until the mid-nineteenth century that British universities began constructing research-focused physical laboratories, largely for the production of accurate electrical standards for the telegraph industry. See Simon Schaffer, 'Physics Laboratories and the Victorian Country House', in Crosbie Smith and Jon Agar (eds.), *Making Space For Science: Territorial Themes in the Shaping of Knowledge* (Basingstoke: Macmillan Press Ltd, 1998), 149–80, at 150–1; Graeme Gooday, 'Precision Measurement and the Genesis of Physics Teaching Laboratories in Victorian Britain', *British Journal for the History of Science*, 23/1 (Mar. 1990), 25–51, at 29; on the production of electrical standards for the telegraphy industry, see Bruce J. Hunt, 'Doing Science in a Global Empire: Cable Telegraphy and Electrical Physics in Victorian Britain', in Bernard Lightman (ed.), *Victorian Science in Context* (Chicago: University of Chicago Press, 1997), 312–33.

[45] Graeme Gooday asserts that, rather than limit our analysis to the accepted sites of modern scientific activity, we need to take a more inclusive look at 'what kinds of spaces have served as venues of experimental epistemology, whether called laboratories or otherwise'. See Graeme Gooday, 'Placing or Replacing the Laboratory in the History of Science', *Isis*, 99/4 (Dec. 2008), 783–95, at 785.

and incalculable uncertainties that threatened to undermine any conclusions drawn from such experimental work. As Steven Shapin has demonstrated, the validation of scientific knowledge has historically been marked by visual regimes of social and experimental organization.[46] In the Cornish mine, however, this regulation was difficult. This was a site which offered unique encounters of natural phenomena and rich resources for fashioning new understandings of Nature, but it was also a profoundly difficult space to discipline. Unlike the visual organization that Shapin identified as crucial to the production of knowledge, the mine defied such ordering regimes. The audiences of these places of labour were socially diverse, transcending class boundaries and established political hierarchies, including mine owners, captains, and working-class miners. The subterranean world was, therefore, a philosophically troubled world of uncertain epistemological worth.[47] At stake here was the question of what constituted an 'experiment': was it an arrangement to be ordered within a disciplined space, or was it the testing of Nature beyond the laboratory? Whatever the answer, this alterative experimental space placed Fox in a unique position from which to contribute to the broader study of terrestrial magnetism at a crucial moment for this science. As Chapter 2 shows, Fox's work received mixed reactions from wider national and international audiences. Taking the knowledge produced underground, and transferring this to metropolitan centres of science like Paris and London, was a difficult task. This was not just a matter of producing experimental data: Fox developed a series of strategies, including the mobilization of working miners' know-how and an industrious social networking campaign, to secure credibility for his philosophical claims.

While these first two chapters are regionally focused, they demonstrate how Cornish natural philosophers, like Fox and his collaborators, perceived of their work as global in remit: their geological, electrochemical, and magnetic claims were, they believed, applicable to the entire world. Unknown to Fox, however, it was not his philosophical assertions regarding the Earth's subterranean heat and electric properties that were to prove significant, but his instrumentation. Fox's primary concern was to understand terrestrial magnetism as part of a broader

---

[46] Steven Shapin, 'The House of Experiment in Seventeenth-Century England', *Isis*, 79/3, special issue on Artifact and Experiment (Sept. 1988), 373–404, at 374; For comparison, in seventeenth- and eighteenth-century Europe, it was the home, specifically the kitchen with all its mundane culinary instruments, that was the predominant place of philosophical investigation. These private sites were the location of complex physical and social regimes: the organization of space was crucial to the fashioning and reception of new claims over nature. See Alix Cooper, 'Homes and Households', in Katherine Park and Lorraine Daston (eds.), *Cambridge History of Science*, iii: *Early Modern Science* (Cambridge: Cambridge University Press, 2006), 224–37; Anita Guerrini, 'The Ghastly Kitchen', *History of Science*, 54/1 (2016), 71–97; Simon Werrett, *Thrifty Science: Making the Most of Materials in the History of Experiment* (Chicago: University of Chicago Press, 2019), 42–63.

[47] Explored in Jenny Bulstrode, 'Men, Mines, and Machines: Robert Were Fox, the Dip-Circle and the Cornish System', Part III diss., History and Philosophy of Science Department, University of Cambridge, 2013.

explanation for the formation of subterranean mineralogical veins, rich in tin and copper, and on which Cornwall's mining industry depended. To measure the strength and direction of magnetic forces at work within a mine involved the use of robust instruments, capable of suspending magnetic needles, which Fox worked with local miners and mechanics to design and build. Beyond Cornwall, this question of the Earth's magnetic properties was subject to increased philosophical attention. In many respects, the planet exhibited the characteristics of an immense magnet but, since the late fifteenth century, European mariners had known that its qualities varied over time and space, with the north magnetic pole constantly changing location. Since Alexander von Humboldt's study of terrestrial magnetism during his travels in the Americas between 1799 and 1804, and a series of naval expeditions to the Arctic that followed the end of the Napoleonic Wars in 1815, European natural philosophers were increasingly aware of how transient the Earth's magnetism was, and how radically its influence varied around the globe.

The remainder of *An Empire of Magnetism* examines how Fox's Cornish science contributed to an experimental regime that extended far beyond its immediate local contexts, transitioning from the production of regional observations to the establishment of general laws and universal truths. To explain this escalation, the book makes two claims: first, that the key to Fox turning his local experiments into a worldwide investigation was his ability to contribute to the BMS's systematic collection of data. And second, that it was the nineteenth-century British state that delivered the material means for realizing this system. The great difficulty, however, in examining terrestrial magnetism at a global level was of how to measure it and, as Chapter 3 shows, it was to this problem that Fox's magnetic needles promised a solution. During his subterranean experiments, Fox developed instruments to measure terrestrial magnetic phenomena: his dipping needle, the Fox type, was a product of Cornwall's mines, and while some questioned its accuracy, it secured a reputation of robustness. The very characteristics that made Fox's work epistemologically troubling concurrently placed it in a strong position to be incorporated within the greatest philosophical venture of the age. A crucial part of achieving this involved the demonstration that his experimental programme could be expanded beyond Cornwall. Fox's exhibition of his instrument's potential during the magnetic survey of the British Isles in the 1830s and during his own travels across Europe helped in this endeavour.

Above all, it was for the on-board experiments of oceanic voyages that Fox would establish his instrument as the foremost option for magnetic survey work. Chapter 4 analyses how Fox, from Falmouth, directed limited trials in Canada and on transatlantic voyages to achieve this ambition. Between 1838 and 1839, he demonstrated that his devices were capable of obtaining magnetic data on a large scale. At the same time as Fox was crafting a reputation for robustness, a network of the nation's leading scientific elites was campaigning for the government to support a global magnetic survey. By the time this succeeded, Fox's dipping needles were well established as the choice instruments of magnetic survey work. The departure of James Clark Ross's 1839 Antarctic Expedition, equipped with two Fox-type dipping needles, inaugurated over a decade of government-funded naval expeditions, tasked with fulfilling the BMS's aims. But it also represented a triumph for Fox, who had won his apparatus the approval of the BMS's promoters.

Throughout the 1840s, Fox oversaw the use of his experimental practices and apparatus on the Atlantic, Pacific, and Indian oceans, as well as across Canada, central West Africa, and in the polar regions. As Chapter 5 argues, this increasingly took on the character of a system. To measure the Earth's magnetic properties involved the formation of a cohort of trained naval and army officers, working with unified instruments and to routine practices. Throughout the 1830s and 1840s, Fox drafted several manuals on the use of his dipping needles, and provided experimental instruction for officers in his subtropical gardens at Falmouth. In these locations, free from disturbing magnetic influences, Fox delivered guidance on what expeditionary officers could expect from their needles and the Earth's magnetism. Once they departed, these officers corresponded with Fox, updating him on their progress and seeking instrumental advice. They returned broken needles and requested replacement parts that were despatched to colonial ports for collection. Fox also deployed mathematical formulae to adjust incoming magnetic results for human error and instrumental failings, as well as local magnetic and temperature variations. Through manuals, instruction, unified instruments, regulating calculations, a ready supply of new apparatus and repairs, and correspondence, Fox orchestrated a system capable of obtaining magnetic data from around the world, via a network of practitioners of varying experimental experience. By synchronizing ordered experiments over considerable geographical space and long durations of time, Fox was able to exert a systematizing influence over the great scientific enterprise of the day.

This notion of a 'system' as a means of knowledge production builds on histories of nineteenth-century empire that have placed technology at the centre of their analysis, notably in terms of oceanic steamships, railways, and telegraphy.[48]

---

[48] For instance, Britain's oceanic steamship companies represented the most iconic of these engineering enterprises. The operation of shipping lines, capable of ordering travel around the world, depended on a sophisticated network of shipyards, workshops, coaling stations, administrative offices, repair facilities, and hotels, as well as complex ships, consisting of carefully designed hulls, steam engines, effective propulsion, and reliable navigation apparatus. The projectors and engineers of these systems also had to build confidence into their lines, appearing in control of human and technological components while managing relations with the media, stock investors, and officials in government and at the Admiralty. See Crosbie Smith, *Coal, Steam and Ships: Engineering, Enterprise and Empire on the Nineteenth-Century Seas* (Cambridge: Cambridge University Press, 2018), 11 and 14; see also Daniel R. Headrick, *The Tools of Empire: Technology and European Imperialism in the Nineteenth Century* (Oxford: Oxford University Press, 1981), 3–5.

The conception of engineers, inventors, and scientific authorities as builders both of empires and systems develops Thomas Hughes's analytical approach which encouraged a historical focus on the organization of physical, social, institutional, and legal artefacts towards a unified goal. System builders mobilize diverse networks of component parts, including social practices, rules, material objects, and specialist knowledge, to sustain regularity and eradicate uncertainty. As such systems are often intended to operate over time and great geographical space, constant management is required to overcome environmental challenges, as well as those arising from human error. System builders must therefore ensure that those responsible for the system's operation have the freedom to respond at moments of crisis, while being suitably rationalized so as to minimize interference.[49] As Crosbie Smith has argued, the great strength of this historical methodology is that it emphasizes the role of human agents in terms of choices, cultural values, and personal failings: projectors and managers of these systems work to maximize order and eliminate capriciousness from their technological networks.[50] In the nineteenth century, processes of standardization represented an especially important technique for achieving this goal, regulating technical, practical, and social behaviour. Efforts to impose standardized cultures of measurement, map disorganized geographical space, or reduce natural variations to charts, were essential to the effective operation of imperial engineering systems: steam engines were measured in units of power, trains ran on unified gauges of track, and telegraphy cables were organized with precise measures of electrical resistance.

> As Ben Marsden and Crosbie Smith put it, these were attempts to standardise or, better, to make routine, aspects of the skills necessary to give technologies stability and expansive power—often through an interplay of disciplined human agent and the specialist instrumentation that embodied skill[.][51]

These standardizing programmes were central to the orchestration of nineteenth-century engineering empires, promising to instil reliability and predictability over space, and to rationalize the operation of technological systems.

---

[49] Thomas Hughes, 'The Evolution of Large Technological Systems', in Wiebe E. Bijker, Thomas Hughes, and Trevor Pinch (eds.), *The Social Construction of Technological Systems: New Directions in the Sociology and History of Technology* (Cambridge, MA: MIT, 2012), 45–76, at 45–9; see also Thomas Hughes, *Human-Built World: How to Think about Technology and Culture* (Chicago: University of Chicago Press, 2004), 82–7.
[50] Smith, *Coal, Steam and Ships*, 1.
[51] Ben Marsden and Crosbie Smith, *Engineering Empires: A Cultural History of Technology in Nineteenth-Century Britain* (Basingstoke: Palgrave Macmillan, 2005), 9.

This was exactly what Fox envisaged for his contribution to the BMS. By developing standardized magnetic instruments and a routine of experimental practices through disciplining training and instruction, he oversaw the formation of a system capable of obtaining magnetic data from around the world and unified so that such information could be reduced to charts, substantiating philosophical knowledge of the Earth's magnetic phenomena. It was a programme intended to make the experimental practices of those producing magnetic measurements consistent and discipline their behaviour. Each magnetic experiment performed with a Fox dipping needle was, in this way, the end result of an extensive network of social and technical components, managed from Falmouth, and constituting an elaborate system. His standardizing regime allowed him to exert influence over immense geographical space in an age without direct communication: it marked an escalation from local to global scientific investigation. And while Fox drew on Cornwall's resources in the design and construction of his instruments, the key to this transformation was his ability to win government support through his collaboration with Sabine and the BMS. As much as the British state conscripted Fox's scientific resources within its magnetic survey, Fox effectively secured the naval, military, colonial, and financial means required to extend his Cornish observations to the entire world.

For all these systematizing efforts and Fox's promises of robustness, it becomes very clear in Chapter 6 that the BMS remained an immensely fragile enterprise. The production of magnetic data relied on disciplined practices and standardized routines and apparatus, but it was under constant threat of compromised integrity. Human error, technological failings, bad weather, accidents, the wear and tear of travel, local magnetic variations arising from geological anomalies, the influence of iron on-board ships, pirates, and breakdowns in relations between British naval crews and local inhabitants, all risked disturbing magnetic experiments. A concatenation of disruptions conspired to undermine the accuracy of measurements and the authority of the scientific claims that depended on them. Chapter 6 unpacks the ways in which Fox worked to perfect his instruments and techniques, focusing on the successful expeditions of HMS *Samarang* and HMS *Rattlesnake*, as well as the disastrous Franklin Expedition of 1845, but the real story here is the enduringly capricious nature of magnetic survey work. For the BMS's director, Sabine, there was a constant struggle to maintain the credibility of this system of magnetic data collection. Fox's continuing support, instruction, and development of dipping needles were crucial to this project. But this chapter also emphasizes how, by the mid-1840s, the BMS's ambitions had extended far from the limits of formal empire. With expeditions taking Fox types to the Arctic, Korea, the Far East, Japan, and the Pacific, this was a project that had transcended the imperial to become a global science. By the 1850s, however, the BMS had lost much of its early zeal, which becomes apparent in Chapter 7. After a decade of expeditions and observatory measurements, and a wealth of magnetic data, the task now fell to Sabine and his military bureaucrats to transform the thousands of magnetic observations into publications and charts. Here, the data that Fox had done so much to help collect would be organized into maps of the Earth's magnetic properties, effectively transforming the planet into a vast magnetic system. Throughout these chapters, we effectively 'follow' Fox's dipping needles from their deep subterranean origins within Cornwall's mines, to trials in London and the British Isles, and then on voyages and overland expeditions around the world.

1
Steam-Engine Economy and the Heat of the Mine in Early Nineteenth-Century Cornwall

The origins of Fox's magnetic instruments that would prove so crucial to Britain's worldwide magnetic investigation between the 1830s and 1850s, were to be found in the deep mines of early nineteenth-century Cornwall. These were places of tremendous heat and labour. A grim industrial spectacle awaited anyone brave enough to venture down into the deep dark bowels of the Earth. After visiting a Cornish mine in the late eighteenth century, physician James Forbes described how, on arrival, he was equipped with a woollen shirt, trousers, nightcap, jacket, old shoes, and a candle. In these 'greasy and filthy' clothes, pungent, and 'stocked with a republic of creepers', the mine's captain guided him down the central shaft. A hot and damp descent followed, in which Forbes became 'soon wet through, weak from want of proper respiration and half-stifled with the fumes of sulphur'. Eighty fathoms down, he came across a vein of copper ore where miners 'were busied in the process of their miserable employment—with hardly room to move their bodies, in sulphurous air, wet to the skin and buried in the solid rock... pecking out the hard ore by the glimmering of a small candle'.[1] Descending to 130 fathoms, the heat became intoxicating. This would have been the overwhelming sensation in a deep Cornish mine during the intensity of the Industrial Revolution. 'The heat of the mine is excessive', reported Forbes, the 'miners are quite naked when engaged at their work, and they told me that the change of climate and the revolutions of winter and summer were not to be perceived at their great depth'. Nature's seasons wielded no influence here: all was unbearable, constant heat. Weakened by this atmosphere, a visitor feeling faint might find relief from the sprayed water of the pumping steam engine. As Forbes put it, 'when the heat of the mine combines with the fumes of sulphur to fatigue and oppress you', that was the only remedy.[2] Had you wanted to experience the raging industry of late eighteenth- and early nineteenth-century Britain, it was to these claustrophobic

---

[1] Quoted in A. K. Hamilton Jenkin, *The Cornish Miner* (London: George Allen & Unwin, 1962), 102–3; thanks to Michael Carver for directing me to this source.
[2] Ibid. 103.

regions that you would have had to descend to feel the heat of the mine and vapour of the steam engine.

For all the satanic connotations of the mine, this was, throughout the early nineteenth century, a place of extensive scientific investigation and philosophical experiment. Here, deep below ground, knowledge of Nature could be produced, with the labours of miners providing unique opportunities for natural philosophers. Along with revealing the hidden geological structure of the Earth, most mines also boasted a specimen of that most celebrated of all human inventions, the steam engine, which constantly drove the great oscillating cast-iron beams as they pumped water from the depths of the Earth to keep Cornwall's subterranean regions from flooding. What Nature could boast, man could match through Cornish engineering and steam power. From the introduction of the first steam-powered pumping engines in the 1720s until the extremely economical machines of the mid-nineteenth century, Cornwall's mines were at the centre of Britain's coal-fuelled industrialization (Figure 1.1). Around the intensive mining district of West Cornwall, a rich engineering culture developed, tasked with the most urgent challenge of the day: how to make the coal-hungry steam engines, on

Figure 1.1 Botallack Mine during the 1850s, the most iconic of West Cornwall's mines, stretching out for miles beneath the Atlantic Ocean (Author's image, 2023).

which mining relied, economical. Through years of experiment and trial, a network of celebrated Cornish engineers pushed the limits of steam-engine technology to new heights and set global standards for fuel efficiency.

Cornwall was not just an engineering centre. Amid questions over coal consumption and mining, a rich scientific culture developed, committed to the philosophical investigation of Nature, which took shape through a powerful network of societies and individuals eager to extend and promote Cornish knowledge of the natural world across Britain and the wider world. This was inseparable from the region's industrial expansion. As the mine entrepreneur and spokesman of Cornish science John Taylor put it in 1839, reflecting on Cornwall's recent escalation of steam-engine efficiency, more had 'been done by practical experiences than by scientific research'.[3] This chapter explores the relationship between Cornish engineering and the expansion of the county's scientific culture, but it also argues that the region's shared industrial and philosophical concerns shaped new understandings over the internal workings of the Earth. In early nineteenth-century Cornwall, it was the experience of great heat within mines, which miners asserted increased with depth, that particularly aroused the interest of natural philosophers, engineers, and experimentalists. However, beyond Cornwall's mining communities, the existence of heat within the Earth's interior was a controversial question which troubled Europe's leading mathematical and scientific authorities. If the Earth's deepest regions were hot and molten, then this suggested that the planet's outer crust had once also been fluid and had cooled over time. Such a slow process to the Earth's formation clearly had worrying implications for biblically guided accounts of Creation, taken from Genesis. Cornish natural philosophers, taking up claims of underground heat from local miners and captains, ventured deep below the surface to find an empirical answer to this pressing geological concern. They sought to mobilize the mine as a powerful place of experimentation to resolve a leading philosophical controversy but, in doing so, they provided answers informed by the regions' industrial contexts. Their evidence for, and account of, subterranean heat was fashioned through discussions with leading steam-engine builders and mine captains, and drew on Cornwall's engineering culture which was focused on the economic management of steam power. Just as steam engines transformed heat into motive power, so the Earth's subterranean heat acted like a vast boiler: a place of heat and pressure akin to an engine. Unsurprisingly, the engineers leading the county's drive to mobilize high-pressure steam for mining helped shape new theories of the planet's subterranean heat. Fox's earliest philosophical interventions were at the forefront of these discussions.

---

[3] Taylor, quoted in Thomas Lean & Brother, *Historical Statement of the Improvements Made in the Duty Performed by the Steam Engines in Cornwall, from the Commencement of the Publication of the Monthly Reports* (London: Simpkin, Marshall, and Co., 1839), 5.

1.1. Economies of Coal

In March 1768, a disturbing discovery was made on Parys Mountain in Anglesey, which sent a chill throughout Cornwall. Until then, Cornwall's copper mines had known little competition for their valuable ore. On Parys Mountain, however, there lay a deposit of copper so rich, so easily workable and extensive, that Cornwall's monopoly over the metal was instantly lost. For twenty years, with cheap Welsh copper flooding the world's markets, the Cornish economy fell into depression. The problem was that Anglesey copper could be quarried from a depth not exceeding 7 feet but, in Cornwall, centuries of excavation meant that ore had to be extracted from deep below the surface, which required expensive pumping machinery to prevent water flooding excavations.[4] It was steam that drove the pumps which kept Cornwall's mines open; powerful beam engines sustained deep mining. Although the potential of steam power had been known since ancient times, it was not until 1720 that a steam engine was applied practically to pump a Cornish mine. Devonian military engineer Thomas Savery (c.1650–1715) had, in 1698, patented a machine for raising water by the power of fire, but this invention was not commercially viable. Instead, it was Thomas Newcomen's (1663–1729) engine that made steam power a practical proposition, consisting of a cylinder in which a piston moved up and down, connected by a chain to the end of a 'beam' which oscillated on a central pivot, to work a bucket pump. By filling the cylinder with steam, produced in a boiler, the piston was raised. When the cylinder was full of steam, the supply was cut off and cold water sprayed in, forming a vacuum as the steam condensed. The greater weight of the external atmosphere in relation to this vacuum then drove the piston down to the bottom of the cylinder, creating the source of motive power from this difference in pressure between the air inside and outside the cylinder. On turning off the cold-water spray, and opening the supply of steam, which forced its way into the cylinder and pushed the piston back up, the cycle repeated.[5] Newcomen first employed this engine practically in a colliery near Dudley Castle in Staffordshire, pumping the coal mine from 1712. Word of this soon reached Cornwall and, in 1720, a Newcomen engine was in operation at Wheal Fortune.[6] For half a century, Cornwall's engines remained largely unmodified according to Newcomen's original design.

---

[4] D. B. Barton, *A History of Copper Mining in Cornwall and Devon* (Truro: D. Bradford Barton Ltd, 1968), 26; for an economic history, see John Rowe, *Cornwall in the Age of Industrial Revolution* (St Austell: Cornish Hillside Publications, 1993), 114–64.

[5] Donald S. L. Cardwell, *From Watt to Clausius: The Rise of Thermodynamics in the Early Industrial Age* (London: Heinemann, 1971), 15–16.

[6] D. B. Barton, *The Cornish Beam Engine* (Truro: D. Bradford Barton Ltd, 1965), 15–16; for an account of building a Cornish beam engine in 1922, see J. H. Trounson, *Cornish Engines and the Men Who Handled Them* (St Ives: J. H. Trounson & the Trevithick Society, 1985).

As with Anglesey's copper, however, it was the Welsh who again proved the root of Cornwall's troubles. The difficulty with the Newcomen engines was their ravenous consumption of coal, which Cornish miners had to import from south Wales. With Cornwall lacking coal deposits of its own, Welsh mine owners charged a premium for the commodity. Concomitantly, until 1831, the Cornish had to pay tax on seaborne coal. The relentless hunger of the Newcomens and the high cost of fuel was just about manageable with a virtual monopoly over the copper trade, but Anglesey's cheap ore made the extraction of minerals from deep below ground unprofitable. Cornwall's salvation was to be found in Scotland, in the form of an unknown instrument maker at the University of Glasgow. Born at Greenock on the River Clyde, James Watt (1736–1819) was to make a colossal, if sometimes controversial, contribution to Cornwall's mining fortunes, with an invention so radical that it would unleash a revolution in British industry and allow Cornish mines to compete with Anglesey's copper. Watt identified that the problem with the Newcomen was wasted heat. Injecting cold water into the steam-filled cylinder was essential to creating a vacuum, but this cooled the cylinder, making each cycle of the engine inefficient. Such extreme changes in temperature were essential to producing power, but tremendously wasteful: a reduction of temperature resulted in a fall in pressure. Watt recognized that to employ hot condensing water, preserving the cylinder's temperature, would cost an engine power by diminishing the opposing weight of atmospheric air which acted to drive the piston down. In short, Watt could not keep the walls of the cylinder hot, preserve a pool of near-boiling water at its bottom, and simultaneously produce a vacuum inside: the hot water would boil, creating steam, thus destroying the vacuum.[7] For maximum economy, the cylinder had to be kept hot, but for maximum power, it required cooling each cycle. As a cylinder could not be simultaneously hot and cold, Watt's solution was to have two separate cylinders, and it was this idea of a 'separate condenser' that was central to the conception of his new engine. Watt's design consisted of a cylinder with a moving piston, just as in a Newcomen, but with a pipe at the bottom, controlled by a valve, connected to a second cylinder, for condensing. As steam in the first cylinder cooled, the valve was opened, so the steam escaped into the new condensing cylinder, where it was sprayed with cold water (Figure 1.2). Although there was no condensing of steam in the hot cylinder, a vacuum was formed in both, with the second chamber kept cold. Seeing that the atmospheric air pressing the piston down, being cooler than steam, reduced the temperature of the first cylinder's walls, Watt added a connecting pipe to the top of the cylinder, through which steam at normal atmospheric pressure could enter to drive the piston down, while maintaining the cylinder's high temperature. Watt proposed this in 1765, before

---

[7] Cardwell, *From Watt to Clausius*, 46.

Fig. 59. — Machine à balancier de Watt.

v. Tuyau de prise de vapeur ; T, tiroir ; J, cylindre ; H, condenseur ; PE pompe d'épuisement ; WY pompe alimentaire de la chaudière UX pompe d'alimentation de la bâche R ; *p* Z régulateur ; *dd* excentrique ; ABCD parallélogramme ; GM bielle et manivelle ; V volant.

Figure 1.2 Watt's steam engine, with 'H' marking the separate condenser, receiving a spray of cold water (Author's image, 2021).
Source: Amédée Guillemin, *La Vapeur*, (Paris: Hachette, 1873).

securing a patent in 1769 'for a method of lessoning the consumption of steam and fuel in fire engines'.[8]

Through his association with the Birmingham manufacturer Matthew Boulton (1728–1809), Watt was able to turn his engine into a commercial triumph. In correspondence from 1766, and first meeting at Boulton's Soho works in Birmingham in 1768, the industrialist soon recognized the potential of Watt's new engine to power his manufacturing and convinced the Glaswegian instrument maker to move to Birmingham six years later. They finally went into partnership in June 1775, establishing one of the most successful business collaborations in history, with Boulton having succeeded in getting Parliament to extend Watt's original patent in May 1775 on the grounds that it had taken time to transform the design into a profitable application.[9] It was not long before Cornwall's mine proprietors heard news of Watt's engine. In 1776 a Cornish deputation visited Boulton's Soho

---

[8] Richard L. Hills, *Power from Steam: A History of the Stationary Steam Engine* (Cambridge: Cambridge University Press, 1994), 51–5; Cardwell, *From Watt to Clausius*, 48–9; Ben Marsden and Crosbie Smith, *Engineering Empires: A Cultural History of Technology in Nineteenth-Century Britain* (Basingstoke: Palgrave Macmillan, 2005), 48–51; see also Ben Marsden, *Watt's Perfect Engine: Steam and the Age of Invention* (Cambridge: Icon Books, 2002), 43–69.

[9] Hills, *Power from Steam*, 58–9; Marsden and Smith, *Engineering Empires*, 56–7.

works and witnessed Watt's first engine at work in the Staffordshire colliery near Tipton. Boulton and Watt offered to pay to build and install an engine in Cornwall to demonstrate its advantages. In January 1778, parts began arriving at Falmouth for this machine, to be erected at the Ting Tang Mine, although by then another Watt engine had already been in operation at Chacewater Mine since September 1777. Together, these two engines exhibited increased power and reduced coal consumption over Cornwall's existing Newcomen engines. By late 1783, about twenty Watt types were at work in Cornwall and, when his twenty-five-year patent expired in 1800, his engines dominated Cornwall's mining industry.

Across Britain, Watt's machines powered a revolution in steam, from the collieries of South Wales and Newcastle to the booming manufacturing towns of Manchester and Birmingham. Yet nowhere did they take on the same significance that they did in Cornwall. In Britain's coal mines there was obviously no shortage of fuel. Manufacturers, on the other hand, relied largely on water power but, when they did invest in steam, they prioritized reliability, favouring engines that would not break down and threaten production schedules.[10] In Cornwall, however, there was only ever one question on the minds of mine proprietors: how could you get the most work from a piece of coal? West Cornwall was unusual in being an intensive mining district without any native coal reserves and this shaped a constant anxiety over steam-engine economy. Around the mines and Watt engines a unique engineering culture developed, almost completely fixated on the challenge of fuel consumption, or the problem of what engineers called 'duty', this being the measure of a steam engine's performance in terms of the weight of water, in millions of pounds, that it could lift 1 foot high for the consumption of a bushel of coal. Watt had first introduced this concept, which became central to the county's engineering practices.[11] The best duty that Newcomens could achieve was approximately 9,450,000 lb of water per bushel of coal. Watt boasted a duty of 23.4 million in 1779 and, by 1800, claimed to have trebled the duty of even the most economical Newcomens.[12]

Yet Watt's patent often stifled innovation. Most famously, it was the engineering dynasty of the Hornblower family who found Watt's patent an impasse to engine development. Born in Shropshire as the son of the engine builder Joseph Hornblower (c.1696–1762), Jonathan Hornblower (1717–80) constructed Newcomens across Derbyshire, Shropshire, and Wales, before settling in Cornwall at Salem, near Chacewater. Between 1748 and 1775, he constructed over twenty Newcomens, constituting the majority of the forty or so in use in Cornwall at the time. He remained prominent after the arrival of the Watt engines, cooperating in the erection of the first seven or eight of these new engines and serving as Boulton

---

[10] Hills, *Power from Steam*, 68–71 and 113–19.   [11] Lean & Brother, *Historical Statement*, 2.
[12] Barton, *A History of Copper Mining in Cornwall and Devon*, 44; Lean & Brother, *Historical Statement*, 6–8.

and Watt's resident engineer. Of Jonathan's six sons, at least two became respected engineers, the most significant of which was his youngest, Jonathan Hornblower junior (1753–1815), who in 1781 developed and patented a new form of steam engine which would overshadow all his father's achievements, consisting of two cylinders and the expansive use of steam. This 'compound' engine, first built in 1782, had a small cylinder which was kept at high pressure, and a second larger cylinder of a lower pressure. In this process, a fixed amount of steam moved from the boiler into the first cylinder, driving the engine's piston, before then moving into the lower pressure cylinder where it expanded, producing another stroke of the piston. This extracted more work out of a given amount of steam than a traditional low-pressure Watt engine. Unfortunately, the use of high-pressure steam was hard to manage and undermined the physical integrity of Hornblower's machines. From 1791, he began installing these in mines, about nine or ten in total, but they produced few savings over Watt's engines.[13] This programme was brutally curtailed when Boulton and Watt took out legal proceedings against Hornblower, successfully claiming that he had infringed Watt's patent.

When the patent finally expired, Boulton and Watt recalled their resident engineer in Cornwall, William Murdock, back to Birmingham and effectively abandoned their investments in the region. Yet this departure did not initiate an immediate increase in innovation. On the contrary, Cornish engineering stagnated with mine captains content to profit from rising copper prices and the end of royalty payments due for Watt's engines.[14] As one reviewer put it in 1839, with Boulton and Watt's withdrawal, 'the county was left in a manner stripped of men possessing science and experience sufficient to maintain the engines in the improved state to which these eminent engineers had carried them'.[15] But despite this, Cornish engineers gradually stoked up fresh interest in engine duty and the potential economies of high-pressure steam. Although Jonathan Hornblower had failed to successfully manage high-pressure steam, the potential of such expansive working had been well noted among Cornwall's steam-engine builders. With the expiry of Watt's patent came the chance to retry Hornblower's experiment and it was Richard Trevithick (1771–1833), the doyen of Cornish engineering, who would lead this drive for economy.

Born in Illogan in 1771, Trevithick grew up in Camborne where his father had been appointed captain of Dolcoath Mine with responsibility for installing and improving a Newcomen. At school, Trevithick demonstrated a mix of disobedience and mathematical ability, before cutting his teeth amid Camborne and Redruth's intensive practical mining culture, largely under his father's guidance.

---

[13] Cardwell, *From Watt to Clausius*, 78.
[14] Barton, *A History of Copper Mining in Cornwall and Devon*, 45; Barton, *The Cornish Beam Engine*, 28–9.
[15] Lean & Brother, *Historical Statement*, 8.

In 1790, Trevithick's father employed him at Stray Park Mine with a monthly salary of 30 shillings, reflecting that the young engineer was already considerably experienced with steam engines. He remained there until 1792, when he moved to Tincroft Mine at Pool to report on the comparative efficiency of a Watt engine and a high-pressure Hornblower double-cylinder compound. Three years later he introduced fuel-saving measures to an engine in use at Wheal Treasury, securing the post of engineer at the Ding Dong Mine near Penzance in 1797.[16] With Boulton and Watt's departure, Trevithick set about introducing high-pressure steam, developing Jonathan Hornblower's principles. High-pressure engines were harder to maintain, but had the potential to be more efficient. This was an idea that Trevithick's leading patron, Davies Giddy (1767–1839; later Davies Gilbert), eagerly encouraged. Giddy was a firm advocate of high-pressure steam, having earlier promoted Hornblower's work and provided a mathematical expression to theoretically explain the expansive power of steam. Giddy showed that the relation between pressure and volume was proportional to the power generated by a steam engine.[17] In 1802, Trevithick built a small engine at Coalbrookdale with what was, at the time, a staggeringly high pressure of 145 lb per square inch (psi), compared to Watt's machines that usually operated at between 5 lb and 10 lb psi. Following an explosion of a high-pressure pumping engine in Greenwich in 1803, and Watt's warnings that such engines were likely to blow up, Trevithick conceived of a series of safety measures. Working throughout the 1800s, he continued to enhance the performance of Watt's pumping engines until, in 1811, he was declared bankrupt. That same year, he took employment as consulting engineer at Dolcoath Mine, where he installed his first high-pressure Cornish steam engine. The key feature to this was the addition of what was to become known as a 'Cornish boiler', which was in operation from 1812. Trevithick's new boiler had a large cylinder running through it, connected to a fire, which produced steam of increased heat and was an effort to work an engine by pressure alone, without condensation. These boilers were soon employed across Cornwall, enabling the expansive use of steam.[18]

Throughout the 1810s, Trevithick remained at the forefront of an engineering culture which was to make Cornwall the world's centre of economical steam power, employing new boilers and high-pressure steam to drive up the region's duty figures.[19] Along with Trevithick, it was the nefarious Arthur Woolf (bap. 1766, d. 1837) who, following his 1804 patent for a two-cylinder compound expansive engine, led this Cornish endeavour to mobilize the principles of

---

[16] Francis Trevithick, *Life of Richard Trevithick, with an Account of His Inventions*, 2 vols. (London: E. & F. N. Spon, 1872), i. 56–7; on Trevithick's early life, see Anthony Burton, *Richard Trevithick: Giant of Steam* (London: Aurum Press, 2000), 14–37; L. T. C. Rolt, *The Cornish Giant: The Story of Richard Trevithick, Father of the Steam Locomotive* (London: Lutterworth Press, 1960), 27–43.
[17] Cardwell, *From Watt to Clausius*, 79–81 and 83.
[18] Lean & Brother, *Historical Statement*, 151.   [19] Hills, *Power from Steam*, 103.

high-pressure steam. The eldest son of Jane Newton and Arthur Woolf, a carpenter at Dolcoath Mine, young Woolf took an apprenticeship in carpentry at Pool and then worked as a millwright in London. By 1795, he had risen to the position of master engineer, erecting a steam engine at a colliery in Newcastle upon Tyne. Convinced of the advantages of high-pressure steam, he initially took out a patent in 1804 for 'an improved apparatus for converting water and other liquids into vapour or steam for working steam engines'.[20] Three years later, he became a partner in a steam-engine factory at Lambeth, where he revived Hornblower's original compound engine, in which steam passed first into a small high-pressure cylinder, performing a stroke of the piston, before then moving into a second larger, lower-pressured cylinder where the steam expanded and delivered a second stroke of the piston. In 1811, armed with this design, Woolf returned to Cornwall where, for over twenty years, his engines would set new standards for duty. He built his first double-cylinder high-pressure engine, applied to winding, at Wheal Fortune, followed by his first pumping engine, in October 1814, at Wheal Abraham. This initially registered a duty of 34 million, rising to 52.2 million in 1815 and 55.9 million in May 1816, representing an almost 100-per-cent improvement on a low-pressure Watt engine.[21] A second Woolf pumping engine followed in 1815 at Wheal Vor, boasting a duty of 45 million in its first month. Compared with Trevithick's engine at Dolcoath, which registered a duty of 35 million, Woolf's engines appeared unbelievably efficient.

After his compounds at Wheal Abraham and Wheal Vor, Woolf designed two huge engines with single 90-inch cylinders, celebrated as the most powerful machines on earth, for the Consols copper mine in 1821.[22] Woolf's steam empire expanded across Cornwall, with engines installed at Wheal Busy in 1824, Wheal Alfred and Wheal Sparnon in 1825, and Consols in 1827. Consulting engineer to over thirty Cornish mines, resident engineer at Consols from 1818 until 1830, and superintendent of Harvey & Co's engine foundry at Hayle between 1816 and 1833, Woolf was at the centre of the boom in Cornish engine economy.[23] Nevertheless, for all the excitement that surrounded high-pressure steam, it was difficult to keep pistons steam-tight in high-pressure cylinders over long periods of time. Instead, improvements to single-cylinder low-pressure engines of Watt's design proved a more practical approach to enhancing duty.[24] The savings in coal and rise in duty were not as dramatic or immediate as with high pressure, but by

---

[20] E. Carlyle, rev. P. Payton, 'Woolf, Arthur (bap. 1766, d. 1837)', Oxford Dictionary of National Biography, https://www.oxforddnb.com/view/10.1093/ref:odnb/9780198614128.001.0001/odnb-9780198614128-e-29953, 23 Sept. 2004, accessed 30 Nov. 2019; Cardwell, From Watt to Clausius, 155–6.
[21] Hills, Power from Steam, 107; Cardwell, From Watt to Clausius, 156.
[22] Barton, The Cornish Beam Engine, 41.
[23] Carlyle, rev. Payton, 'Woolf, Arthur (bap. 1766, d. 1837)', Oxford Dictionary of National Biography; Barton, A History of Copper Mining in Cornwall and Devon, 46–7.
[24] Barton, The Cornish Beam Engine, 38–40.

the gradual perfection of low-pressure Watt engines, Cornish engineers were able to extract greater efficiency from their engines in a far more sustainable fashion. By the 1820s, Cornish mine captains were returning to steam engines of lower pressure.

What really set Cornwall's steam-engine culture apart from its rivals, however, was not just its focus on economy and experimental trials with high-pressure steam, but also its systematic reporting of duty statistics. This unique practice provided an unprecedented comparison between high- and low-pressure engines. Between 1811 and 1904, these monthly reports provided not only statistics for the duty of each engine, but also their cylinder sizes, dimensions, and the names of their engineers. Originally established as the *Monthly Duty Paper*, it was renamed *Lean's Engine Reporter* in 1838, reflecting the familial character of the publication.[25] Its first editor, Joel Lean (1749–1812), born in Gwennap to Wesleyan parents, had an impeccable reputation for integrity. From 1801 until 1806, Lean managed the Crowan and Oatfield mines in Crowan Parish, where he set about improving the existing dilapidated engines, halving their coal consumption.[26] After this success, he sought to extend his influence for improving steam-engine duty throughout Cornwall's mining regions. As his sons later recalled, he had been 'convinced that it would be attended with much good, if the public generally, but more especially all who were adventurers in mines, had the means of comparing the different engines with each other'.[27] According to his son Thomas Lean (1784–1847), Joel subsequently secured the support of Captain John Davey (1757–1819) of Gwinear, who provided the figures for his engines at Wheal Alfred in the first report, which Lean published in July 1811. A month later, Lean's *Reporter* gave figures for eight engines, including Wheal Alfred's, three at Dolcoath, one at Cook's Kitchen, and one at Wheal Fanny. These engines produced an average duty of 15.7 million and, by December, twelve engines featured in the journal, averaging 17 million. Joel senior died in September 1812, but his sons Thomas and John (1791–1851) took over the business. After three years of reporting, their publication included thirty-two engines with an average duty of 20.6 million.[28] From 1813 until 1815, Trevithick's 63-inch-cylinder engine at Stray Park led duty figures. But Trevithick's lead was broken when Woolf built his first patent compound engine at Wheal Abraham in September 1814. In April 1815, Woolf's second engine, at Wheal Vor,

---

[25] Bridget Howard, *Mr. Lean and the Engine Reporters* (Penryn: Trevithick Society, 2002), 7 and 44; Barton, *The Cornish Beam Engine*, 32.
[26] Howard, *Mr. Lean and the Engine Reporters*, 13–15; Barton, *The Cornish Beam Engine*, 29; Lean & Brother, *Historical Statement*, 8–9.
[27] Lean & Brother, *Historical Statement*, 10.
[28] Barton, *The Cornish Beam Engine*, 32; Howard, *Mr. Lean and the Engine Reporters*, 18; Lean & Brother, *Historical Statement*, 10–11.

initially recorded 45 million, rising to 50 million in June. Compared to the other thirty-eight machines which averaged 20.5 million, Woolf's engines seemed to be making a comprehensive demonstration of the advantages of double-cylinder compound high-pressure steam engines.[29]

Nevertheless, the Cornish practice of duty reporting clearly demonstrated that Woolf's engines could not sustain their high levels of economy. It was hard to keep the two cylinders steam-tight and the cast-iron water-tube boilers in order under such relentless heat and pressure, while the engine's repeated heating and cooling caused cracking.[30] At the same time, Woolf's influence was not quite so benign as Thomas Lean would have his readers believe. In her 2002 *Mr. Lean and the Engine Reporters*, Bridget Howard made a compelling argument that Woolf himself had established the *Reporter* as a publicity coup for his own engines and, with the aid of Thomas Lean, manipulated figures to enhance the performance of his own engines. Given that the finances of the *Reporter* remain unclear, Howard speculated that Woolf probably owned the *Reporter* from 1811 until 1829. With Joel senior's death in 1812, Woolf began manipulating the figures, dominating the two young sons. In May 1815, the *Reporter* stated that thirty-four engines had performed an average monthly duty of 20.4 million, but this did not include Woolf's new 53-inch engine at Wheal Vor, which achieved 49.9 million. Were this added to the calculation of average duty, the figure would be 35 million but, by omitting it, Woolf's engine appeared to be doing more than double the average performance.[31] Similar discrepancies continued until, in October 1827, Woolf claimed his engine at the Consols Mine had produced a duty of 63.7 million, followed by 67 million in November. Local engineers were outraged and demanded a public trial in which the engine achieved just 63.6 million. Critics were quick to point out that trials usually produced figures above those reported monthly, so to have underperformed in this manner was clearly suspicious. Yet despite the *Reporter*'s shady origins, the publication of duty figures was central to Cornwall's engineering culture of steam engine economy and power. The combination of experimental high-pressure steam, engine refinement, and competition that Lean's *Reporter* actively encouraged, put Cornwall at the centre of Britain's steam-powered Industrial Revolution. By the 1850s, thanks to the engines of Newcomen, Watt, the Hornblowers, Trevithick, and Woolf, Cornwall had been at the centre of a dramatic cultivation in steam power for almost a century.

---

[29] Barton, *The Cornish Beam Engine*, 34–7.
[30] Alessandro Nuvolari and Bart Verspagen, '*Lean's Engine Reporter* and the Development of the Cornish Engine: A Reappraisal', *Transactions of the Newcomen Society*, 77 (2007), 167–89, at 174.
[31] Howard, *Mr. Lean and the Engine Reporters*, 21–5.

1.2. Cornwall's Societies and Patrons of Science

Nineteenth-century Cornish engineering was inexorably bound to the business of mining. It depended on understanding the relationship between heat and pressure and how they could be applied for the production of steam power. This engineering activity extended beyond the mine. Trevithick was renowned for his development of high-pressure steam engines for mine pumping, but the expansive use of steam also provided opportunities for steam-powered locomotion. High-pressure engines meant that Watt's separate condenser could be done away with and, given that a high-pressure cylinder was smaller than a normal low-pressure cylinder but gave the same power, it was practical to mount such an engine on wheels.[32] Trevithick's first successful model locomotive was complete by 1797, running on high-pressure steam and, with Gilbert's support, he built a full-scale road engine in Camborne, displaying it publicly for the first time on Christmas Eve 1801. He continued to experiment with steam-powered locomotion until 1808, when successive financial failures exhausted his interest in the project.[33] Trevithick was not the only Cornishman interested in the potential high-pressure steam for locomotive power. His steam-powered transportation experiments proved inspirational for aspiring young engineers, perhaps most famously for the gentleman inventor Goldsworthy Gurney (1793–1875) who, as a child, met Trevithick and witnessed his steam locomotive experiments three years later. After schooling at Truro Grammar, young Gurney took an apprenticeship in medicine at Wadebridge and moved to London in 1820.[34] Having secured appointment as lecturer of chemistry and natural philosophy at the short-lived Surrey Institution, where he performed experiments on heat, gases, and electricity, Gurney published *The Elements of Science* in 1823, before turning to the application of steam power to road transportation during the late 1820s. Like Trevithick, Gurney abandoned his 'steam drags', consisting of an engine coupled to a carriage, on financial grounds, when Parliament legislated for heavy tolls on such transportation to protect the horse-coach industry.[35] Nevertheless, such entrepreneurialism marked Cornwall out as a place of bustling engineering innovation.

Importantly, however, Cornwall's mining industry and development of steam-engine technology encouraged not only a rich engineering culture, but also considerable scientific activity. Along with the mastery of steam and fuel economy, access to the deepest parts of the Earth's interior, a depressed agricultural sector, and increasingly global trade links via the port of Falmouth provided opportunities

---

[32] Cardwell, *From Watt to Clausius*, 84.
[33] Ben Marsden and Crosbie Smith, *Engineering Empires: A Cultural History of Technology in Nineteenth-Century Britain* (Basingstoke: Palgrave Macmillan, 2005), 135.
[34] Dale H. Porter, *The Life and Times of Sir Goldsworthy Gurney: Gentleman Scientist and Inventor, 1793–1875* (Bethlehem: Lehigh University Press, 1998), 25, 29, and 35.
[35] Ibid. 71–132; Marsden and Smith, *Engineering Empires*, 132–3.

to investigate geology and chemistry, and to acquire exotic botanical specimens from ships returning from all over the world. In 2010, Simon Naylor emphasized the importance of location and local setting to the formation of knowledge in Cornwall. As he has shown, 'Cornish science' was shaped by the region's industry, its coast and agriculture, and its temperate climate in which tropical plants could be grown outdoors.[36] This expanding scientific culture took shape in a series of philosophical societies throughout the county. Formed in 1814, the Cornwall Geological Society in Penzance claimed to be the second oldest organization in the world dedicated to the study of geology, preceded only by the Geological Society of London, established in 1807. Keen to emulate its London rival, the Cornish society soon acquired royal patronage of its own, becoming the Royal Geological Society of Cornwall (RGSC), and set up its own geological museum.[37] The *Transactions of the Royal Geological Society of Cornwall* provided a respected venue for local scientific papers. Its first printed geological map, of the Lizard District, appeared in 1818. Though a socially elite society, the focus of the majority of papers published in the RGSC's *Transactions* had strong industrial connotations.[38]

The foundation of the Cornwall Philosophical Society in 1818 represented the region's second scientific society. Originally meeting in the County Library at Truro, with the intention of cultivating geological, mining, engineering, and agricultural knowledge, it soon specialized in natural history and antiquarianism, becoming the Royal Institution of Cornwall (RIC) in 1821. Just as socially elite as the RGSC, the RIC secured Admiral Edward Pellew (1757–1833) as its first president, followed by Penryn's Whig MP, Charles Lemon (1784–1868), who remained in this position between 1818 and 1830.[39] Along with the organization of these prestigious institutions, more specialist associations constituted a broader cultivation of the natural sciences, including zoology, botany, and meteorology, such as Penwith Agricultural Society and the Penzance Natural History and Antiquarian Society. These new organizations took advantage of local know-how surrounding questions of Nature. For instance, in 1832 the formation of a horticultural society drew on the county's flourishing gardening culture, especially

---

[36] Simon Naylor, *Regionalizing Science: Placing Knowledge in Victorian England* (London: Pickering & Chatto, 2010), 7–10.

[37] Ibid. 30–1.

[38] Ibid. 59–71; Simon Naylor, 'Geological Mapping and the Geographies of Proprietorship in Nineteenth-Century Cornwall', in David N. Livingstone and Charles W. J. Withers (eds.), *Geographies of Nineteenth-Century Science* (Chicago: University of Chicago Press, 2011), 345–70; Paul Weindling, 'The British Mineralogical Society: A Case Study in Science and Social Improvement', in Ian Inkster and Jack Morrell (eds.), *Metropolis and Province: Science in British Culture, 1780-1850* (London: Hutchinson, 1983), 120–50, at 126–32. This was equally apparent for the Geological and Polytechnic Society of the West Riding of Yorkshire, established in 1837 amid Yorkshire's coal mines; see Jack Morrell, 'Economic and Ornamental Geology: The Geological and Polytechnic Society of the West Riding of Yorkshire, 1837–1853', in Inkster and Morrell (eds.), *Metropolis and Province*, 230–56.

[39] Naylor, *Regionalizing Science*, 31–2.

subtropical gardens, like Trebah, Glendurgan, and Trelissick. In May 1832 the industrialist-politician John Vivian (1785–1855) of Pencalenick called together an initial gathering in Truro, with the first general meeting of what became the Royal Horticultural Society of Cornwall held a month later. Although short-lived, collapsing in 1861 due to a decline in finances and interest, its early success in the 1830s demonstrated the very real interest in botany among Cornwall's leading families.[40]

The influence of this flourishing scientific culture extended beyond Cornwall. Thanks to an industrious network of natural philosophers, engineers, mathematicians, and experimentalists, supported by powerful political patrons, the county was increasingly well represented in the natural sciences across Britain, especially in London. Perhaps most famously, it was Humphry Davy (1778–1829) who made a name for Cornish natural philosophy in the capital. The son of a Penzance woodcarver, Davy pursued his private interests in chemistry, anatomy, and botany while apprenticed to a local surgeon from 1795. After reading French chemist Antoine Lavoisier's (1743–94) *Traité élémentaire de chimie*, published in 1789, Davy began his own chemical researches by investigating the properties of heat, melting ice cubes through friction. In 1799, he won fame for his attack on Lavoisier's description of heat as a substance, instead contending that it was a motion. Subsequently taking a post as a medical assistant in Bristol, Davy experimented on the effects of oxygen and nitrogen, or 'laughing gas', writing an account in 1800 where he advocated its use as an anaesthetic. The following year his investigations into electricity and battery construction led to his first publication in the prestigious *Philosophical Transactions of the Royal Society*. These early successes earned Davy an invitation to lecture on the chemistry of tanning at London's Royal Institution on Albemarle Street. Ideally suited to the institution's early focus on agricultural improvement, by 1802 he was lecturing more broadly on chemistry. A brilliant speaker, he was awarded the Royal Society's Copley Medal for his course on tanning and became a secretary and an editor of its *Philosophical Transactions*. Davy thus secured honours and position at London's two leading scientific venues. Throughout the 1800s, he continued experimenting at the Royal Institution, using electricity to isolate potassium from caustic potash. This use of electrolysis to split compounds into their comprising elements allowed him to identify new metals, including sodium, magnesium, calcium, barium, and strontium, earning him a reputation as the 'Newton of chemistry'. Knighted in 1812, his eminence expanded rapidly until, in 1820, he was elected president of the Royal Society, marking the culmination of a precocious scientific career.[41]

---

[40] A. Pearson, 'The Royal Horticultural Society of Cornwall', *Journal of the Royal Institution of Cornwall*, NS 7/2 (1974), 165–73, at 165–7; on botany and meteorology, see Naylor, *Regionalizing Science*, 101–24 and 149–70.

[41] D. Knight, 'Davy, Sir Humphry, baronet (1778–1829)', *Oxford Dictionary of National Biography*, https://www.oxforddnb.com/view/10.1093/ref:odnb/9780198614128.001.0001/odnb-9780198614128-e-7314, 23 Sept. 2004, accessed 1 Dec. 2019; on his elemental discoveries, see Humphry Davy,

For his meteoric rise, Davy owed much to his Cornish social connections. Early on, he had secured the patronage of the county's leading promoter of up-and-coming natural philosophers, Davies Gilbert. Born at St Erth, the son of a curate, Gilbert attended Penzance Grammar School, before studying mathematics in Bristol and at Pembroke College, Oxford, where he matriculated in 1785. Originally Davies Giddy, but taking the name Gilbert on his marriage to Mary Ann Gilbert (1776–1845) as a condition for inheriting her family's estates in Sussex, he went on to champion several promising young Cornishmen. Gilbert had promoted Jonathan Hornblower's work between 1791 and 1800 and Trevithick's from 1814 until 1833, along with Davy's from 1797.[42] Gilbert's support of Davy helped win him presidency of the Royal Society in 1820. Gilbert had been the favoured successor of the incumbent president, Joseph Banks (1743–1820), but turned down the nomination in favour of his protégé. However, he had a second opportunity seven years later. Although Davy was re-elected in October 1826, his presidential address was a strained affair. Two months later, he had a stroke and resigned the following summer, before retiring to Europe. In Rome he suffered a second, this time fatal, stroke in 1829.[43] With Davy having vacated the presidency, Gilbert finally ascended to the chair, holding the position for three years. The election of two successive Cornish presidents, Davy and Gilbert, marked a considerable coup for the region's scientific reputation: between 1820 and 1830 Cornish science effectively had two immensely influential patrons at the very centre of London's scientific community. At the same time, the county also benefited from an extensive network of scientifically interested politicians, eager to promote Cornish science in the capital. Along with Gilbert, Charles Lemon, William Molesworth (1810–55), and Richard Vyvyan (1800–79) were all Cornish MPs who, at various times, advanced the careers of local natural philosophers and engineers. Indeed, local MPs were particularly important for the movement of philosophical knowledge from the regional to the national, being present both in their constituencies and at Westminster. They were the ideal facilitators in communicating knowledge of Nature attained from around the British Isles, notably Cornwall, to London.[44] With sympathetic MPs at Westminster and presidents at the Royal Society, Cornish science had a rich cohort of patrons in

'Electro-Chemical Researches, on the Decomposition of the Earths; with Observations in the Metals Obtained from the Alkaline Earths, and on the Amalgam Procured from Ammonia', *Philosophical Transactions of the Royal Society*, 98 (1808), 333–70.

[42] For his patronage of Hornblower, Trevithick, and Davy, see A. C. Todd, *Beyond the Blaze: A Biography of Davies Gilbert* (Truro: D. Bradford Barton Ltd, 1967), 57–78, 96–111, and 23–31; D. Miller, 'Gilbert [formerly Giddy], Davies (1767–1839)', *Oxford Dictionary of National Biography*, https://www.oxforddnb.com/view/10.1093/ref:odnb/9780198614128.001.0001/odnb-9780198614128-e-10686, 23 Sept. 2004, accessed 1 Dec. 2019.

[43] D. Knight, 'Davy, Sir Humphry, baronet (1778–1829)', *Oxford Dictionary of National Biography*, https://www.oxforddnb.com/view/10.1093/ref:odnb/9780198614128.001.0001/odnb-9780198614128-e-7314, 23 Sept. 2004, accessed 1 Dec. 2019.

[44] Argued in Edward Gillin, 'Cornish Science, Mine Experiments, and Robert Were Fox's Penjerrick Letters', *Notes and Records of the Royal Society*, 76/1 (Mar. 2022), 49–65.

the capital. These networks for the patronage of early nineteenth-century Cornish science delivered powerful resources for advancing the county's philosophical investigations.

1.3. Pressure, Steam, and Subterranean Heat

Cornwall's industrial preoccupation with the mine and the extraction of work from coal shaped new understandings over the workings of the Earth, particularly in terms of its internal heat. In 1812, Joel Lean the younger (1779–1856) took out a patent with a young Quaker from Falmouth by the name of Robert Were Fox (Figure 1.3). Although not specifying precisely the nature of their invention, their fourteen-year patent announced that they had 'invented certain Improvements on Steam Engines and the apparatus needful or expedient to be used with the same' and that they were 'the first and true inventors' of these measures.[45] Considering the timing of this invention, just a year after Woolf's return to

Figure 1.3 Robert Were Fox (Author's image, 2021).

[45] KKW, X114/1, 'Patent to Robert Were Fox and Joel Lean, Improvements on Steam Engine' (1812).

Cornwall and coinciding with Trevithick's application of high-pressure steam at Dolcoath Mine, Fox and Lean's improvements almost certainly involved the manipulation of the heat and pressure within a steam engine. Given the Leans' prominence with the *Reporter* and collection of data on duty, Joel junior was at the very centre of Cornwall's recently instigated practice of duty reporting and would have been profoundly aware of Trevithick and Woolf's development of high-pressure steam. Although the precise nature of the patent is unclear, one obituary of Fox described how 'Watt's improvement in the steam-engine…drew the attention of active minds to the study of the conditions of steam at various temperatures, and, in 1812, we find Mr. Fox associated with Joel Lean, making costly experiments on the laws regulating the elasticity of high pressure steam; and subsequently he aided Trevithick in his progress with his ingenious machines'.[46] In 1827, on receiving a copy of his Royal Society paper on the steam engine from his patron, Gilbert, Fox recalled that at the time of his patent he had been experimenting on high-pressure steam power. He had been 'convinced by many experiments on vapour, which I made in 1812, that its density, when in contact with the liquid producing it, increased nearly in proportion to its elasticity, and that the advantages of using high pressure steam, is almost entirely mechanical'.[47]

With his mutual interest in natural philosophy and steam power, Fox was typical of Cornwall's industrial science. Born in 1789 in Falmouth, he had worked with his father in the family's ship-brokering, mining, smelting, and foundry enterprises.[48] Excluded from a university education due to his Nonconformist Quaker faith, Fox's upbringing amid the natural beauty of Cornwall cultivated 'a love of nature', while his mining connections sustained an occupation with practical knowledge from a young age.[49] Natural philosophy was very much a family concern. His brother, Charles Fox (1797–1878), was engaged in scientific research, as well as being a partner in the family shipping business and general manager of Perran Iron Foundry at Perranarworthal from 1825 until 1842, while

---

[46] Anon., 'Mr. Robert Were Fox', *The Athenaeum*, 2597 (4 Aug. 1877), 153–5, at 153.

[47] Robert Were Fox to Davies Gilbert, 27 July 1827; thanks to Rachel Morin and Hilary Watson for the loan of this letter, currently in the author's possession.

[48] Susan E. Gay, *Old Falmouth: The Story of the Town from the Days of Killigrews to the Earliest Part of the 19th Century* (London: Headley Brothers, 1903), 149–60; P. Payton, 'Fox, Robert Were (1754–1818)', *Oxford Dictionary of National Biography*, https://www.oxforddnb.com/view/10.1093/ref:odnb/9780198614128.001.0001/odnb-9780198614128-e-42083, 23 Sept. 2004, accessed 5 Dec. 2019; Anon., *People from Falmouth* (London: Books LLC: 2010); for a business history of the Fox family, see Charles Fox, *On the Brink: The Story of G. C. Fox and Company: A Quaker Business in Cornwall through Eight Generations* (London: Zuleika, 2019), 155–77; the Fox family's personal and business papers are held at Kresnow Kernow (the Cornwall Centre) in Redruth and the Library of the Royal Polytechnic Society of Cornwall in Falmouth.

[49] J. H. Collins, *A Catalogue of the Works of Robert Were Fox, F.R.S, &c. Chronologically Arranged, with Notes and Extracts, and a Sketch of His Life* (Truro: Lake & Lake, 1878), 3–4; D. Crook, 'Fox, Robert Were (1789–1877)', *Oxford Dictionary of National Biography*, https://www.oxforddnb.com/view/10.1093/ref:odnb/9780198614128.001.0001/odnb-9780198614128-e-10062, 23 Sept. 2004, accessed 5 Dec. 2019. For Robert Were Fox's scientific papers, see Library of the Royal Polytechnic Society of Cornwall, Falmouth, Fox family archives: Robert Were Fox, 1789–1877'.

Figure 1.4 Robert Were Fox's subtropical garden at Rosehill, located centrally in Falmouth (Author's image, 2020).

brother Alfred Fox (1794–1874) established a subtropical garden at Glendurgan. Robert Were Fox's own tropical gardens at Rosehill (Figure 1.4) and Penjerrick (Figure 1.5) benefited from Falmouth's tropical microclimate, becoming celebrated sites of experimentation on botanical acclimatization. In 1814, he had married Maria Barclay (1785–1858), a Quaker from Surrey, with whom he had two daughters, Anna Maria (1816–97) and Caroline (1819–71), and a son, Barclay (1817–55). They too shared this love of Nature, with Barclay cultivating botanical interests of his own and both daughters becoming leading patrons of Cornish science.[50] The Fox family's scientific endeavours went well beyond the creation of botanical wonders. They were also extremely well connected in local mining circles, including with Captains William and John Davey of Wheal Alfred, as well as Woolf. In the spring of 1816, for instance, Fox, his father, and Alfred toured Devon's manganese mines. Leaving Falmouth on 2 April, they travelled up to

---

[50] Collins, *A Catalogue of the Works of Robert Were Fox*, 5–6; Anon., 'Memorial of Maria Fox, Wife of Robert Were Fox, of Falmouth, England', *Friends' Review: A Religious, Literary and Miscellaneous Journal*, 12/22 (Feb. 1859), 337; Anon., *People from Falmouth* (London: Books LLC, 2010); V. Chancellor, 'Fox, Caroline (1819–1871)', *Oxford Dictionary of National Biography*, https://www.oxforddnb.com/view/10.1093/ref:odnb/9780198614128.001.0001/odnb-9780198614128-e-10019, 23 Sept. 2004, accessed 5 Dec. 2019. Fox and his wife are both buried at the Quaker Burial Ground in Budock, near Falmouth.

Figure 1.5 Located just outside Falmouth, Fox's Penjerrick garden became a place of dense subtropical growth and rich botanical specimens (Author's image, 2020).

Redruth, before overnighting at St Austell. At their inn, they met Captains Hodge, Teague, and Woolf, and spent the evening with this eminent mining company before continuing their journey to Devon the following morning.[51] The Fox family, in fact, boasted significant investments in mines employing Woolf's engines. In April 1814, Fox received a letter from his father, explaining how he was about to purchase a 1/58th share of Wheal Vor which, in 1815, was where Woolf had built his second high-pressure steam engine.[52] In December 1815, Joseph Tregelles Price (1784–1854), a managing partner in the Fox family's iron foundry at Neath Abbey, testified that Woolf's compound engines performed the same duty as a Watt engine for half the consumption of coal. Price was, unsurprisingly, employed at the ironworks that were building Woolf's engines.[53] The Foxes were, therefore, business associates of the county's foremost high-pressure steam engineer.

Joel Lean and Robert Were Fox's 1812 patent made evident that they understood the principles of steam engine technology and the expansive power of steam, especially concerning the relationship between heat and pressure. But in

---

[51] KKW, FOX/B/2/63: 'Part of A. Fox's Diary' (1816).
[52] KKW, FOX/B/2/60: R. W. Fox senior to R. W. Fox junior (1 Apr. 1814).
[53] Howard, *Mr. Lean and the Engine Reporter*, 27.

the early nineteenth-century Cornish mine, the steam engine was not the only source of heat. Miners had long known that as they descended deep below the surface of the Earth, a perceptible increase in temperature was encountered. This fact was not lost on Cornwall's engineers; dealing with questions of heat and pressure on a daily basis, they were naturally curious at the Earth's thermal qualities. Early in 1815, Joel Lean had emphasized to Fox the philosophical importance of understanding subterranean heat, suggesting that he conduct scientific research into the relationship between depth and temperature. Fox later recalled how Lean was convinced 'that the high temperature observed in our mines existed in the earth itself, increasing with the depth; and shortly afterwards his brother Thomas Lean, at our joint request, kindly made many experiments in Huel [Wheal] Abraham Copper Mine, of which he was the manager, in order to test the correctness of this view'.[54] Thomas Lean descended the mine's shaft armed with a thermometer, taking readings of temperature as he went deeper. He found that at the surface his thermometer registered 59°F in the shade. At 20 fathoms below the surface this increased to 64½°, rising to 68½° at 100 fathoms, 70° at 160, and 79° at 190 fathoms, this being the bottom of the mine. After these measurements, made in June 1815, he repeated his observations in December, finding 50° at the surface, 57° at 20 fathoms, 66° at 100, and 78° at 200.[55] Lean's figures seemed to confirm what Cornish miners had long known to be true. In the same year, Fox asked that similar observations be made at Dolcoath Mine, which John Rule junior subsequently provided.

Thomas Lean published his results in 1818 in the *Philosophical Magazine*, before Fox combined them with his own measurements and presented them to the RGSC in September 1819 and October 1820. Published in the society's *Transactions* in 1822, these two papers appeared directly after an introduction in which geologist John Hawkins (1761–1841) outlined some of the advantages which Cornwall possessed for the study of geology. An owner of considerable mining property and graduate of Trinity College Cambridge, Hawkins drew attention not only to the rich ore and mineral deposits in the county, but also to the exhibition of heat that could be experienced deep below the Earth's surface. Invoking the language of the steam engine, Hawkins described how this was a geological landscape forged through heat and pressure.[56] Fox's two RGSC papers illustrated Hawkins's point perfectly. Throughout 1815, he had collected observations from the mines of Wheal Abraham, Dolcoath, Cook's Kitchen, Tincroft, and

---

[54] Robert Were Fox, 'Report on Some Observations on Subterranean Temperature', *Report of the Tenth Meeting of the British Association for the Advancement of Science; Held at Glasgow in August 1840* (John Murray, 1840), 309–19, at 309.

[55] Thomas Lean, 'On the Temperature of the Mines in Cornwall', *Philosophical Magazine: A Journal of Theoretical, Experimental and Applied Physics*, 52 (1818), 204–6.

[56] John Hawkins, 'On Some Advantages Which Cornwall Possesses for the Study of Geology, and on the Use Which May Be Made by Them', *Transactions of the Royal Geological Society of Cornwall*, 2 (1822), 1–13, at 7.

the United Mines. His initial results suggested a similar expansion of heat with depth as found in Thomas Lean's observations.[57] He elaborated on this general observation in his second paper, presenting more results thanks to the assistance of Captain Thomas Lean at Wheal Abraham, John Rule junior at Dolcoath, and Captain William Teague of Treskerby. Fox also thanked Joel Lean for drawing his attention to the subject. Comparatively, Fox had found the temperature in Dolcoath at 130 fathoms to be 63°, 75° at 160 fathoms at the United Mines, and 64° at 80 fathoms in Ting Tang Mine, rising to 68° at 110 fathoms.[58] He recognized that workmen, candles, and blasting might have an influence on the temperature of the mine, but planted his thermometers 6 to 8 inches into the rock and packed them with soil to help keep warm air from corrupting the readings. Fox believed his findings to be convincing evidence that the Earth's subterranean regions possessed great heat, concluding that while the sun was 'a powerful agent in producing evaporation', it was likely that the Earth's 'internal heat has also a share in this very curious and important part of the economy of nature'.[59]

Fox and the Leans were, in fact, wading into what was, in early nineteenth-century Europe, an urgent philosophical concern with implications over how the Earth had formed: was the globe's interior hot or cold, and what were the ramifications of its temperature for the deep history of the Earth? While Cornwall's deep mines provided a unique experimental space to observe phenomena related to this question, it was in Paris that it received the greatest attention. In 1820, the French mathematician Joseph Fourier (1768–1830) proposed that the present stability of the Earth was the result of gradual cooling and that the conduction of heat through the Earth's interior was so slow that after an initial period of rapid cooling at its surface, resulting in the Earth's solid crust, there might still be great heat trapped underground. Were Fourier's central heat to be proven, as residue evidence of the planet's slow cooling, his theory over the Earth's geological formation would be established. Fourier continued to expand his theory, publishing his influential 'Remarques générales sur les températures du globe terrestre et des espaces planétaires' in the *Annales de chimie et de physique* in 1824, again arguing that the Earth's interior was a source of immense heat. Pierre Louis Antoine Cordier (1777–1861), a geologist and mineralogist to Napoleon Bonaparte's ill-fated expedition to Egypt in 1798, took up Fourier's physical explanation of a cooling Earth. Were the principle of increased temperature within deep mines established, this could be taken as empirical evidence in support of Fourier's

---

[57] Robert Were Fox, 'On the Temperature of Mines', *Transactions of the Royal Geological Society of Cornwall*, 2 (1822), 14–18, at 14 and 18.
[58] Robert Were Fox, 'On the Temperature of Mines', *Transactions of the Royal Geological Society of Cornwall*, 2 (1822), 19–28, at 19–20.
[59] Ibid. 27.

mathematical understanding of the Earth's formation.[60] While Paris was an intellectual centre with a strong mathematical tradition and the ideal location for the production and reception of Fourier's theoretical work, the observable evidence to support his hypothesis required mines of great depth and, in the 1810s and 1820, it was Cornwall that boasted the most extensive excavations in which to experiment. Fox and the Leans were mobilizing their access to the world's deepest regions to contribute to one of Europe's leading philosophical dilemmas. For all that Cornwall appeared somewhat isolated from mathematical and academic centres like Paris and Cambridge, it provided unique spatial resources for the investigation of Nature.

1.4. The Earth's Great Heat

Unsurprisingly, given their philosophical and geological implications, Fox's papers on the temperature of mines did not escape critical evaluation from Cornwall's scientific community. Fox helped persuade several prominent natural philosophers of the existence of heat beneath the surface of the Earth, including the Prussian polymath Alexander von Humboldt (1769–1859), who met Fox in 1814 while the Cornishman was honeymooning in France. Humboldt had originally rejected the claims of Saxon miners that temperature increased with depth.[61] However, Fox's investigations aroused significant opposition. Appearing in the RGSC's 1822 *Transactions*, a local physician, John Forbes, contributed his own observations on subterranean heat, drawing attention to the warming influence of labouring miners on a mine's temperature.[62] He produced figures for the temperature at different depths in six mines, including Wheal Abraham, Ding Dong, and Botallack. Unlike Fox, he also gave statistics for the number of men at work in each mine and the monthly expenditure on candles and gunpowder. For instance, in July 1819, Botallack Mine gave a temperature of 62° at its bottom, 570 feet below the surface, with 150 men at work and a monthly consumption of 1,200 lb of candles and 600 lb of gunpowder. When Forbes returned to repeat his measurements in May 1822, the mine had been excavated down to 672 feet, where he found a temperature of 67°. In comparison, Wheal Abraham, at 1,440 feet, gave 71° with 4,800 lb of candles and 1,500 lb of gunpowder used per month.[63] Forbes's results demonstrated a clear increase in heat with depth, but he emphasized

---

[60] Martin J. S. Rudwick, *Worlds Before Adam: The Reconstruction of Geohistory in the Age of Reform* (Chicago: University of Chicago Press, 2008), 124–6; Joseph Fourier, 'Remarques générales sur les températures du globe terrestre et des espaces planétaires', *Annales de chimie et de physique*, 27 (1824), 136–67.

[61] Collins, *A Catalogue of the Works of Robert Were Fox*, 3.

[62] John Forbes, 'On the Temperature of Mines', *Transactions of the Royal Geological Society of Cornwall*, 2 (1822), 159–215, at 162.

[63] Ibid. 175 and 191.

the significant influence miners had on temperature, through their animal heat, candle combustion, friction from digging, and use of gunpowder. Wheal Vor, for example, had a monthly expenditure of 3,000 lb of candles and 3,500 lb of gunpowder, along with 548 men at work. Clearly, Forbes argued, such a huge production of 'artificial caloric' would contribute to the mine's recorded temperature of 67°. Reflecting on Fox's results, Forbes claimed that his own records concurred with the claim that there was heat deep below the Earth's surface, increasing with depth, but warned that Fox had overestimated the extent of this rise, having failed to deduct between 6° and 8° which could be accounted for by artificial causes of heat.[64]

This qualification aside, Forbes agreed with Fox's suggestion that heat was conveyed from the Earth's innermost regions by the movement of steam, formed under heat and pressure. He thought it likely that there was a 'perpetual flow of aqueois vapour from the centrical part of the earth to the surface'.[65] Fox's assertion rested on the observation of high-pressure jets of water which could be found flooding into mines, which he reasoned resulted from the condensing of subterranean vapours. In this way, the Earth was like a high-pressure steam engine, with steam formed under pressure in the Earth's hot interior, before forcing its way towards the surface, and condensing in the cooler regions through which mines were excavated. This understanding of the movement of steam from an area of high pressure to a cooler one of low pressure, clearly invoked the high-pressure Cornish steam engines of Woolf and Trevithick, with which Fox was well acquainted. It was all too easy for Cornish natural philosophers and engineers like Fox and Forbes to think of the Earth as a vast steam engine, with a boiler at its core.[66]

Not all found such analogies so compelling. Throughout 1822, Matthew Moyle, a physician from Helston, took up Forbes's concerns over the influence of miners on temperature to completely reject Fox's suggestion of a great source of subterranean heat. Moyle had seen that Fox had sent his observations on subterranean heat to the *Annales de chimie et de physique* and that the French journal had published Fox's results along with an extract of Fourier's geometrical researches on the Earth's heat. Although Fox did not directly invoke Fourier's work, his observations were presented to French audiences as supporting evidence for the mathematician's theory. In Paris, Fox's systematic observations of ten different mines, including an increase from 50.18° at 10 fathoms to 82.04° at 240 fathoms in Dolcoath, appeared compelling. But, responding in the *Annals of Philosophy*, Moyle claimed to have performed similar experiments at Dolcoath with different

---

[64] Ibid. 202 and 208.   [65] Ibid. 213.
[66] As argued in Jenny Bulstrode, 'Men, Mines, and Machines: Robert Were Fox, the Dip-Circle and the Cornish System', Part III diss., History and Philosophy of Science Department, University of Cambridge, 2013, 21. I am grateful to Jenny Bulstrode for providing me with a copy of her thesis.

results. As to Fox's claim that he had found a jet of water of 80.04° at 240 fathoms below the Earth's surface, Moyle thought this little basis for a theory of the entire globe. Moyle had found a rise in temperature with depth at Dolcoath, but that was in the working part of the mine, where the labours of miners had clearly interfered with the mine's temperature. With 'the presence of so numerous a body of workmen in different parts of the said mine', often over 400 miners, and most deployed at the bottom of a mine, the rise in temperature was hardly surprising.[67] Fox's error was in not recording the temperature in the parts of mines remote from active workings, as Moyle had done at Wheal Unity. In a gallery at a depth of 150 fathoms, unworked for a year, he found a temperature of just 65°, compared to 74° in a working section at the same depth. Such trials produced similar results at Wheal Trumpet and the Old Trevenen Tin Mine.[68] A month later, Fox responded that his experiments had been performed on mineral veins and at cross-levels away from workmen. He acknowledged the difficulty of precise measurements in such a chaotic experimental site, but felt his results to be substantiating and alleged that only a small proportion of a mine's workforce laboured in its deepest regions at any given time.[69]

Undeterred, Moyle renewed his critique in June, with new results obtained from a mine not worked in over a year. The owners of Wheal Trenowith had kept the mine pumped thanks to their engine being powered by a waterwheel, rather than coal, presenting a unique opportunity to examine an unflooded mine that was not under excavation. After 34 fathoms of descent, Moyle found no increase in temperature and at 10 fathoms the temperature was a cool 54°, compared to 62° at the surface. Yet, importantly, on returning to the surface, he found the temperature to have risen by more than 1°, due, Moyle contended, to his presence, along with two assistants.[70] Moyle presented his arguments to the RGSC in October 1822, publishing an account in the same issue of the society's *Transactions* as Fox's original two papers, as well as Forbes's results. There was, Moyle accepted, an increase of about 1° per 10 fathoms of descent in most of the mines he inspected, but he again asserted that such a slight escalation was attributable to human influence. There was no evidence for subterranean heat: if it existed, Moyles surmised that 'there must be a mass of melted matter existing at the centre', but there were no facts to prove this.[71]

---

[67] M. P. Moyle, 'Observations on the Temperature of Mines in Cornwall', *Annals of Philosophy*, NS (Jan.–June 1822), 308–10, at 309; a point well explored in Bulstrode, 'Men, Mines, and Machines', 16 and 21–2.

[68] Moyle, 'Observations on the Temperature of Mines in Cornwall', 309–10.

[69] Robert Were Fox, 'Observations on the Temperature of Mines in Cornwall', *Annals of Philosophy*, NS (Jan.–June 1822), 381–3.

[70] M. P. Moyle, 'On the Temperature of Mines in Cornwall', *Annals of Philosophy*, NS (Jan.–June 1822), 415–16.

[71] M. P. Moyle, 'On the Temperature of the Cornish Mines', *Transactions of the Royal Geological Society of Cornwall*, 2 (1822), 404–15, at 415.

The solution, as Fox saw it, to the problem of temperature arising from miners was to secure more evidence testifying to the existence of subterranean heat. Throughout the 1820s, he continued his observations on temperature in mines with increasing rigour, producing an industrious body of measurements. In 1827, Fox added his latest findings to those published in 1822, offering further proof of increasing heat with depth. Here he presented supporting data from the Consolidated, Poldice, and Wheal Unity mines in Gwennap; South Wheal Towan in St Agnes; East Liscombe, Wheal Friendship, and Bere Alston mines in Devon, expanding the geographic range of mines investigated from his initial experiments. Several results were of particular interest. An accident to the pumping engine at the Ting Tang Mine in Gwennap resulted in flooding with water at 63.5°, but when this was pumped out, water 10 fathoms above the mine's bottom was found to be 65°. This suggested 'that the currents of water which supplied the pumps from the deepest level, were at least 1½° warmer than the water in the shaft 10 fathoms above'. Likewise, when the United Mines at Gwennap flooded up to a depth of 30 fathoms, the miners were evacuated, leaving Fox to record a high temperature of 87° at 170 fathoms of depth after the mine was pumped. This provided a measurement in which workmen could not have overtly influenced the mine's temperature.[72] Mine floods and pumping engine failures clearly presented fortuitous opportunities for temperature measurements free of human influence. Overall, Fox's 1827 figures produced an average temperature of 73.5° in mines at 150 to 160 fathoms, and 66.5° between 110 and 120 fathoms. Fox drew attention to the raised temperature of water jets flowing into mine excavations, which were generally warmer than the temperatures encountered at these depths. In Dolcoath, at the great depth of 240 fathoms, Fox observed water springs of 82° and 80°. This was in a shaft far deeper than the lowest levels of works, where no miner had laboured for several years. Here, a 'thermometer four feet long, having its bulb buried three feet deep in a hole in the rock, which was closed up with clay, indicated the constant temperature of 75.5°, for nearly 20 months'.[73] This was not only a measure that Fox felt to be free from Moyle's claims of human interference, but also evidence of the greater heat of water springs from regions below the mine's deepest depths.

Not all were confident in these fresh results. Although publishing his 1827 temperature measurements in the RGSC's *Transactions*, Fox had originally wanted these results to appear in the more eminent *Philosophical Transactions of the Royal Society*, bringing his work to national audiences. However, London's leading scientific journal declined this submission, with Gilbert writing to notify

---

[72] Robert Were Fox, *Further Observations on the Temperature of Mines; Communicated to the Royal Geological Society of Cornwall and Published in Their 3rd Volume of Transactions* (Penzance: T. Vigurs, 1827), 4–5; on the importance of 'stopped mines' in temperature readings, see Bulstrode, 'Men, Mines, and Machines', 22.
[73] Fox, *Further Observations on the Temperature of Mines*, 8.

Fox of the rejection. In response, Fox thanked Gilbert for his 'encouragement not to be deterred from sending other papers to the Royal Society'. Nevertheless, he regretted that he did not know how he could 'ever obtain a stronger impression of any of my experiment being new, than I had before I presented my paper on heat'.[74] Having known each other since at least 1807, Gilbert proved a valuable ally for Fox in London.[75] Regardless of the Royal Society's scepticism, Fox was convinced that his results constituted proof of the relatively high temperature of springs gushing directly into excavations from mineral veins. These water jets could not, he felt, be subject to the influences of labourers and candles, but indicated the great subterranean heat beneath the mines. Having analysed water samples from hot springs in the United Mines, Wheal Unity, Poldice, and the Consolidated Mines, and finding that each contained common salt, Fox suggested that seawater must have been penetrating 'into the fissures of the earth'. He speculated that it then filtered down to the warmer regions of the globe, where the water was heated, evaporating beneath the Earth's surface, and then being forced up as steam, before condensing as it burst into the mines as jets of hot water.[76] This account of warm water jets from deep below ground clearly drew on Fox's earlier experiences of the expansive use of steam power.

Indeed, just a year after securing his 1812 patent with Joel Lean, Fox discussed the relationship between water jets and pressure with none other than Jonathan Hornblower, whose 1781 compound engine had been the first to work with high-pressure steam. The two engineers had a mutual interest in steam engines and hot-water springs, discussing these subjects during the early 1810s. In 1813, Hornblower informed Fox that he had, at their last meeting, been impressed by his doctrine concerning the 'Principle of Spouting Water'. Hornblower subsequently trialled experiments with high-pressure jets of water that seemed to agree with Fox's speculations, before conceiving of 'a practical method of asserting the Absolute Force of any given size column of water' under various pressures.[77] With his brother Alfred, Fox later conducted experiments on water jets and the effects of pressure on their rates of discharge.[78] They found that water discharged at the same speed, whether underwater, into the atmosphere, or under mercury, suggesting that 'the force with which a moving or spouting fluid recoils is not affected by the surrounding medium'.[79] The study of high-pressure water jets fit neatly

---

[74] Robert Were Fox to Davies Gilbert (27 July 1827); letter in author's possession.
[75] RS, Robert Were Fox Papers, MS/710/36: Robert Were Fox to Davies Giddy (2 May 1807).
[76] Fox, *Further Observations on the Temperature of Mines*, 9 and 14.
[77] RS, Robert Were Fox Papers, MS/710/55: Jonathan Carter Hornblower to Robert Were Fox (3 May 1813).
[78] Robert Were Fox, 'On the Discharge of a Jet of Water Under Water', *Journal of the Royal Institution of Great Britain*, 1 (Oct. 1830–May 1831), 368.
[79] Robert Were Fox, 'On the Discharge of a Jet of Water Under Water', *Journal of the Royal Institution of Great Britain*, 1 (Oct. 1830–May 1831), 599–600, at 600.

alongside understandings of how steam and water might interact within a steam engine, as well as in the innermost regions of the Earth.

Likewise, while he investigated subterranean heat in Cornwall's mines, Fox continued to take an interest in steam-engine design and the industrial use of steam power throughout the 1820s. At St Katharine Docks in London in 1827 he witnessed a novel arrangement of fused-iron tubes 'for generating steam, but the engine was not working. The fire was not advantageously applied, being by the side of, & not under the iron case, which contained the fused metal & tubes'.[80] Fox thought this a troublesome contrivance, given the thickness of the metal between the fire and water, which was detrimental to the 'transmission of heat' and would make it hard to sustain a uniform temperature 'if a brisk fire be used, which I believe to be important in an economical point of view'. He also speculated that 'the cases containing the fused metal at a high temperature of 500 to 600°' would soon wear out.[81] Not long after this, Fox witnessed a working steam engine which Goldsworthy Gurney had designed for locomotive purposes, judging that

> it bids fair to answer for many purposes at least, as it combines lightness with strength in a superior degree, & must I conceive, be economical as it respects fuel. It may perhaps be questioned whether it will generate steam with sufficient rapidity for large engines & whether its numerous joints may not subject it to frequent leaks, but Gurney says he has found it to answer fully after fourteen months trial.[82]

Gurney had not yet tasked his boiler with mine pumping, but gave Fox 'a very flourishing report' of his steam-carriage trials around Regent's Park. Similarly, in June 1830, Fox's clerk and scientific assistant, William Jory Henwood (1805–75), reported 'that some of the Government Boat Engines... lift upwards of 30 millions of lbs one foot high by the use of each buschel of coal'. Considering this application of the measure of duty to steam navigation, Henwood concluded that if 'this be true it fully confirms me in the belief I have long had of the superiority of Watt's Engines... over those now made in Cornwall'.[83] Fox and his family continued to discuss steam engines with Lean well into the 1840s. After attending a monthly meeting of the RGSC on 21 August 1839, Barclay Fox dined at Joel Lean's house in Penzance. Here Barclay examined Lean's 'patent (now expired) to produce motion by compressed air acting on 2 sets of valves in a cylinder', which the young Fox thought 'Very ingenious'. In April 1841 the Foxes returned the favour, hosting Lean and his family. Barclay recorded an enjoyable night on their arrival,

---

[80] Robert Were Fox to Davies Gilbert (27 July 1827); letter in author's possession.
[81] Ibid. [82] Ibid.
[83] RS, Robert Were Fox Papers, MS/710/52: William J. Henwood to Robert Were Fox (22 June 1830).

where the Foxes and Leans 'Talked of steam engines all the evening'.[84] Robert Were Fox and Joel Lean evidently shared a mutual interest in steam power long after their 1812 patent. Fox's mine experiments were conducted amid his daily encounters of Cornish steam engineering.

1.5. Conclusion

What is particularly interesting about Fox and the Leans' measurements of subterranean temperature is that they are still, in the 2020s, of industrial relevance. These historic data sets constitute some of the oldest in existence on subterranean heat and are valuable to present-day geologists in modelling the historic qualities of Cornwall's geothermic gradient. In the present age of escalating fuel prices and demand for low-carbon energy, these temperature records are proving useful for the private and public sectors in calculating the energy potential of Cornwall's now-flooded mines, such as at the Camborne School of Mines, part of the University of Exeter, where the hope is that such geothermal energy will help heat commercial, residential, municipal, and industrial buildings on a large scale, alleviating fuel poverty, while containing the ruinous impacts of anthropogenic climate change.[85] This was, of course, not the point of Fox's nineteenth-century investigation, but it was equally connected to questions over fuel economy and power. Then, as now, mines provided mysterious places in which to investigate the natural world. For all that they are dark, the extraction of precious minerals and excavation of the Earth's interior has long promised knowledge of Nature and Creation. As early as the sixteenth century, the mine offered powerful moral lessons for Protestant Reformation ministers, such as Johannes Mathesius (1504–65), who conscripted these places of toil into his sermons on the Resurrection. In the Bohemian mining town of Joachimsthal, famous for its silver, ore provided evidence of the resurrection of the body in a fashion that was familiar to Mathesius' listeners. Just as the body would one day rise from the earth, the soil hid silver and gems, waiting to be uncovered. The Earth was a place not just of decay but also of rebirth, and while the body would one day be transformed through its resurrection, silver required smelting to be of value. Along with the growth of seeds, mines could be knowledge-imparting metaphors, consistent with Martin Luther's insistence that Nature showed evidence of both decay and resurrection, which was so central to Reformed thought in sixteenth-century Europe.[86] By the

---

[84] Barclay, quoted in R. L. Brett (ed.), *Barclay Fox's Journal* (London: Bell & Hyman, 1979), 159 and 225.
[85] Thanks to Dr Nick Harper, Geology Impact Fellow for Deep Digital Cornwall, for bringing this to my attention.
[86] Erin Lambert, *Singing the Resurrection: Body, Community, and Belief in Reformation Europe* (Oxford: Oxford University Press, 2017), 68–9; thanks to Dr Silke Muylaert for drawing my attention to this.

nineteenth century, mines not only promised to be valuable sites of knowledge production, both of God's Creation and the order of Nature, but were also now sites of steam-engine innovation and the economic management of heat.

Cornwall's industrial use of steam power and rich mining culture gave rise to distinctive interpretations of the Earth's structure and the history of its formation. It was not that mine engineers like Joel and Thomas Lean, or natural philosophers like Forbes and Fox, were alone in conceiving of the Earth's interior as a place of great heat; Fourier drew similar conclusions in Paris, far from Cornwall's rich seams of copper and tin. Rather, it was the form that such understandings took that demonstrated how the contexts of steam-engine design and deep mining shaped the character of Cornish science. Instead of mathematical theory, it was the experiences of temperature increase with depth that provided evidence and direction for Fox's subterranean investigations. And when conceiving of the manner in which heat circulated below ground and the operation of hot-water jets, it was from the steam engine that Fox drew inspiration. With the Earth's core acting as a colossal boiler, the discharge of heated high-pressure water in Cornwall's mines could be likened to the expansion and condensation of high-pressure steam within the cylinder of a steam engine. Here was not only a valuable analytical tool for theorizing about the Earth's structure, but also a beautiful comparison between the works of man and the works of Nature. Fox brought his own expertise in steam engineering to the philosophical question of the Earth's heat. But he was also able to mobilize Cornwall's rich industrial resources to this experimental pursuit. He not only had access to deep mines, but also drew on a valuable network of engineers, including the Lean brothers and Hornblower, as well as local mine captains who could assist in experiments, to fashion new scientific knowledge. Fox would, however, take these inquiries much further in the 1830s and 1840s. In his 1827 RGSC *Transactions* paper on subterranean heat, Fox first suggested that it was likely 'that electricity may be a primary exciting cause of the high temperature of the earth'.[87] This speculation represented the start of a sustained effort on Fox's behalf to link subterranean phenomena of heat, electricity, and, eventually, magnetism within a grand vision of Nature.

---

[87] Fox, *Further Observations on the Temperature of Mines*, 15.

2
The Earth's Laboratory
Underground Experiments, Philosophical Miners, and Knowledge from the Mine

The underground laboratory of London's Royal Institution was the site of Michael Faraday's most famous experiment. In August 1831, the celebrated experimentalist solved a problem that had, for over a decade, troubled natural philosophers across Europe: how to combine motion and magnetism in a way that would produce electricity. Since the Danish chemist Hans Christian Ørsted's (1777–1851) observation in 1820 that a magnetic compass needle could be made to jump by passing electricity through a closely positioned wire, it had been known that magnetism and electricity could produce movement. The following year, Faraday found that when a current passed through a wire freely suspended over a magnet, the wire oscillated around the magnet's pole. This laboratory experiment established the principle of electromagnetic rotation. With movement produced from electricity and magnetism, the remaining challenge was to reverse this process and induce electricity from magnetism and movement. Faraday eventually solved this problem in the autumn of 1831 by repeatedly passing a magnet through a coiled helix of copper wire: he observed the needle of a galvanometer, employed to detect an electric current, twitch at each transit of the magnet through the helix. By continually moving the magnet, Faraday sustained a current, successfully converting motion and magnetism into electricity in what was effectively the first electric generator.[1] The experimental production of electromagnetic induction initiated over twenty-five years of investigation into electrical phenomena, published at regular intervals as his *Experimental Researches in Electricity*. In the first of these papers, originally read to the Royal Society in November 1831, Faraday announced his observation of electromagnetic induction. However, while he could describe how he had induced electricity from motion and magnetism, it was difficult for Faraday to explain why this actually worked. Electromagnetic induction was an unknown phenomenon and, in the months

---

[1] Geoffrey Cantor, David Gooding, and Frank A. J. L. James, *Michael Faraday* (New York: Humanity Books, 1991), 47, 52, and 11–12; on Faraday's experiments, see David Gooding, '"In Nature's Schools": Faraday as an Experimentalist', in David Gooding and Frank A. J. L. James (eds.), *Faraday Rediscovered: Essays on the Life and Work of Michael Faraday, 1791–1867* (Basingstoke: Macmillan Press Ltd, 1985), 105–35.

following his discovery, Faraday wrestled to reconcile his celebrated experiment with established scientific research.

At the Royal Society on 12 January 1832, Faraday delivered the institution's prestigious Bakerian Lecture, opting for the subject of 'Terrestrial Magneto-Electric Induction'. In what would constitute the second of his *Experimental Researches in Electricity* papers, Faraday used this opportunity to connect the electromagnetic action produced in his helix experiment with the magnetic behaviour of the Earth. Specifically, he invoked Robert Were Fox's experiments on the movement of electrical currents through copper veins deep below the Earth's surface. Faraday speculated that these trials might enhance understanding of electromagnetic phenomena and informed his august audience that 'Mr. Fox of Falmouth has obtained some highly important results respecting the electricity of metalliferous veins in the mines of Cornwall'. Fox had recently published the results of his underground electrical experiments in the *Philosophical Transactions of the Royal Society*, and it was here that Faraday had an opportunity of examining the Cornishman's work 'with a view to ascertain whether any of the effects were probably referable to magneto-electric induction'. For Fox, this publication in such an eminent journal represented a real breakthrough. 'When parallel veins running east and west were compared, the general tendency of the electricity *in the wires* was from north to south', reported Faraday, and 'when comparison was made between parts towards the surface and at some depth, the current of electricity in the wires was from above downwards'.[2] In conceiving of the globe as a giant magnet, Fox claimed to have measured the movement of electric currents through metalliferous veins of copper ore to determine the direction of the Earth's magnetic pole. Although conducted deep within a mine, this experiment appeared comparable to how the passing of electricity through a copper wire suspended over a magnet revealed the direction of the magnet's pole. Fox's own work built directly on Faraday's principle of electromagnetic rotation, only instead of trialling refined copper wire in a laboratory, he used veins of copper ore in its natural state as it lay in the Earth.

As interesting as Fox's experiments were, Faraday was uncertain of their implications. Attempting to draw out any possible connections between the results of Fox's subterranean trials and his own recent observation of electromagnetic induction, Faraday surmised that

> If there should be any natural difference in the force of the electric currents produced by magneto-electric induction in different substances, or substances in different positions moving with the earth, and which might be rendered evident

---

[2] Michael Faraday, *Experimental Researches in Electricity* (London: Richard and John Edward Taylor, 1839), 54–5.

by increasing the masses acted upon, then the wires and veins experimented with by Mr. Fox might perhaps have acted as dischargers to the electricity of the mass of strata included between them, and the directions of the currents would agree with those observed as above.[3]

The existence of subterranean electrical currents, nonetheless, remained unproven. Faraday was sceptical, concluding that there was no discernible relation between electromagnetism and the direction of underground electrical currents. Yet what is revealing about this reference to Fox's work is how, in the early 1830s, scientific knowledge of electric and magnetic phenomena was as much the product of experiments deep below ground in the mines of Cornwall as it was a rarefied laboratory construct.

Between the 1820s and 1850s, Fox endeavoured to put the Cornish mine at the centre of philosophical debates over the Earth's structure and the nature of terrestrial magnetic phenomena. Through subterranean experiments and the experiences of Cornwall's miners, he constructed knowledge in which it was not the laboratory, but the mine, that was the crucial site of investigation. This resulted in a distinctive and localized account of Nature that eminent scientific authorities, like Faraday, engaged with, albeit cautiously. Geological and electromagnetic phenomena were critical subjects of philosophical concern in mid-nineteenth-century Britain and, by mobilizing Cornwall's industrial resources, Fox was able to contribute privileged, highly provincial, knowledge of both. In doing so, he revealed the unique philosophical opportunity the mine presented for scrutinizing Nature. However, this experimental space was of paradoxical scientific value. On the one hand, mine experiments and tacit mining knowledge seemed to provide knowledge straight from Nature but, concurrently, the mine was an uncontrollable place of industrial labour and danger. Evidence collected in this space appeared extremely prone to error. Similarly, without access to a mine, Fox's experimental evidence could be neither completely rejected nor absolutely validated. At stake here were wider tensions between the provincial mine and the metropolitan laboratory as reliable sites of experiment. Such epistemological dilemmas were at the centre of Fox's efforts to connect his earlier study of subterranean heat with broader, and increasingly fashionable, questions over electromagnetic phenomena. In this work, we find his earliest experimental use of magnetized needles: these experiences would shape his later development of dipping-needle instruments. As this chapter shows, this was a matter not just of building apparatus and obtaining data, but also of cultivating social connections and increasing his own credibility with the nation's leading scientific authorities.

[3] Ibid. 55.

2.1. Mine Experiments

After fifteen years of measuring the heat of mines, Fox turned his attention to the subject of how seams of ore developed. This had long been a troubling geological problem. Back in 1805, during his Royal Institution lecture on mineral veins, Humphry Davy had expressed doubts over existing theories of how these geological features formed, including James Hutton's (1726–97) speculations that they resulted from molten fluid from below the Earth's crust seeping into fault lines and then cooling.[4] Fox wanted to resolve this question: he believed that the escalation of subterranean heat with depth was related to the formation of the metalliferous veins that yielded Cornwall's valuable copper and tin. Specifically, he thought that the combination of underground heat and moisture generated electrical currents that caused saline water, collected in fissures and natural faults, to deposit its mineral content as ore. On 30 March 1830, Fox wrote to Charles Lemon with news of his 'successful experiments on the electricity of copper veins with a galvanometer, in Huel Jewel, Dolcoath, & Tresavean mines'. A fellow of London's Royal Society since 1822 and soon-to-be Whig MP for Penryn, Lemon combined the powerful social, political, and scientific networks that were so crucial for aspiring Cornish natural philosophers to advance their work. Fortunately for Fox, he and Lemon were already close associates. Hoping to secure Lemon's patronage, Fox explained how

> Slips of sheet copper two feet long, & 3 inches wide, were fastened by copper nails to different parts of the veins, and an electrical communication was established between two of these & the galvanometer, by copper wire coated with sealing wax. The distance of the copper slips from each other was generally less than 20 or 30 fathoms.—The deviation of the needle was in some cases considerable, its oscillations extending over more than half the circle, in others, the action was comparatively slight.[5]

These were, Fox declared, encouraging results, suggesting mine veins were capable of conveying electrical currents.

Accompanying these experiments, Fox had also employed a magnetic needle, suspended by silk, to measure the intensity of the Earth's magnetic force within the mine. Hoping to determine that the direction of electric currents passing through veins was relative to terrestrial magnetism, he had recently 'made some experiments in mines with the magnetic needle[.] ... The needle I used was seven

---

[4] Davy, in Robert Siegfried and Robert H. Dott, Jr (eds.), *Humphry Davy on Geology: The 1805 Lectures for the General Audience* (Madison: University of Wisconsin Press, 1980), 114–15.
[5] Robert Were Fox to Charles Lemon, 30 Mar. 1830, 1–5, at 1; letter in author's possession.

inches long, suspended through a glass tube, by very fine untwisted silk'.[6] Along with measuring the dip, or angle, of mineral lodes, and the direction of subterranean electricity currents, Fox was using the mine to measure the Earth's magnetic force, or intensity. Observing the Earth's magnetism at different depths, Fox found that there was no increase of intensity with depth, even 1,000 to 1,200 feet below sea level, compared to that at the surface. Fox had intended to write to Faraday about these experiments and their implications for understanding terrestrial magnetic phenomena. It would, Fox informed Lemon, be of interest

> to know whether any of my views might obtain the sanction of such an authority, or if not, I might have the benefit of his corrections. I cannot but feel very diffident on the subject of my opinions, till I am more acquainted with the recent...conclusions of this Philosopher. Perhaps thou wilt have the kindness to mention my experiments to him, if a suitable opportunity should occur.[7]

Fox was eager to bring his experiments before Britain's foremost natural philosophers, and there was no better place to do this than at London's Royal Society, where Fox had a powerful patron in its president, Davies Gilbert. Together, Lemon and Gilbert secured Fox's 'On the Electro-Magnetic Properties of Metalliferous Veins in the Mines of Cornwall' a reading at a society meeting on 10 June 1830, instantly bringing the Cornishman's work to the attention of leading British scientific audiences. Here, Fox delivered an account of his experiments to conduct electrical currents through mineral veins deep below ground. His investigations suggested 'that the existence of electricity in metalliferous veins similarly circumstanced, and capable of conducting it, will prove to be as universal a fact, as the progressive increase of temperature under the earth's surface is now admitted to be'.[8] As with subterranean heat, it was the Cornish mine that provided the experimental space in which to investigate this philosophical claim.

Deep below ground, Fox had fixed small plates of sheet copper to veins of copper ore, using copper nails, connecting plates at various locations throughout the mine with copper wire. Copper ore veins were in this way fashioned into electrical circuits which could then be measured when connected to a galvanometer. Some of these circuits involved almost 300 fathoms of copper wire. By including a metalliferous vein within a circuit in this manner, Fox contended that he was able to directly measure the intensity of an electromagnetic impulse passing through ore within a mine. At different locations, he observed varying deviations of the galvanometer's needle, noting that a vein conducted a greater impulse when it contained a higher amount of copper ore. Fox detected less electrical action

---

[6] Ibid.   [7] Ibid. 4.
[8] Robert Were Fox, 'On the Electro-Magnetic Properties of Metalliferous Veins in the Mines of Cornwall', *Philosophical Transactions of the Royal Society*, 120 (1830), 399–414, at 399.

where there was little ore. From this, he suggested that 'electro-magnetism may become useful to the practical miner in determining with some degree of probability at least, the relative quantity of ore in veins, and the directions in which it most abounds'.[9] Fox noticed that if the copper plates were positioned between a horizontal vein that was rich in copper ore and undisrupted by any non-conducting substances, the galvanometer registered no motion. However, if a cross-vein of quartz or clay interrupted the path of the vein, the galvanometer displayed a significant electrical action, due to the current's irregularity. A very decisive electrical action was recorded between two plates at different depths of the same vein, as well as between varying veins at the same level. The direction of positive electricity was both from east to west and from west to east. At the Great St George Mine, he found a very productive copper vein, running through soft ground, conducting a powerful electrical current. Here he remained at the surface and dropped a wire down through the mine's shaft, finding that as it descended it caused a connected galvanometer's needle to flicker. In Wheal Jewel, Fox sustained an electrical current between a plate fixed at a heap of copper at the surface and ore in the same vein at different depths below.[10]

While Fox found that the conducting powers of different metallic ores seemed to have no relation to the electrical properties of their pure metals, what was revealing was the manner in which electricity passed through non-metallic matter. Significantly, Cornish 'killas', a clay-slate, seemed to convey electricity to a very limited extent, yet only in the direction of the rock's cleavage. Fox reasoned that this was due to retained moisture in the clay. This conduction led the Cornish experimentalist to speculate that moisture, retained in softer rocks, conducted electrical currents which, over time, formed the metalliferous veins on which Cornish mining depended. Veins appeared to form along channels through which subterraneous water and vapour circulated. It also seemed that horizontal veins bearing tin and copper tended to run from east-north-east to west-south-west, with few deviations.[11] Reports from miners of veins in Mexico, Guatemala, and Chile resembled the character and direction of those in Cornwall, implying that this was a global phenomenon. Already, Fox was looking to link up his Cornish observations with comparative reports from overseas.

Having established the conducting capacity of metalliferous veins, as well as moisturized clay and rock, Fox sought to connect this phenomenon to broader questions of terrestrial magnetism and the Earth's temperature. Reflecting on the parallels between electromagnetic observations in a laboratory and the passing of electrical currents through veins, Fox surmised that 'analogies become highly interesting when regarded in connection with terrestrial electricity, magnetism, and heat; for if it be granted that the two latter increase in intensity at great depths

---

[9] Ibid. 400. [10] Ibid. 401. [11] Ibid. 403 and 405.

in the earth, they are evidently so connected with electrical action that the augmentation of it also, in the interior of the globe, may be reasonably inferred.'[12] Just as electrical currents influenced a magnetic needle at the Earth's surface, so they might have a similar effect on metalliferous veins deep underground. The Earth's rotation was, Fox thought, connected with this formation of veins in a way that was comparable to the electromagnetic rotation of a wire around a magnet. He concluded that the 'rotation of the earth on its axis from west to east seems moreover to harmonize with the idea of oblique electrical currents; since rotation in the same direction may be produced by corresponding electro-magnetic arrangements'.[13] Here was knowledge straight from the mine that was consistent with Ørsted's and Faraday's laboratory experiments.

Fox's paper was evidently well received at the Royal Society in June 1830, securing publication in *Philosophical Transactions* later that year. One anonymous member of the meeting's audience subsequently reported that Fox's findings would 'be of practical utility to the miner in discovering the relative quantity of ore in veins, and the directions in which it most abounds'.[14] Similarly, Gilbert lavished praise on Fox's work. 'Your paper was read last evening at the Royal Society,' wrote Gilbert; 'I believe that no paper has been read for many years which excited so great a degree of attention.' The mine experiments were, he declared, 'an epoch in Geological Science'.[15] High praise indeed for a Royal Society paper: this was a venue accustomed to scientific epochs. Given the Royal Society's rejection of Fox's work on subterranean heat in 1827, his publication in *Philosophical Transactions* was nothing short of a coup. Fox's continued networking with Gilbert and Lemon had paid off.

Fresh from his first successful Royal Society publication, Fox was keen to secure a second. This time it was his experiments with a magnetized needle that he wanted to promote, having measured the strength and direction of the Earth's magnetic force both within a mine and at surface level. Originally read at a Royal Society meeting on 17 March 1831, Fox's 'On the Variable Intensity of Terrestrial Magnetism, and the Influence of the Aurora Borealis upon It' provided insights on magnetic action at the Earth's surface. At Falmouth, Fox had made observations on 'the variable force of the magnetic attraction, and the action of the aurora borealis on the direction and intensity of the [magnetic] needle'.[16] Throughout 1830, he had recorded the movements of a magnetic needle to ascertain if it was affected by changes of the Earth's distance from the Sun or in its relation to the

---

[12] Ibid. 406.   [13] Ibid. 407.
[14] Anon., 'Review: *On the Electro-Magnetic Properties of Metalliferous Veins in the Mines of Cornwall*, By Robert Were Fox (Read June 10, 1830)', *Journal of the Royal Institution of Great Britain*, 1 (Oct. 1830–May 1831), 345–6, at 345.
[15] RS, Robert Were Fox Papers, MS/710/43: Davies Gilbert to Robert Were Fox (11 June 1830).
[16] Robert Were Fox, 'On the Variable Intensity of Terrestrial Magnetism, and the Influence of the Aurora Borealis upon It', *Philosophical Transactions of the Royal Society*, 121 (1831), 199–207, at 199.

equator. He had taken two needles, one possessing an excessive northern pole and the other in which the south pole was predominant. With silk thread, he suspended each of these in two identical slate boxes, mounted on bricks, and employed an 8-inch-long magnetic bar to control the needles' movement from below. A slip of glass at one end of the boxes allowed the movements of the two needles to be seen. Over the course of a year, Fox measured the direction of force of the needles with their opposing polarizations. Generally, he found that the yearly mean magnetic intensity was reasonably equal between the two poles; one needle was attracted as strongly as the other was repelled. However, the intensity of this force varied over time. This led Fox to connect his needle observations to those he made earlier on metalliferous veins, asserting that it 'seems most reasonable to refer the phenomena of the earth's magnetism to the urgency of electrical currents existing under its surface, as well as above it'.[17] Given the variations in magnetic intensity discernible through his suspended-needle arrangement, such phenomena could not be attributable entirely to meteorological causes, as these were so transitory: a constant subterranean influence on the magnetic needles was, he contended, the only explanation.

Having taken magnetic needle readings throughout 1830, Fox's location at Falmouth presented a privileged position to observe the aurora borealis, which appeared frequently and with unusual intensity over Cornwall during the winter of 1830–1. This striking natural marvel 'sent up streams of red and white light' that coincided with unusual magnetic activity from the suspended needles. At 7 p.m. on evenings that streams of light occurred, both needles initially rested at 0°, but after the streams reached their zenith, just after dusk, the north ends of the needles moved east. By 8.30 p.m. they had produced an easterly variation of 1° 15′, returning west at 8.45 p.m., and being stationary in their original magnetic meridian by 10 p.m. Fox claimed this regular movement to the east was an impact of some 'electrical phenomenon' (Figure 2.1).[18] He appealed to the Royal Society for similar experiments to be made across the world, preferably on remote islands, far from any interferences from land, where delicate influences on needles would be made visible. A replication of his needle observations at Falmouth promised to throw light 'on the hypothesis of electrical currents under and above the surface of the earth'.[19]

Fox's observations on the aurora borealis's magnetic influence subsequently secured publication in the *Philosophical Transactions*, marking his second contribution to the journal. This article attracted Faraday's attention and he cited it the following year during his 1832 Bakerian Lecture, 'Terrestrial Magneto-Electric Induction', later published in the second series of his *Experimental Researches in Electricity*. Faraday explained that Fox had observed deflections of a magnetic

---

[17] Ibid. 201.   [18] Ibid. 202.   [19] Ibid. 203.

A TABLE showing the Times of Vibration of two Magnetic Needles, described in the annexed paper, in the neighbourhood of Falmouth.

Needle No. 1.					Needle No. 2.				Barometer.	Thermometer in room with needles.	Thermometer out, exposed to the north.	Lunar phases.	Extraordinary variations of the needles by night.	Winds.	General Remarks.
Time of Observation.			Arcs 80° to begin, ended as under.	Number of seconds in making 80 vibrations.	Time of Obs.		Arcs 80° to begin, ended as under.	Number of seconds in making 80 vibrations.							
1830.	A.M.	P.M.			A.M.	P.M.									
Aug. 2	......	10	......	1183†.00											
3	10	......	39.00	1179†.00											
4	......	10	39.00	1184†.00	......	......	......	....	....	......	......	○	....	......	Clear.
5	10	......	39.00	1186†.00											
6	......	10	39.00	1181†.00	......	......	......	....	....	......	......	......	....	......	Fine weather.
9	......	10	39.00	1183†.00											
Mean for the month		....		1182.66											
[a] Oct. 7	......	10	38.30	1180†.00	......	10.30	43.00	1137.00	30.30	60.00	......	☾ 8th	....	......	[a] Suspended nearly two months in consequence of absence from home.
9	......	3.30	38.30	1180†.00	......	4	43.00	1132.00	30.40	......	......	......	....	......	Fine clear weather.
13	......	10	39.00	1177†.00	......	10.30	43.00	1131.00							
15	......	10	39.00	1180†.00	......	11	43.00	1133.00							
Mean for the month		....		1179.25	......	......	......	1133.25	....	......	......	● 16th			
[b] Nov. 8	......	8	39.00	1177†.00	......	8.30	42.00	1127.50	....	......	......	......	....	......	[b] Again suspended from absence.
12	......	8	39.00	1178.00	......	8.15	43.00	1130.00							
13	10.30	......	39.00	1180.00	11	......	44.00	1134.50	....	......	......	......	....	S	A great storm with rain.
15	......	7.30	39.00	1179.00	......	8.30	43.00	1134.00	29.20	58.00	......	●	....	SW	A great storm without rain.
16	10	......	40.00	1182.00	10.30	......	44.00	1136.00	28.50	58.00	......	......	....	"	Calm throughout the day, but a great storm at night.
19	......	7.30	39.00	1176.00	......	9.45	44.00	1133.00	30.00	54.00	......	......	....	"	Fine.
22	......	8	39.00	1176.00	......	9	44.00	1131.00	29.90	55.00	......	......	....	"	Nearly calm and clear weather.
23	10	......	39.30	1176.00	10.30	......	44.00	1130.00	30.00	54.00	......	☽	....	"	Nearly calm and clear weather.
24	......	2.15	40.00	1177.00	......	2.30	44.00	1126.50	30.20	55.00	......	......	....	SE	A strong dry wind.
25	11.30	......	40.00	1177.00	10	......	44.30	1125.00	30.10	55.00	......	......	....	"	A strong dry wind, very clear.
26	11.30	......	40.30	1175.00	noon		44.30	1123.00	29.80	52.00	......	......	....	"	Clouded, and wind stronger—some rain.
27	......	11	40.00	1176.00	......	11.30	44.30	1124.00	28.90	52.00	......	......	....	ESE	Rain in morning, evening clear.
29	10.30	......	39.30	1180.00	11	......	44.00	1128.00	29.40	56.00	......	......	....	......	Strong wind and dry weather.
Mean for the month		....		1177.61	......	......	......	1129.38	29.60	54.90		○ 30th			
Dec. 1	......	10	39.30	1178.00	......	10.30	45.00	1127.00	29.60	54.00	......	......	....	SE	Wind strong—clouded.
2	10	......	39.30	1178.00	10.20	......	45.00	1127.00	29.30	54.00	......	......	....	ESE	Wind not strong—rain.
3	......	10	40.00	1177.00	......	10.30	44.30	1126.00	29.30	54.00	......	......	....	NE	Clouded, and light wind.
4	10.30	......	39.30	1176.00	10.50	......	45.30	1126.00	29.50	54.00	......	......	....	SE	Clouded, and light wind.
6	......	9.30	40.00	1176.00	......	10	45.00	1128.00	28.10	54.00	......	......	....	"	A great storm to-day, which abated in the evening.
7	9.30	......	41.00	1175.50	10	......	46.00	1128.00	28.40	55.00	......	☾	....	NE	Rain.
8	......	8.30	41.00	1174.00	......	7.30	45.30	1127.00	28.70	54.00	......	......	....	SE	Rain.
9	......	10	41.30	1173.00	......	10.20	45.30	1125.00	28.60	52.00	......	......	....	NE	Clouded.
10	9	......	41.30	1173.00	......	......	46.00	1125.50	28.80	51.00	......	......	....	NW	Showers.
11	......	7.30	41.00	1175.75	......	7	46.00	1126.50	29.20	49.50	......	......	....	W	Strong squalls with hail showers.
"	......	11.15	42.00	*1177.30	......	11.40	45.30	*1129.00	*29.*20	*49.00	......	......	....	"	Aurora visible this evening; the variation not observed.
12	......	......	......	....	......	......	......	....	....	......	......	......	6 P.M. 45′ E.	NNE	Aurora again visible in N.N.E. to N.W. from 6 to 11 P.M.
13	......	4	41.00	1172.00	......	4.30	46.00	1122.50	30.30	47.50	......	......	....	SW	Hard frost in the morning, but now gone.
14	......	11	41.30	1172.00	......	......	46.00	1123.50	30.50	48.00	......	......	....	SSW	Sunshine.
15	......	10	40.00	1172.00	......	......	45.00	1124.00	30.45	49.00	......	●	....	S	
16	9	......	41.00	1174.00	9.30	......	44.30	1126.00	30.25	50.00	......	......	....	N	Clouded, and small rain.
"	......	9.30	40.45	*1171.00	......	10	44.30	*1124.00	"	"	......	......	9 P.M. 20′ E. 10½ P.M. 15′ E.	ESE	Strong wind, and clear.
17	9	......	41.15	1172.00	9.30	......	44.30	1124.50	29.30	49.00	......	......	....	NNE	Nearly calm.
18	......	10.30	41.15	1169.50	......	11.30	45.00	1122.50	30.30	46.00	......	......	....	W	
20	9	......	41.30	1173.50	9.30	......	45.15	1126.00	29.50	47.50	......	......	10 P.M. 20′ E.	N	Showers.
21	......	10	41.15	1174.00	......	10.45	45.15	1126.50	29.30	49.00	......	......	....	NNW	Clouded.
22	......	7	42.00	1173.50	......	7.30	45.30	1126.00	29.40	49.00	......	☽	....	WNW	Strong gusts of wind, and rain.
23	9	......	41.30	1171.50	9.30	......	46.00	1125.00	29.45	46.50	......	......	....	NW	Hail showers.
"	......	11	40.30	*1168.50	......	11.30	45.30	*1123.00	*29.50	46.00	30.00	......	9½ P.M. 30′ E.	N	Snow showers.
24	......	11	40.30	1167.00	......	11.30	45.30	1128.00	29.50	41.00	27.00	......	....	"	Mostly clouded; deep snow on the ground.
25	......	2	40.15	1167.00	......	3.15	45.30	1128.50	29.30	41.00	25.00	......	....	"	Clear weather, and deep snow on the ground.
"	......	9.30	40.00	*1173.50	......	10	46.00	*1134.00	*29.20	*41.00	*33.50	......	9¼ P.M. 35′ E. 10 P.M. 50′ E. 12 P.M. 30′ E.	"	Snow and sleet falling.
27	9.15	......	40.30	1171.50	10	......	46.30	1131.00	28.70	41.50	36.00	......	....	W	Clouded.
"	......	11	40.00	*1172.00	......	11.30	46.00	*1132.00	*28.70	*42.00	*39.00	......	9 P.M. 15′ E.	SW	Clouded, and rain occasionally.
28	9	......	40.30	1171.00	10	......	45.30	1132.00	28.90	42.50	39.00	......	....	W	Clouded, and showers of hail and rain.
29	9	......	40.15	1172.00	9.30	......	44.45	1132.00	29.50	43.00	39.00	○	....	SW	Rather hazy.
"	......	10.45	40.40	*1174.00	......	10	45.00	*1133.00	*29.35	*44.00	*44.00	......	9 P.M. 25′ E. 9½ P.M. 50′ E. 10½ P.M. 15′ E.	S to E	Just commencing to blow strongly—showers.
30	9.30	......	40.30	1174.50	10	......	44.30	1134.50	28.85	44.00	47.00	......	....	E	Strong wind and much rain.
31	9	......	41.30	1176.00	10.30	......	45.00	1134.00	29.00	46.50	45.00	......	....	W	Clouded and damp.
"	......	11.30	41.00	*1173.00	11	......	45.30	*1131.00	*29.50	*45.50	*38.50	......	....	W	Clear, and fine night.
Mean for the month		....		1173.30	......	......	......	1126.72	29.39	48.61	36.60				

Figure 2.1 Fox's tables, published in *Philosophical Transactions*, detailing the movements of two magnetic needles during the aurora borealis at Falmouth between August 1830 and February 1831 (Author's image, 2023).

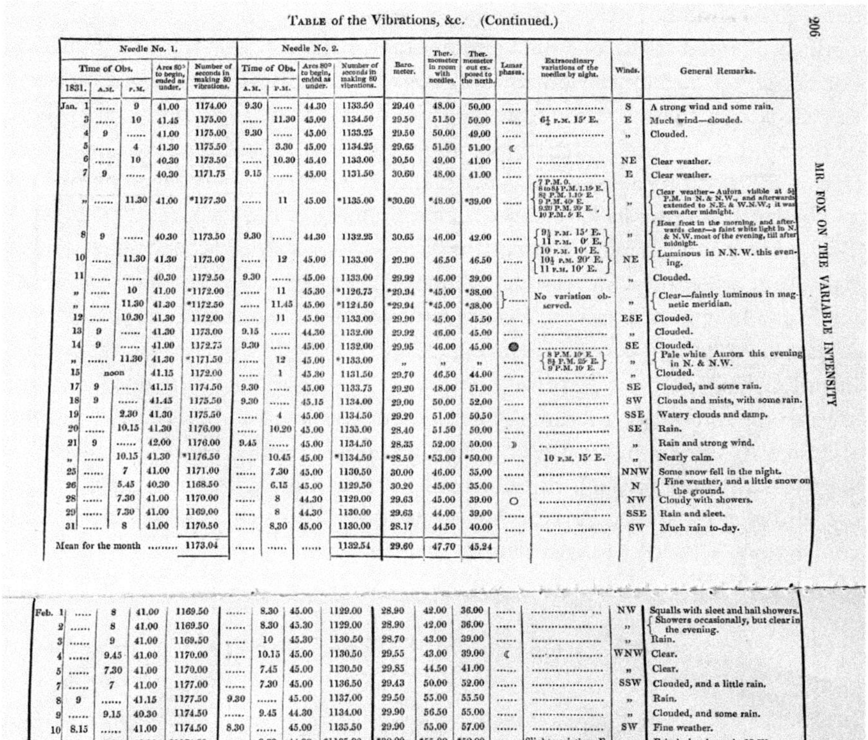

206

MR. FOX ON THE VARIABLE INTENSITY OF TERRESTRIAL MAGNETISM.

TABLE of the Vibrations, &c. (Continued.)

Needle No. 1.					Needle No. 2.				Barometer.	Thermometer in room with needles.	Thermometer out exposed to the north.	Lunar phases.	Extraordinary variations of the needles by night.	Winds.	General Remarks.
Time of Obs.			Arcs 80° to begin, ended as under.	Number of seconds in making 80 vibrations.	Time of Obs.		Arcs 80° to begin, ended as under.	Number of seconds in making 80 vibrations.							
1831.	A.M.	P.M.			A.M.	P.M.									
Jan. 1	......	9	41.00	1174.00	9.30	......	44.30	1133.50	29.40	48.00	50.00	......	....	S	A strong wind and some rain.
3	......	10	41.45	1175.00	......	11.30	45.00	1134.50	29.50	51.50	50.00	......	6¼ P.M. 15′ E.	E	Much wind—clouded.
4	9	......	41.00	1175.00	9.30	......	45.00	1133.25	29.50	50.00	49.00	......	....	,,	Clouded.
5	......	4	41.30	1175.50	......	3.30	45.00	1134.25	29.65	51.50	51.00	☾			
6	......	10	40.30	1173.50	......	10.30	45.40	1133.00	30.50	49.00	41.00	......	....	NE	Clear weather.
7	9	......	40.30	1171.75	9.15	......	45.00	1131.50	30.60	48.00	41.00	......	....	E	Clear weather.
,,	......	11.30	41.00	*1177.30	......	11	45.00	*1135.00	*30.60	*48.00	*39.00	......	7 P.M. 0. 8 to 8¼ P.M. 1.15′ E. 8½ P.M. 1.10′ E. 9 P.M. 40′ E. 9.20 P.M. 20′ E. 10 P.M. 5′ E.	,,	Clear weather—Aurora visible at 5½ P.M. in N. & N.W., and afterwards extended to N.E. & W.N.W.; it was seen after midnight.
8	9	......	40.30	1173.50	9.30	......	44.30	1132.25	30.65	46.00	42.00	......	9½ P.M. 15′ E. 11 P.M. 0′ E.	,,	Hoar frost in the morning, and afterwards clear—a faint white light in N. & N.W. most of the evening, till after midnight.
10	......	11.30	41.30	1173.00	......	12	45.00	1133.00	29.90	46.50	46.50	......	10 P.M. 10′ E. 10½ P.M. 20′ E. 11 P.M. 10′ E.	NE	Luminous in N.N.W. this evening.
11	......	......	40.30	1172.50	9.30	......	45.00	1133.00	29.92	46.00	39.00	......	....	,,	Clouded.
,,	......	10	41.00	*1172.00	......	11	45.30	*1126.75	*29.94	*45.00	*38.00	......	No variation observed.		Clear—faintly luminous in magnetic meridian.
,,	......	11.30	41.30	*1172.50	......	11.45	45.00	*1124.50	*29.94	*45.00	*38.00			,,	
12	......	10.30	41.30	1172.00	......	11	45.00	1133.00	29.90	45.00	45.50	......	....	ESE	Clouded.
13	9	......	41.30	1173.00	9.15	......	44.30	1132.00	29.92	46.00	45.00	......	....	,,	Clouded.
14	9	......	41.00	1172.75	9.30	......	45.00	1132.00	29.95	46.00	45.00	●	....	SE	Clouded.
,,	......	11.30	41.30	*1171.50	......	12	45.00	*1133.00	,,	,,	,,	......	8 P.M. 10′ E. 8½ P.M. 25′ E. 9 P.M. 10′ E.	,,	Pale white Aurora this evening in N. & N.W.
15	noon		41.15	1172.00	......	1	45.30	1131.50	29.70	46.50	44.00	......	....	,,	Clouded.
17	9	......	41.15	1174.50	9.30	......	45.00	1133.75	29.20	48.00	51.00	......	....	SE	Clouded, and some rain.
18	9	......	41.45	1175.50	9.30	......	45.15	1134.00	29.00	50.00	52.00	......	....	SW	Clouds and mists, with some rain.
19	......	2.30	41.30	1175.50	......	4	45.00	1134.50	29.20	51.00	50.50	......	....	SSE	Watery clouds and damp.
20	......	10.15	41.30	1176.00	......	10.20	45.00	1135.00	28.40	51.50	50.00	......	....	SE	Rain.
21	9	......	42.00	1176.00	9.45	......	45.00	1134.50	28.35	52.00	50.00	☽	....	,,	Rain and strong wind.
,,	......	10.15	41.30	*1176.50	......	10.45	45.00	*1134.50	*28.50	*53.00	*50.00	......	10 P.M. 15′ E.	,,	Nearly calm.
25	......	7	41.00	1171.00	......	7.30	45.00	1130.50	30.00	46.00	35.00	......	....	NNW	Some snow fell in the night.
26	......	5.45	40.30	1168.50	......	6.15	45.00	1129.50	30.20	45.00	35.00	......	....	N	Fine weather, and a little snow on the ground.
28	......	7.30	41.00	1170.00	......	8	44.30	1129.00	29.63	45.00	39.00	○	....	NW	Cloudy with showers.
29	......	7.30	41.00	1169.00	......	8	44.30	1130.00	29.63	44.00	39.00	......	....	SSE	Rain and sleet.
31	......	8	41.00	1170.50	......	8.30	45.00	1130.00	28.17	44.50	40.00	......	....	SW	Much rain to-day.
Mean for the month				1173.04	......	......	......	1132.54	29.60	47.70	45.24				
Feb. 1	......	8	41.00	1169.50	......	8.30	45.00	1129.00	28.90	42.00	36.00	......	....	NW	Squalls with sleet and hail showers.
2	......	8	41.00	1169.50	......	8.30	45.30	1129.00	28.90	42.00	36.00	......	....	,,	Showers occasionally, but clear in the evening.
3	......	9	41.00	1169.50	......	10	45.30	1130.50	28.70	43.00	39.00	......	....	,,	Rain.
4	......	9.45	41.00	1170.00	......	10.15	45.00	1130.50	29.55	43.00	39.00	☾	....	WNW	Clear.
5	......	7.30	41.00	1170.00	......	7.45	45.00	1130.50	29.85	44.50	41.00	......	....	,,	Clear.
7	......	7	41.00	1177.00	......	7.30	45.00	1136.50	29.43	50.00	52.00	......	....	SSW	Clouded, and a little rain.
8	9	......	41.15	1177.50	9.30	......	45.00	1137.00	29.50	55.00	55.50	......	....	,,	Rain.
9	......	9.15	40.30	1174.50	......	9.45	44.30	1134.00	29.90	56.50	55.00	......	....	,,	Clouded, and some rain.
10	8.15	......	41.00	1174.50	8.30	......	45.00	1135.50	29.90	55.00	57.00	......	....	SW	Fine weather.
,,	......	10	40.30	*1174.00	......	9.30	44.30	*1135.00	*30.00	*55.00	*52.00	......	Slight variation, E.	,,	Faintly luminous in N.W.
11	10.15	......	41.15	1175.50	10.30	......	45.00	1136.50	30.15	55.00	52.00	......	....	WSW	Clouded and damp.
12	9	......	41.15	1176.50	9.30	......	45.00	1137.00	30.20	53.00	53.50	●	....	,,	Clouded and damp.
14	......	6.45	41.00	1175.00	......	7	45.00	1135.50	30.00	51.00	50.50	......	....	S	Clouded, and some rain.
15	9.15	......	41.30	1175.50	10	......	45.00	1136.00	29.85	50.50	50.00	......	11 P.M. 25′ E. 11¼ P.M. 0′ E.	,,	Clouded, and some rain—at 11 P.M. horizon in N.N.W. a little luminous, but much increased at 11½ to 12.
16	8.15	......	41.30	1176.00	8.30	......	45.00	1136.00	29.80	51.00	48.00	......	....	NW	Fine morning, afterwards clouded.
17	8	......	41.15	1173.75	8.30	......	45.30	1134.50	30.10	48.50	48.00	......	7 P.M. 30′ E. 7¼ P.M. 15′ E. 7½ P.M. 0′ E.	,,	Fine weather, slightly luminous in N.N.W. at 7 P.M.
18	8	......	41.00	1174.00	8.20	......	45.30	1133.00	30.25	48.00	45.00	......	....	W	Clear weather.
19	9.45	......	41.15	1174.00	10.30	......	45.30	1134.50	30.20	49.50	51.00	☽	....	NW	Clouded.
21	......	7.15	41.00	1173.00	......	7.45	45.30	1133.20	30.05	47.00	47.00	......	....	WNW	Fine day—some clouds.
22	8	......	41.15	1174.50	8.30	......	45.45	1135.40	30.00	48.00	51.00	......	....	N	Some rain and damp.
23	8	......	41.00	1174.50	8.20	......	45.15	1134.60	30.37	47.00	44.00	......	....	NE	Clouded, but dry weather.
24	8	......	40.30	1174.00	8.30	......	45.30	1135.00	30.15	47.50	50.00	......	....	SW	Clouded—some showers.
25	8	......	41.00	1174.00	8.20	......	46.00	1135.50	29.87	48.00	47.00	......	....	NW	Clouded—damp.
26	......	6	41.30	1175.00	......	6.20	46.00	1136.50	29.35	49.00	48.00	○	....	WSW	Much rain.
28	......	9.30	41.00	1173.50	......	10	46.00	1134.50	29.73	48.00	42.00	......	....	NNW	Clear this evening and dry.
Mean for the month				1173.75	......	......	......	1134.17	29.78	48.83	47.37				

Note 1.—I have added 8″ to each of the numbers marked thus †, in consequence of a slight alteration having been made in the dip of needle No. 1, on the 8th of November, which appeared to add at least eight seconds to the time of its performing 80 vibrations.

2.—The numbers in the above Table marked thus *, are not comprised in the monthly averages, as I thought it best to include in the mean, only one observation for each day, and most of the results so marked are rather anomalous, or were obtained about the time when the needle varied considerably to the eastward. I may here remark that I had four needles, from which I noted the variations.

3.—There is a considerable difference in the arcs described by the two needles: this I attribute to the circumstance of No. 1 being not so heavy as No. 2, the latter having of course required a more considerable counterpoise, in order to give an elevation to its south pole equal to the depression of the north pole of No. 1.

No. 2 weighs, including the support and counterpoise	898 grains.
No. 1 weighs, including the support and counterpoise	854 ——
Difference	44 grains.

On adding 202 grains to the weight of No 1, and retaining it at the same dip, instead of 41° 15′, it described an arc of 50°, after having performed eighty vibrations. The difference in the times of the two needles, I attribute to No. 2 being more strongly magnetic than No. 1.

207

Figure 2.1 Continued

needle at Falmouth which appeared to be caused by the aurora borealis: the evidence seemed credible to the London experimentalist, who reported that all of Fox's observed variations were towards the east 'and this is what would happen if electric currents were setting from south to north in the earth under the needle, or from north to south in the space above it'.[20] It appeared that Fox's investigations were revealing a profound order of electromagnetic phenomena both above and below the Earth's surface. Given the prominence of Fox's work in *Philosophical Transactions* and Faraday's subsequent interest, this Cornish science had secured significant attention with national audiences. Fox continued to chart the movement of a magnetic needle in Falmouth long after the aurora borealis of 1831. As late as 1838 he was still measuring the effects of varying climatic conditions, including hail, rain, and storms, observing how a magnetized needle swung towards the east during mornings, before wandering almost 15' west by lunchtime, and then 10' east in the evenings.[21] Through both his underground conduction of electrical currents through metalliferous veins and his use of magnetic needles suspended at the Earth's surface, Fox's experimental arrangements demonstrated connections between a range of magnetic phenomena.

2.2. Rejection at the Royal Society

In 1832, Fox looked to complete a hat-trick of *Philosophical Transactions* publications. Despite successful articles in 1830 and 1831, Fox was about to have a painful lesson on the philosophical limits of the mine. At first, all went well, with Gilbert himself presenting Fox's paper on the effects of temperature variation on a magnetic needle and its implications for the study of terrestrial magnetism to a meeting of the Royal Society on 3 May. A week later, Fox's patron resumed this account of his fellow Cornishman's experiments to 'discover the cause of the irregularities in the indications of the intensity of terrestrial magnetism given by the vibrating magnetic needle' when suspended in a box around which water of different temperatures was circulated.[22] Through his inquiries on the 'intensity of terrestrial magnetism', Fox claimed to have demonstrated that 'the vibrating needle affords at best but a tedious means of investigating the subject; and that its indications are often anomalous in different places, however inconsiderable their distance from each other; it appeared to me highly desirable to discover the cause of these irregularities'. Fox therefore had

---

[20] Faraday, *Experimental Researches in Electricity*, 57.
[21] CUL, Add.9942/19: 'Variation of Magnetic Needle at Falmouth, March 22nd 1838'.
[22] Robert Were Fox, 'On Certain Irregularities in the Magnetic Needle, Produced by Partial Warmth, and the Relations Which Appear to Subsist between Terrestrial Magnetism and the Geological Structure and Thermo-Electric Currents of the Earth', *Proceedings of the Royal Society*, 10 (1831–2), 123–5, at 123.

a box made of sheet copper, having one side open, which was afterwards covered with glass for the purpose of observing the vibrations of a magnetic needle delicately suspended in it by unspun silk. This box, the glass side excepted, was enclosed in a copper case of much larger dimensions, with sufficient space between them to admit of my surrounding the former with water at any given temperature.[23]

On trialling how long it took a 6-inch magnetized needle to complete forty vibrations at varying temperatures, he found that the application of warm water at different positions around the oscillating needle produced disturbances to its vibrations. When heat was applied equally on all sides of the box, the needle's vibrations remained constant, but when he applied heat to only one part of the box, the oscillations became irregular. He supposed that temperature inequalities within the box produced currents of air that interfered with a magnetic needle's movements.

Having observed the flaws due to temperature variation, Fox proposed 'a substitute for the vibrating needle, which seems to me no less adapted for making experiments on the magnetic intensity at sea when circumstances permit'. This apparatus consisted 'of a needle, suspended by a few filaments of unspun silk from one & a half to two inches in length in a wooden box; or if it be intended for nautical purposes, it will perhaps be necessary that the needle should rest on a pivot to insure steadiness'.[24] On the base of the box, magnets should be placed enclosed in copper cases in which water of a uniform temperature could circulate to keep up a constant magnetic presence on the needle. This would ensure that 'the needle will become an index of intensity of the earth's magnetism, its relation to the former as well as to the latter, being equally affected by changes of temperature'. The magnets could be moved nearer to, or further from, the needle, so as to always be at a constant angle from the magnetic meridian, 'and thus the distance of one or both magnets, will be the measure of the magnetic intensity,—a scale being provided to ascertain it. A graduated circle...should be fixed immediately beneath the glass, in order to insure accuracy in observing the place of the needle'.[25] Fox suggested that his new instrument be used in mines and on top of mountains, and that it could be helpful in obtaining 'information on the profound oceans of the globe, which hitherto we have desired in vain'. Over the previous twelve months he had performed continuous experiments on the vibration of two horizontal needles in one of his underground cellars, finding 'that their intensity

---

[23] RS AP/15/20: Robert Were Fox, 'On Certain Irregularities in the Magnetic Needle Produced by Partial Warmth, and the Relations Which Appear to Subsist between Terrestrial Magnetism and the Geological Structure, and Thermo:electric Currents of the Earth' (1832), 1.
[24] Ibid. 4.   [25] Ibid. 4.

has been diminishing'.[26] While these experiments were local, his instrumental developments were conceived of as a means of extending them on a worldwide scale.

Fox also provided new results from his mine experiments, reporting that he had found no evidence of 'any increase of magnetic intensity at the depth of 1,000 or 1,200 feet below the level of the sea; but if any thing, rather the reverse; but, on the whole, the discrepancy in the results was so great, that no dependence can be placed on them as establishing a general fact of this importance'.[27] Although the direction of subterranean electric currents appeared diverse, Fox asserted the general tendency was of positive currents running from east to west. Gilbert concluded Fox's Royal Society address by reporting that the Falmouth philosopher was convinced that these electric currents exerted great influence over all phenomena of terrestrial magnetism and were the result of the great temperature of the globe's interior. This had repercussions for broader understandings of magnetic phenomena, surmised Fox. Until recently, the general view was that 'the earth's magnetism is owing to a central magnet', but having shown the 'existence of intense heat in the interior of the globe' and that a magnet could not retain its force at a high temperature, the cause of the Earth's magnetism was better explained through an understanding of the 'circulation of electrical currents around it'.[28] Terrestrial magnetism should, Fox argued, be understood as the result of a combination of the Earth's rotation from west to east and the impact of solar rays on the planet, acting so differently on its polar and tropical regions. This global disequilibrium in temperature, united with the conductive properties of rock, Fox claimed, sustained subterranean electric currents, the direction of which was determined by the Earth's rotation. He explained that the

> prevalent direction of the electrical currents, taken in the aggregate, acting on the needle in Europe, Africa, parts of Asia & North America, as well as most of the Atlantic & Indian oceans, seems at present to be to the southward of west: while the currents producing magnetism in South America, the Pacific ocean, & parts of North America, and the eastern coasts of Asia, seem to take a more westerly course, verging to the northward of west. It is said that in Asia, there are two lines of no variation, and as they are comparatively near each other, they may probably afford examples of the two cases I have suggested, in one of which, the needle on either side such line or meridian, must point towards it; and in the other from it.[29]

---

[26] Ibid. 6.
[27] Fox, 'On Certain Irregularities in the Magnetic Needle', *Proceedings of the Royal Society*, 10 (1831–2), 124.
[28] RS AP/15/20: Robert Were Fox, 'On Certain Irregularities in the Magnetic Needle', 7.
[29] Ibid. 11.

In this way, Fox linked his geological work on the formation of metalliferous veins and subterranean heat with the causes of terrestrial magnetism.

The difficulty Fox faced was in converting this Royal Society paper into a *Philosophical Transactions* publication. His treatise went out for peer review in May 1832, with the specialist in terrestrial magnetism Samuel Hunter Christie (1784–1865) appointed to evaluate Fox's submission. In late June, Christie reported on Fox's work with a devastating critique, leaving little doubt that the treatise was far below the Royal Society's expected standards. The Cornishman's style appeared to Christie 'so discursive, and the subjects to which the author, in different parts of it refers, so various, that it is difficult to say what is the precise object of the communication'.[30] Fox claimed 'opinions and views, as new' which were well known, and seemed poorly acquainted with the subject of terrestrial magnetism. That needles should be kept in wooden boxes of a consistent temperature was obvious, as were the daily effects of the Sun on a magnetic needle, which had been established in 1759, and discussed in *Philosophical Transactions* as recently as 1825. As for the magnetic influence of the Earth's heat, Christie recalled that the Royal Society of Edinburgh had published an article in 1824, arguing that the varied action of the Sun's rays on the planet, between tropical and polar regions, effectively converted the Earth into a 'vast magnetic apparatus'.[31] Christie himself had written a paper in *Philosophical Transactions* in 1827 that experimentally demonstrated how the Sun determined the diurnal variation of a needle in London. As for Fox's references to links between the aurora borealis and a magnetic needle's wanderings, these had been confirmed in 1756.[32] While stopping short of accusing Fox of plagiarism, Christie thought the Cornishman profoundly naïve. He also regretted that Fox had a tendency to contradict himself. At the start of one paragraph, for instance, Fox asserted that the cause of terrestrial magnetism must be due to phenomena at the Earth's surface. But by the end of the same paragraph, Fox had concluded 'that the principal cause of terrestrial magnetism must exist at a great depth in the Earth'.[33] Fox's reasoning appeared muddled. Even his skill as an experimentalist was in doubt. While reviewing the submitted manuscript, Christie scribbled, 'I doubt greatly these results, especially the last that heated the needle accelerated it... These results I cannot... be convinced of the accuracy.'[34] Surely, Christie surmised, the observation that a magnetic needle's vibrations accelerated with an increase in temperature should have led any competent experimentalist to 'repeat the experiment in order to determine the cause of this anomaly'.[35] He was forced to 'doubt the accuracy' of Fox's results and concluded that the paper was 'quite unfit for publication in the Transactions

---

[30] RS RR/1/73: Samuel Christie, 'Report on Mr Fox's Paper, 27th June 1832' (1832), 1.
[31] Ibid. 2–3.   [32] Ibid. 6.   [33] Ibid. 9–10.
[34] RS AP/15/20: Robert Were Fox, 'On Certain Irregularities in the Magnetic Needle', 2. Christie has annotated this.
[35] RS RR/1/73: Samuel Christie, 'Report on Mr Fox's Paper, 27th June 1832' (1832), 10.

of the Royal Society.'[36] Still, were Fox to repeat his experiments 'with care' and extend them 'to form a series from which definite results might be obtained', the research might, Christie supposed, prove of some value. What was wanted was the 'collection of new and interesting facts', not continued 'theoretical views unsupported by such evidence'.[37]

Christie's review must have been a painful blow to Fox but he made little effort to revise his manuscript and, in late 1832, secured its publication as two articles in the *Philosophical Magazine*. That the same content that Christie had rejected appeared so rapidly in print revealed something of the challenge of publishing with *Philosophical Transactions* as opposed to a less eminent journal. Fox's first article presented his experimental results on the vibrations of a magnetic needle in relation to different temperatures. Here he explained his arrangement of a suspended needle within a box of sheet copper and circulation of water of variant temperatures about the box. Again, Fox asserted that when heat was applied below the box, the needle's vibrations increased but its arcs of movement diminished. When the box's heat was consistent, there was less disturbance to the needle. He continued to attribute these observations to air currents. This was, word for word, the same text that Christie had rejected back in June. The one edit Fox did make was to remove his contradiction over terrestrial magnetism being a phenomenon that resulted from the planet's surface, as well as its interior. Fox instead asserted clearly that the source of this phenomenon 'must be far removed from us, so far indeed as to require powerful currents to produce the effects observable at the surface'.[38]

The second part of Fox's Royal Society submission appeared as a separate study of his experiments on the application of heat to igneous rocks. In his original manuscript, the inclusion of these inquiries in a paper on terrestrial magnetism had confused Christie. In the *Philosophical Magazine*, however, Fox focused specifically on how the size of small pieces of various igneous rock could be expanded when raised to extreme temperatures, before contracting on cooling. Granite, for example, increased in size when raised to a dull red heat, but broke up when at a white heat. Fox speculated from this that if igneous rocks had been caused through heat and subsequently cooled, they would be riddled with fissures. In Cornish mines, however, this was not so. As Fox observed, 'on the contrary, our mineral veins...traverse all the rocks without any necessary change in their size or direction'.[39] The fissures which contained mineral deposits could not, he concluded,

[36] Ibid. 11.   [37] Ibid. 11–13.
[38] Robert Were Fox, 'On Certain Irregularities in the Vibrations of the Magnetic Needle Produced by Partial Warmth; and Some Remarks on the Electro-Magnetism of the Earth', *Philosophical Magazine*, 1 (1832), 310–14, at 314.
[39] Robert Were Fox, 'Some Facts Which Appear to Be at Variance with the Igneous Hypothesis of Geologists', *Philosophical Magazine*, 1 (1832), 338–40, at 338.

be the result of the expansions and contractions of rocks with extreme temperature variations: rather, there must be some other law at work.

Despite rapid publications in the *Philosophical Magazine*, Christie's rejection reflected much wider anxieties surrounding Fox's work. Although a valuable supporter, even Lemon had his doubts over the reliability of experimental knowledge acquired from the mine and Fox's philosophical claims. On 9 April 1834 Lemon, now Whig MP for West Cornwall, received a letter from Fox reporting his continuing experiments on electrical currents in mines and with magnetic needles. Along with thanking Lemon for his account of Faraday's recent investigations into electrical action and providing details of a paper on magnetic attraction he had recently submitted to the *Philosophical Magazine*, he continued to press claims for the existence of subterranean electrical currents.[40] But Lemon was unsure of the credibility of these investigations and, looking for verification, forwarded Fox's letter to Faraday. Responding on 25 April 1834, Faraday expressed doubts over Fox's description of these proposed currents as an 'Electric agent', which seemed a vague claim. 'It is easy to imagine forces with certain directions as a kind of abstract notion of electricity but that is saying little', reflected Faraday.[41]

Nevertheless, Faraday confirmed that Fox's experiments were of some value. He hoped they continued, as the Cornishman, with his unprecedented access to the deepest mines in the country, appeared to be finding out much that was new about Nature and the operation of electricity, magnetism, and heat. Crucially, these were experiments that could not be conducted in a laboratory, and could neither be absolutely verified nor completely rejected from Faraday's experimental space in the Royal Institution. The problem was that Fox's experiments yielded a wealth of observations and measurements, but with little solid analysis or explanation. For all Fox's growing credentials, the credibility of his claims remained subject to evaluations from trusted authorities, among whom Faraday and Christie were both prominent. In Cornwall Fox might share scientific knowledge with local gentry and politicians, such as Lemon, but these in turn looked to Faraday for guidance. Faraday's response must have disappointed Fox, revealing that for all his work on the Earth's temperature, subterranean electrical currents, and metalliferous veins, he had produced no solid 'facts' to substantiate his claims.

Fox's earliest exchanges with Faraday had, in fact, been somewhat frosty. On first writing to him on 24 November 1827, asserting that 'Heat, Light, & electricity appear to have strong analogies & possibly a greater degree of identity than has been supposed', Fox made no impression.[42] In January 1834, Faraday sent the

---

[40] Robert Were Fox to Charles Lemon (9 Apr. 1834), 1–2; letter in author's possession.
[41] 'Letter 712: Michael Faraday to Charles Lemon (25 Apr. 1834), in Frank A. J. L. James (ed.), *The Correspondence of Michael Faraday*, ii: *1832–December 1840: Letters 525–1333* (London; Institution of Electrical Engineers, 1993), 178.
[42] Letter 730: Michael Faraday to Robert Were Fox, 20 June 1834, in James (ed.), *The Correspondence of Michael Faraday*, ii. 195–6, at 195.

Cornish experimentalist two recently published papers and promised that if ever he was in Cornwall he would certainly visit Fox at Falmouth, but remained lukewarm towards his philosophical work.[43] Undeterred, Fox attended Faraday's Royal Institution lecture 'On Combustion as an Electrical Phenomenon' later that year and subsequently wrote to him on 18 June 1834, sharing thoughts on electricity and combustion. Fox told Faraday that he had enjoyed 'thy' lecture, but wanted his opinion as to whether the phenomena of heat, light, and electricity could be combined within a united account of Nature. Likewise, he wondered if Faraday recollected his initial letter of 1827, but Faraday had only 'a faint recollection of your letter of Novr. 1827', principally because he thought the views expressed on the connections between natural phenomena to be 'common property'. Fox's speculation that heat, light, and electricity were related seemed, by 1834, obvious. Faraday observed that all who had studied the phenomena 'have had such notions', but without 'facts' these were but 'suppositions', rather than the 'truth'.[44]

It was a fair point: during the 1830s there was a growing consensus that electricity, magnetism, light, heat, and sound were all connected and, from the late 1820s, several highly-influential scientific treatises attempted to make the very connections that Fox suggested. In many respects, these were in response to the godless unifying treatises of the French mathematician Pierre-Simon Laplace (1749–1827), whose *Exposition du système du monde* (1796) and *Traité de mécanique céleste* (1798–1825) made grand connections between natural phenomena for Revolutionary audiences in France. Following contributions on sound, light, and astronomy to the *Encyclopaedia Metropolitana*, John Herschel's *A Preliminary Discourse on the Study of Natural Philosophy* (1830) connected natural phenomena within a religious framework, as did Mary Somerville's unifying treatise *Mechanism of the Heavens* (1831) and subsequent *On the Connexion of the Physical Sciences* (1834).[45] These emphasized God's determining role in the Universe and set out a vision of Nature as a unified system, regulated by a very limited number of laws. The progress of scientific knowledge had 'been remarkable for a tendency to simplify the laws of nature, and to unite detached branches by general principles', declared Somerville, confident that phenomena such as light and heat would 'ultimately be referred to the same agent'.[46] Certainly, Fox ensured his children were well acquainted with Somerville's vision of the universe, with Anna Maria, Barclay, and Caroline all studying her *Connexion* during January 1835.[47] In 1830s' Britain,

---

[43] 'Letter 697: Michael Faraday to Robert Were Fox, 23 Jan. 1834, in James (ed.), *The Correspondence of Michael Faraday*, ii. 163–4, at 163.
[44] Letter 730: Michael Faraday to Robert Were Fox (20 June 1834), in James (ed.), *The Correspondence of Michael Faraday*, ii. 195–6.
[45] John Herschel, *A Preliminary Discourse on the Study of Natural Philosophy* (London: Longman, Rees, Orme, Brown, and Green, 1830); John Herschel, 'Light', in *Encyclopaedia Metropolitana*, iv (London: Baldwin and Cradock, 1830), 341–586.
[46] Mary Somerville, *On the Connexion of the Physical Sciences* (London: John Murray, 1834), p. vii.
[47] R. L. Brett (ed.), *Barclay Fox's Journal* (London: Bell & Hyman, 1979), 73.

then, claims that heat, electricity, and magnetism were connected were hardly novel. Without trustworthy evidence, Fox's assertions over subterranean currents and the interconnections between heat and electricity appeared hollow. It is unsurprising that scientific authorities like Faraday and Christie were so critical, while local associates such as Lemon remained unconvinced.

2.3. Philosophical Miners

As ever, Fox looked to the mine for a solution to his ongoing challenge of credibility. He urgently required new evidence, beyond that which could be obtained through experiment alone. It was not only as an experimental space that the Cornish mine provided Fox with philosophical knowledge. To support his claims over the formation of mineral veins, Fox also mobilized the practical experiences of local miners. As early as 1831, he had emphasized the central role that these labourers would have to take if a comprehensive geological survey of Cornwall's mining districts was to be realized. He proposed the creation of 'an annual premium or medal, to excite practical and well informed miners to report the prevailing facts observable in the mines under their notice; not laying, as is too often the case, an undue stress on the exceptions, rather than on the rule'. Miners would, in this way, be encouraged to bring 'illustrative specimens of the different rocks and vein stones, with their local names' to the attention of natural philosophers.[48] This was something that Edinburgh University's professor of chemistry, Thomas Charles Hope (1766–1844), had emphasized to Fox when they had discussed the theory of mineral vein formation in the autumn of 1830. Hope believed that it was essential to 'the Science & the Practice of Metalliferous veins and mining' that knowledge of how seams of ore formed was developed, so as 'to furnish data to guide the miner in his operations and the Geologist in his speculations'.[49] It was the experience of working miners that Fox wanted to draw on to support his experimental investigations.

Cornish miners did indeed possess and cultivate valuable knowledge of the geological structure of the Earth. It was not just that they worked a subterranean world of varying rock types, diverse mineral veins, and strata, but that the organization of Cornwall's mines contributed to this expansion of tacit knowledge. Unlike in other British mining regions, it was the Cornish custom for a mine's landowner to offer a lease to a mining consortium, known as 'adventurers', who would cover the costs of extraction and pay dues on ore raised. The adventurers

---

[48] Robert Were Fox, *Observations on Metalliferous Veins and Their Electro-Magnetic Properties* (Penzance: T. Vigurs, 1831), 10.
[49] RS, Robert Were Fox Papers, MS/710/54: Thomas Charles Hope to Robert Were Fox (8 Oct. 1830).

would appoint a mine 'captain' with a good knowledge of mining and capable of organizing the workforce. Under this system it was not just captains who had to know the business of mining, but miners too, who were paid on the amount and quality of ore extracted. A good understanding of exactly what they were mining and how to locate the highest quality ore was essential to securing a reasonable wage. Within the mine, areas were divided into 'pitches' which miners, once every nine weeks, would bid to excavate. Given that miners were paid a fraction of the value of the ore raised, knowing which pitches were likely to be more productive was important: miners learnt to identify ore samples and pitches which were indicative of a rich mineral vein, or 'lode'. With a Cornish miner's earnings based on his experience and judgement as well as his physical labour, workers, captains, adventurers, and landowners were all concerned with the economic success of the mine.[50] Across all social classes, Cornish communities were rich networks of geological knowledge.

Through his investigations into subterranean heat, Fox drew on this practical knowledge. Cornish miners not only knew that mines became warmer with depth, but claimed that veins of tin ore produced lower temperatures than copper-heavy lodes. They also knew that gushing water springs often indicated rich seams of ore. Fox's experiments took this practical knowledge much further, developing it into his theory of terrestrial heat. In his 1857 *Cornwall: Its Mines and Miners*, J. R. Leifchild described how, concerning subterranean heat, Fox had revealed 'the curious and anomalous circumstance, that, at more than 150 fathoms deep, the progression again becomes more rapid, and that the ratio at about 150 fathoms in depth is at a minimum, and increases both at greater and smaller depths'.[51] Miners had identified a general heat beneath the Earth's surface, but Fox's experiments appeared to have demonstrated a precise correspondence between depth and temperature.

The challenge of how to make the best use of Cornwall's mining expertise was not an easy one: miners could be troublesome scientific partners. Fox's great advantage, however, was his prominence within the county's mining community. Over the course of six years, he brought together the region's mining experience through a surveying of Cornish miners to produce a substantial body of knowledge, publishing the results of this inquiry in 1837. Making the most of his social connections, he circulated a series of questionnaires to Cornwall's 'practical miners', recording the characteristics of the county's mines (Figure 2.2). Fox requested information on the name and location of each mine, the number of metallic veins or lodes, their size and direction in relation to a compass, the nature of the rocks each lode traversed, which rocks yielded the most productive

---

[50] Anthony Burton, *Richard Trevithick: Giant of Steam* (London: Aurum Press, 2000), 3–5.
[51] J. R. Leifchild, *Cornwall: Its Mines and Miners; With Sketches of Scenery; Designed as a Popular Introduction to Metallic Mines* (London: Longman, Brown, Green Longmans, Roberts, 1857), 118.

ore, and how tin and copper lay together in the rock. He also wanted to know what other metals were mixed in lodes, if 'gosson' could be found around the lodes, if the walls of tin lodes were harder than those of copper lodes, where lodes were most fruitful, whether they transversed through different types of rock, how lodes intersected with each other, and at what depths were lodes most productive for different ores.[52] Fox's questionnaires exhibited his own considerable mining experience, employing local terms for various subterranean phenomena. When writing of the walls around lodes he used the Cornish term 'capels', while referring to quartz by the miners' term 'spar' when asking if crystal found contiguous to copper ore was more porous than that around tin. He noted how this mix of ore and quartz was locally known either as 'honey-comb' or 'sugar spar'.[53] Fox had the dialect to phrase questions so as to encourage detailed responses from Cornish miners. This surveying did not appear as some elite natural philosopher interrogating mine captains and labourers, but as a Cornishman conversant in local mining practices consulting the opinions of his fellow industrialists. This provided an effective strategy in securing a vast number of responses from miners across Cornwall.

These extensive results allowed Fox to produce an account of Cornwall's subterranean mineral veins with an unprecedented level of detail. This was not just an eclectic description of the county's mineral wealth, but a vision of a grand unified system, subject to consistent laws and characteristics, substantiated by the testimonies of miners. Captain Nicholas Vivian provided details of a very productive vein at Wheal Towan, where a copper lode passed westward into the neighbouring Cliffdown Mine, but on the eastern side yielded tin but no copper. Vivian also reported that at the North Roskear Mine there were copper-tin-producing lodes, but at South Roskear just one, though of superior quantity. At the Providence Mines near St Ives, Vivian described two parallel lodes running in granite and killas, one lode producing only copper, the other a tin-copper mix.[54] Captain William Petherick believed that veins at Dolcoath Mine were most productive of ore near hanging walls, claiming that most lodes had isolated masses of rock near cross-courses, which Cornish miners referred to as 'horses'. At Dolcoath, arsenical cobalt and native bismuth both occurred with copper ore in granite. Captain John Rowse reported a cross-course containing a vein of galena intersected and shifted 2 feet by a copper lode, in Wheal Mary, near Perran.[55]

From the parish of St Just, Captain John Taylor provided evidence of the Levant Mine. Here copper ore was found near the hanging wall, with tin below. The 'guides' followed almost a north-to-south bearing, with tin lodes taking a north-west to south-east direction, many of which were 'comby'. Taylor asserted that if miners encountered a jet of water while working, this was normally a favourable

---

[52] Robert Were Fox, *Observations on Mineral Veins* (Falmouth: J. Trathan, 1837), 57–9.
[53] Ibid. 60.   [54] Ibid. 13.   [55] Ibid. 14.

QUESTIONS RELATIVE TO MINERAL VEINS, SUBMITTED TO PRACTICAL MINERS,
By Robert Were Fox.

1. Name of the Mine, as well as of the Parish or District in which it is situated.
2. Number of Metallic veins or lodes, and the description of ore which each contains.
3. Average *size, direction by compass,* and *underlie* of each lode, and whether very variable or not in these respects; and do the lodes generally increase or diminish in size in descending into the earth?
4. Nature of the rocks or country traversed by each lode, whether *granite, killas, elvan,* &c., or all of them; and the bearings of the different rocks with respect to each other.
5. If any elvan courses, (porphyritic dykes;) their appearance, hardness, sizes, directions, and underlie.
6. In which of the rocks have the respective lodes been found most productive of ore, and has there been any difference in this respect, between those of copper and tin, or of any other metal?
7. If copper lodes; do they consist of yellow or grey ore, or of any other variety, and how are the varieties of the ore situated with respect to each other in the lodes?
8. If Copper and Tin occur in the same lodes, are those Metals in different parts of them, or if near together, are they at, or near the opposite walls of the lodes, or are they intimately mixed?
9. If near the opposite walls of underlying lodes, which of these metals is the nearer to the upper or hanging walls, and which to the lower or foot walls;—are these ores separated from each other by "spar" (quartz,) or other substances; and do the hanging and foot walls differ much in hardness?
10. If other metals exist in the lodes, under what circumstances do they occur; and what minerals have a tendency to crystallize, and how are the crystallized masses situated with respect to the contiguous ores?
11. Was there *"gossan"* or other substance observed resting upon, or above the copper ore in the lodes, or if *strictly tin lodes,* were they found to be without gossan?
12. Are the walls or *"capels"* of the tin lodes harder than those of the copper lodes, and if in the case of a copper lode, one wall is harder than the other, is it the nearer one to the copper ore, or that which is the further from it?
13. Are the lodes, generally speaking, most productive of ore on the side of the hanging, or of the foot walls?
14. Is the rock or country immediately contiguous to the walls of any of the lodes, usually softer or harder than at a distance from them; is there any difference in this respect between tin lodes, and lodes of copper, &c.; and is the hardness or softness of the lodes, in any *direct* ratio, or *inverse* ratio, to the hardness or softness of the rock or country contiguous to the walls.
15. Are not the lodes often contracted into small veins or branches, and have any of these been found to open again into large lodes containing ore? In such cases, do not the opposite small veins or branches sometimes overlap each other, or become *"spliced,"* as I believe it is termed?
16. Do the lodes materially vary in size in traversing different rocks, and in which rocks are they the largest? Moreover, in passing from one rock into another, from killas, to elvan or granite, for instance, do they suffer any interruption, or break in their course, and if so, how much, and in what direction?
17. Are there any marks in the walls of a given lode, showing that one of its walls is at a lower level than the other, and, if so, to what extent; and is it not usually the hanging wall which is so circumstanced?
18. Are all or any of the walls, smooth and well defined, or are they imperceptible or indistinctly marked? In either case, are the lodes more or less hard than the ground in which they occur?
19. Are the hanging or foot walls most indistinct, and which of them are the hardest?
20. When the tin lodes meet other lodes, are they intersected by them, and if the intersections take place in their underlie, are they thrown up by them, and how much?
The same question may be asked as it respects other lodes.

Figure 2.2 A copy of the questionnaire that Fox sent to Cornish mine owners and captains in 1836, including some forty questions and a wealth of local mining terms (Author's image, 2022).

21. Are there smaller veins *having distinct walls or divisions* included between the walls of the lodes, that is, are the lodes "*comby*" near the surface, or at a greater depth, and are such small included veins parallel, or oblique, as it respects the walls of the lodes, and of what do the former consist?

22. Are there any veins of clay, or veins, or portions of the containing rock, or country, in the lodes, and are they respectively near the hanging or the foot walls?

23. Have any masses of rock been found in the lodes, termed "*horses*" by the miners, and did they appear to be completely separated by the branches of any given lode from contact with the outer walls or country?

24. What circumstances or appearances in the lodes are considered the most favorable indications of any given ore, and what the least so?

25. Is not an increase in the underlie of lodes usually less favorable for ore than when they become more vertical, and are they not generally more contracted in size, and more filled with mechanical deposits when their underlie is considerably increased?

26. If any of the lodes have crossed or intersected other lodes, has it occured horizontally, or in their underlie, and at what angles; and have they been found more productive of ore at the intersections, or less so?

27. At what depth below the surface have the different lodes been found most productive of their respective ores;—and have many cavities, or "*coughs*" been observed in them, and at what depths?

28. Have the arseniates of copper, iron, or lead, or much fluor spar, occurred in any of the lodes; and how were such substances situated in relation to the ores?

29. Is the "*spar*," or quartz immediately contiguous to copper ore, often more porous or friable, (locally termed "*honey comb* or *sugary spar*,") than that which accompanies tin ore, and even more so, than the spar which is at a distance from the copper ore in the same lode?

30. Are there any *cross courses* or *flucans* intersecting any of the lodes, and what are the directions by compass, underlie, and average sizes of the former; and are they larger or smaller at the upper than at the lower levels?

31. Are the cross courses "*comby*," or subdivided into smaller veins of clay and quartz, or other earthy matter?

32. How far do the cross courses partake of the nature of the country through which they pass?

33. Do they dislocate or heave the lodes and *elvan* courses, and how much each of them; stating the underlie of the two last, at, or near the places of intersection; andare the heaves greater or less at the upper than at the lower levels?

34. Are there any branches, small veins, or "*leaders*" of ore in any of the cross courses between the dislocated extremities of the lodes, or only detached stones of ore; and are the ores in the cross courses the same, or different in their nature or appearance from those in the lodes?

35. Are the lodes more productive of ore near the cross courses, and on both sides, or only on one side of a given cross course, and on which side?

36. Are there any branch veins of ore nearly at right angles to the bearings of the lodes;—of what ores do such rectangular veins consist;—how far have they been seen to extend;—what is their underlie;—and are they near the hanging walls of any cross courses?

37. Are there any beds, or "*floors*" of tin, copper, or other metal, and under what circumstances do they occur?

38. Have they walls like lodes, or are they interposed between the beds or laminæ of the rocks?

39. Are they connected with other veins, or quite distinct from them; in what rocks or country are they most prevalent;—and have any of the ores been observed to occur disseminated, or diffused in the rocks, not as veins, but at a distance from lodes or beds of ore?

40. What are the directions of the joints, heads, or natural divisions of the granite, killas, elvan, or other rocks, and do such joints agree, or not, with the general directions of the lodes and cross courses?

―――― o ――――

R. W. Fox, takes the liberty to submit the accompanying questions to practical miners, hoping that they will kindly reply to them, or to some of them, and add such general observations on the subject of lodes as may appear to be worthy of notice. He hopes that it will not be inconvenient to them to furnish him with the desired information without much delay; and if the fly sheet will not contain the answers, it will be quite sufficient merely to refer to the numbers marked against each question without returning this paper.

In any cases in which plans and sections of lodes and cross courses have been made, rough sketches of them will be much valued.

Figure 2.2 Continued

indicator of ore.⁵⁶ Indeed, in general, miners regarded 'a copious jet of warm water into their workings as a favourable indication in a lode', while in deeper lodes, an increase in descending and circulating water usually corresponded to an abundance of ore.⁵⁷ At Wheal Vyvyan, Captains James Rowse and Richard Dunstan stated that all the lodes ran entirely through granite, usually to a width of between 2 and 40 feet, with copper normally occurring around the hanging wall and tin near the foot wall. Captain Nicholas Grenfell junior informed Fox that at Botallack Mine, just north of Levant, there were seven tin lodes and four lodes of a tin and grey-ore mix, all with a north-west or west-north-west bearing. Most of the lodes were small, between 4 and 12 inches in width, but when crossed by other lodes running north-north-west, became productive of copper. This valuable ore could be found between 25 and 100 fathoms, but was always more fruitful around where the lodes crossed.⁵⁸ Cornish miners also had international experiences to share. Captain Martin Thomas, who had formerly superintended several mines in Chile, reported how lodes in that country generally traversed from east to west and sometimes yielded gold. Captain Richard Tregaskis provided invaluable assistance to Fox with his report, having much first-hand knowledge of veins. He stressed the productivity of lodes passing through cross-courses, which added much value to a vein. Tregaskis recalled a lode in the Consolidated Mines which had a rich vein of copper close to the eastern side of a cross-course, but was poor to the western side.⁵⁹ Fox was confident that Cornwall's veins would be found to differ little from those of other mining districts in terms of their 'mechanical characters'.⁶⁰ He drew attention to mine engineer Westgarth Forster's (1772–1835) published details on veins in northern England, 'to prove to such of our miners, as may not be aware of the circumstance, how analogous they are to the phenomena which have been noticed in the mines of Cornwall'. Forster's evidence suggested the veins of northern England resembled those of Cornwall.⁶¹

Along with this extensive survey, it was Fox's clerk who provided the most direct practical on-hand knowledge of mining and metalliferous veins. William Jory Henwood (1805–75) had begun his career as a clerk in the Fox family's foundry at Perran Wharf in 1822. After reading some natural philosophy at Charles Fox's home, Henwood grew increasingly interested in metalliferous deposits, becoming an authority on the dip and inclination of lodes through his personal experiences of mines. Based on 296 examples, Henwood asserted the mean dip of lodes in Devon and Cornwall to be about 70° from the horizon, and observed the average direction of veins in the ten principal Cornish mining districts to be 4° south of west. Henwood claimed that the majority of Cornish copper lodes ran in an east to west direction, with those of St Just tending more to the north of west than

---

[56] Ibid. 14–15. [57] Ibid. 29. [58] Ibid. 15–16. [59] Ibid. 17–18.
[60] Ibid. 18. [61] Ibid. 19 and 22.

any others.[62] Working alongside local miners, Henwood soon had an unrivalled knowledge of Cornish and Devonian mines, rising to the position of assay master of tin to the Duchy of Cornwall in 1832. When this post was abolished in 1838, he looked beyond the county for opportunities, travelling to Brazil in 1843. In 1858 he retired back to Penzance, having contributed two respected accounts on metalliferous deposits in Cornwall and Devon which appeared in the fifth and eight volumes of the Royal Geological Society of Cornwall's *Transactions*. These helped earn him election as a fellow of the Royal Society and, in 1875, the RGSC's prestigious Murchison Medal. But it was in his early collaboration with Fox that Henwood began his scientific career.[63]

Fox offered Henwood a chance to help in his research, recognizing 'the singular power of exact observation in matters scientific that made Mr. Henwood the leading authority on the subject of metalliferous deposits'.[64] Although Henwood would eventually go on to accuse Fox of plagiarism, they worked closely throughout the 1820s and 1830s, conducting subterranean experiments. In June 1830, a newspaperman from Falmouth visited Henwood at Perran Wharf, having heard about Fox's paper being read at the Royal Society on 10 June. He promised the journalist a brief abstract of this, with Fox's approval.[65] Henwood managed much of the administration for Fox's experiments during the early 1830s. In March 1831, Fox reported to Gilbert that Henwood had commenced his 'geological inspection of mines beginning with the parish of St Just. I have furnished him with apparatus to make experiments on the electricity of lodes'.[66] That October, Adam Sedgwick wrote to Henwood, conveying his 'best wishes for the progress of your underground expts [sic]' and asserting that he did 'not know of any which interested me so much'. Sedgwick also forwarded a letter from Paris, which Fourier's mathematical ally Pierre Louis Antoine Cordier (1777–1861) had written for Henwood on 3 August, stressing the importance of Fox's experiments on heat and electromagnetism in mines and hoping they would establish links between these phenomena.[67] Fox's research had not escaped the attention of France's leading mathematicians. But it is equally apparent that contemporaries like Cordier and Sedgwick acknowledged the practical assistance Henwood provided.

When Fox eventually announced the results of his mine investigations in 1837, combining his underground experiments with the data collected through his

---

[62] Leifchild, *Cornwall*, 100–4.
[63] Denise Crook, 'Henwood (William) Jory (1805–1875)', *Oxford Dictionary of National Biography*, https://www.oxforddnb.com/view/10.1093/ref:odnb/9780198614128.001.0001/odnb-9780198614128-e-12997, 23 Sept. 2004, accessed 24 Mar. 2020; G. T. Bettany, 'The Late W. J. Henwood, F.R.S.', *Nature*, 12 (1875), 293.
[64] Anon., 'Robert Were Fox, F. R. S.', *Leisure Hour*, 27/1370 (30 Mar. 1878), 197–9, at 198.
[65] RS, Robert Were Fox Papers, MS/710/52: William J. Henwood to Robert Were Fox (22 June 1830).
[66] RS, Robert Were Fox Papers, MS/710/37: Robert Were Fox to Davies Gilbert (3 Mar. 1831).
[67] RS, Robert Were Fox Papers, MS/710/101: Adam Sedgwick to William Jory Henwood (26 Oct. 1831); includes attached letter from Louis Cordier to William Jory Henwood (3 Aug. 1831).

questionnaires, this was aimed at not only privileged scientific readers, but also practical miners. He hoped 'that an extensive circulation of the Report amongst the miners of Cornwall, may induce many of them to record the results of their experiences and observations in our mines'.[68] Fox presented the vast sprawling seams of copper and tin running through Cornwall as inherently ordered and claimed that the gathering of miners' observations promised to draw out general rules about the Earth. Fox observed that copper and tin mines usually were found in junctions between granite and killas and that Cornish miners distinguished mineral veins 'into "*lodes*" or metalliferous veins, and *cross courses*, and *cross flucans*, or veins of quartz'.[69] Copper and tin lodes tended to exist in an easterly, or east-north-easterly, horizontal direction, but there were some exceptions, such as the north-westerly course of the tin lodes of St Just. As for the cross-courses and cross-flucans, these were usually to be found at right angles to the lodes in a north-north-westerly direction, sometimes containing ore. Often this network of veins descended vertically through different strata of rock. The average size of lodes was between 3 and 4 feet, but could range from 1 inch up to 10 feet in width in places. Miners believed it a 'more favourable sign' if veins included elements of tin, which usually indicated an abundance of copper nearby. At Dolcoath, for instance, a mix of red-brown iron and tin revealed the presence of a rich copper lode about 200 fathoms from the surface.[70] As for the copper itself, the ore was usually yellow or bi-sulphuret, but miners reported some black and grey ore, as well as a much rarer purple copper-containing ore that Cornish miners called 'horse flesh'.[71] Armed with the evidence collected through his survey of miners and underground experiments, Fox sought to unite his philosophical investigations with practical mining knowledge in an account of the origins of mineral veins. Fox argued that water, charged with matter, circulated through fissures and, under chemical or electrical agency, deposited mineral which crystallized, forming mineral veins. Cornish miners knew that heat and water were indicators of valuable seams of mineral wealth and this was a crucial part of Fox's theory over the formation of veins through fissures.

2.4. Credibility and Consensus

Fox not only looked to mobilize mining knowledge to build credibility into his philosophical claims, but also undertook a relentless programme of social networking, particularly through Britain's leading scientific institutions. Throughout the 1830s, he travelled extensively around Britain and Europe promoting his theories over the Earth's heat and subterranean mineral vein formation. In particular,

---

[68] Fox, *Observations on Mineral Veins*, 3.   [69] Ibid. 4.   [70] Ibid. 5–7.   [71] Ibid. 8.

the newly established British Association for the Advancement of Science (BAAS) provided an invaluable forum in which Fox promoted his work with audiences beyond Cornwall. Established in 1831, the BAAS defined its principal objective as giving 'a stronger impulse and more systematic direction to scientific enquiry' and, to this end, its organizers decided to hold annual meetings in different cities throughout the country in an effort to bring science to provincial and industrial audiences.[72] The choice of York for the inaugural meeting, supposedly at the geographical centre of the British Isles, was very much in keeping with the association's intention to cultivate scientific knowledge beyond London. Avoiding the capital, the second and third meetings were held in Oxford and Cambridge respectively, before Edinburgh in 1834, Dublin in 1835, and Bristol in 1836.[73] Fox used this new organization to enhance his philosophical credentials. Making his BAAS debut at the 1834 meeting in Edinburgh, he delivered an account of experiments made on the electricity of a copper vein in Wheal Jewel, performed just a few days before, on 1 September (Figure 2.3). Fox's son, Barclay, had assisted with these, keeping a detailed journal of the proceedings. Arriving at the mine late in the morning, Barclay and his father,

> with Captain Dee, John Michael, galvanic batteries &c., descended in Michael's shaft, furnished with miners' dresses, hats, candles & all other 'appliances & means to boot'. We stopped at the 70 fathom level to arrange the apparatus, but finding the galvanometers were left 'at grass', we were obliged to send a miner for them & meanwhile descended to the 80 fathom level, where we blasted 2 rocks with sand & surveyed a very fine & promising lode which is yielding the adventurers nearly £3000 per month. We returned to the 70 fathom & meeting the galvanometers there, settled down to experiment on the galvanism in the veins, the results of which were very satisfactory.[74]

In total, Barclay and his father spent over seven laborious hours 'in the lower regions' before returning to the surface.

Less than two weeks later, Fox gave his own account of the investigation to the BAAS at Edinburgh. At the lowest part of Wheal Jewel, he had found a black and vitreous copper vein on top of yellow sulphuret of copper. At 98 fathoms he established an observation station and then set up a series of copper and zinc plates in contact with the vein at different points, connected by copper wires. The point of this arrangement was 'to prove that the electrical action is derived from the vein, and that it is not in any degree excited by the mere contact of the metal with

---

[72] BOD, Ms Dep. Papers of the British Association for the Advancement of Science (BAAS) 5, Miscellaneous Papers, 1831–1869, Folio 39, 'First Resolution of the York Scientific Meeting' (1831).
[73] Jack Morrell and Arnold Thackray, *Gentlemen of Science: Early Years of the British Association for the Advancement of Science* (Oxford: Clarendon Press, 1981), 10, 20–2, and 98–9.
[74] Barclay, quoted in Brett (ed.), *Barclay Fox's Journal*, 69–70.

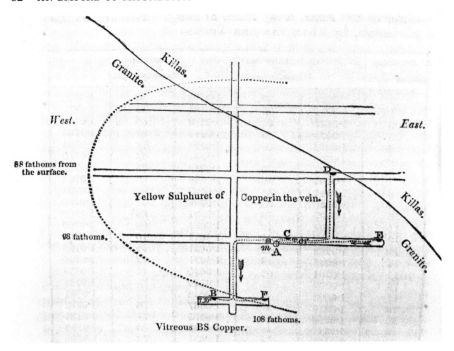

Figure 2.3 Fox's experimental arrangement for measuring the strength and direction of subterranean electrical currents, as published in the report of the BAAS's 1834 meeting (Author's image, 2020).

the ore'.[75] These new experiments were different to those Fox had reported earlier to the Royal Society in that he was trying to establish not just that mineral veins could convey an electrical current, but also that naturally occurring electrical action operated through these metalliferous seams of ore: this was a far more radical philosophical claim to make. To get an 'idea of the electric energy of the vein', Fox used a galvanic trough in the circuit, measuring the deflections made on a magnetic needle. Although noticing the vein lose voltaic activity during the experiments, Fox claimed that it continued to give a violent agitation to the needle on the establishing of a circuit, affording 'strong evidence of the energy of the electricity of the vein'. Comparing his 1834 results with the same experiments conducted at Wheal Jewel in 1830, he determined that 'the direction of the electricity remains unchanged, viz. *positive* from the east'.[76]

Fox's paper evidently made a strong impression on his audience. Cambridge polymath William Whewell was convinced of the worth of Fox's work. In his

---

[75] Robert Were Fox, 'Account of Some Experiments on the Electricity of the Copper Vein in Huel Jewel Mine', in *Report of the Fourth Meeting of the British Association for the Advancement of Science; Held at Edinburgh in 1834* (London: John Murray, 1835), 572–4, at 573.
[76] Ibid. 573–4.

address to the 1835 BAAS meeting in Dublin on the progress in mathematical theories over electricity, magnetism, and heat, Whewell drew attention to the Cornishman's experiments. The Earth's central heat had now been proven, Whewell asserted, with recent observations on the increase of temperature with depth supporting Fourier's analysis of the globe's slow cooling, and, echoing Fox, speculated that this heat was connected to the terrestrial magnetic phenomena.[77] More significantly, Fox's Edinburgh paper won him BAAS funding to purchase more accurate thermometers with which to continue his heat experiments. This marked the first of a series of grants that the society would award him, commencing over two decades of sponsorship. It was not just that BAAS meetings provided opportunities to present research, but also that they were places where Fox could socialize with the most influential figures in British science. Most importantly for Fox's canvassing of BAAS funding was his association with the organization's assistant secretary, John Phillips (1800–74). Phillips had grown up in the care of his geologist uncle William Smith, whose geological map of the British Isles earned him the epithet as the 'father of English geology'. After training as a land surveyor in 1815, Phillips moved to York in 1824, where he was appointed keeper of Yorkshire Museum the following year. He took a leading role in the formation of the BAAS and remained the society's assistant secretary for over thirty years, editing its reports and organizing its first meeting at York.[78] Throughout the 1830s, Fox cultivated a productive relationship with Phillips. He wrote to Phillips in 1836 expressing his delight that the BAAS had established a subcommittee to investigate the Earth's temperature, on which he would play a leading role, but regretted that geologist Henry De la Beche (1796–1855), on learning of this, 'disposed to leave the matter in their [the subcommittee's] hands'. Fox encouraged Phillips to try to persuade De la Beche to change his mind, 'as his authority would be important in such an investigation'.[79] At the same time, their mutual associate Charles Lemon forwarded Phillips a copy of Fox's questionnaire for miners.[80] In Phillips, Fox had an influential ally at the top of the BAAS's organization. With his support, Fox won renewed grants in 1836 and 1838 for further mine experiments.

In September 1837, the BAAS held its meeting in the bustling port of Liverpool, where Fox and his family took every chance to make further social connections.

---

[77] William Whewell, 'Report on the Recent Progress and Present Condition of the Mathematical Theories of Electricity, Magnetism, and Heat', in *Report of the Fifth Meeting of the British Association for the Advancement of Science; Held at Dublin in 1835* (London: John Murray, 1836), 1–34, at 31 and 33.

[78] Jack Morrell, 'Phillips, John (1800–1874)', *Oxford Dictionary of National Biography*, https://www.oxforddnb.com/view/10.1093/ref:odnb/9780198614128.001.0001/odnb-9780198614128-e-22163, 23 Sept. 2004, accessed 24 Mar. 2020.

[79] OUM, Papers of John Phillips, 1800–1874, Phillips 1836/7: Letter from Robert Were Fox to John Phillips (11 Feb. 1836).

[80] OUM, Papers of John Phillips, 1800–1874, Box 101, Item 12: 'Questions Relative to Mineral Veins, Submitted to Practical Miners, by Robert Were Fox' (c.1836).

Fox delivered updates on his BAAS-funded investigation, reporting on his commission from the preceding year to make 'some experiments on the electricity of metalliferous veins on a larger scale than I have yet done, and to have endeavoured to produce changes in the composition of bodies, by long-continued action of electric forces, derived from this source'.[81] Fox replicated his Cornish experiments at Coldberry and Skeers lead mines in Durham, as well as in lead mines in Flintshire. Finding little evidence of electrical currents in any of these, he conceded that 'the electrical action is much more feeble in lead veins when contained in limestone and sandstone than in copper veins included in the lower rocks, such as granite and "*killas*" or clay slate'.[82] Fox also presented new results from his ongoing heat experiments, having found the temperature of 85.3° at 290 fathoms in the Consolidated Mines at Gwennap, compared to 80° at the bottom of the Levant Mine.[83] Fox completed his BAAS report on subterranean temperature in 1840, producing extensive tables of all his observations between 1815 and 1837 (Figure 2.4).[84]

It was not just funding that Fox secured at BAAS meetings: these gatherings allowed him to build invaluable social networks and persuade eminent natural philosophers to visit Falmouth. Along with his properties at Penjerrick and Rosehill, his foundry at Perran, and local aristocratic seats like Carclew House, the town boasted a new learned society at which to host scientific gatherings. In 1833, Fox's 17-year-old daughter, Anna Maria, conceived of a new Falmouth-based organization to promote the works of local miners and artisans. Her siblings, Barclay and Caroline, enthusiastically supported the project, with Caroline suggesting the name of 'Polytechnic'. Mobilizing their father's social connections, including Charles Lemon and Gilbert, they established the Cornwall Polytechnic Society later that year, which secured royal patronage in 1825 to become the Royal Cornwall Polytechnic Society (RCPS).[85] This soon became an invaluable site in which Fox could enhance his philosophical reputation, as he transformed his home town into a leading hub of scientific inquiry and discussion. Notably, in August 1836 Fox took his family to the Bristol meeting of the BAAS. Given the city's proximity to Cornwall, this offered a chance for some of the day's leading natural philosophers to venture further south-west and inspect the region's natural, industrial, and geological curiosities and it was to Fox that they looked for

---

[81] Robert Were Fox, 'Report of Some Experiments on the Electricity of Metallic Veins, and the Temperature of Mines', in *Report of the Seventh Meeting of the British Association for the Advancement of Science; Held at Liverpool in September 1837* (London: John Murray, 1838), 133–7, at 133.

[82] Ibid. 133–4.   [83] Ibid. 136.

[84] Robert Were Fox, *Report on Some Observations on Subterranean Temperature* (London: Richard and John E. Taylor, 1841), 314–15.

[85] Alan Pearson, 'A Study of the Royal Cornwall Polytechnic Society', MA thesis, University of Exeter, 1973, 31; on the RCPS, see Julie Maxted, 'Scientific and Cultural Networks in Nineteenth-Century Falmouth: The Fox Family and the Royal Cornwall Polytechnic Society', BA diss., University of Exeter, n.d., 14–23; Brett (ed.), *Barclay Fox's Journal*, 61.

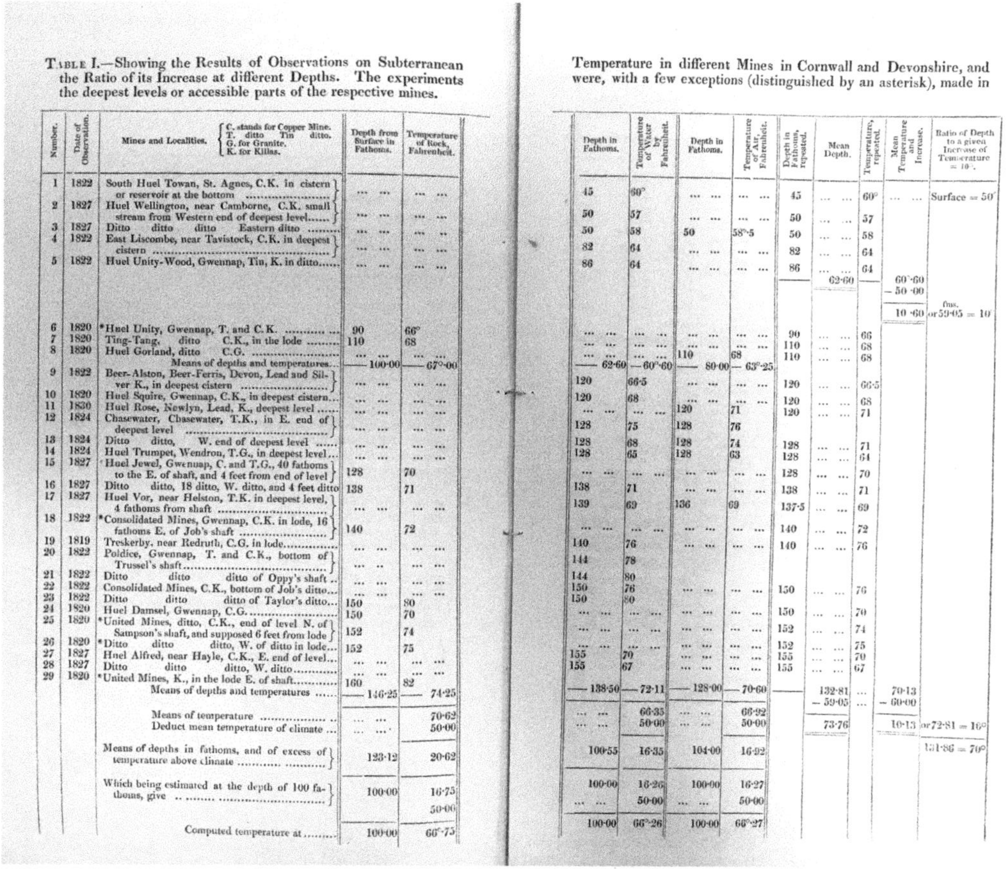

TABLE I.—Showing the Results of Observations on Subterranean Temperature in different Mines in Cornwall and Devonshire, and the Ratio of its Increase at different Depths. The experiments were, with a few exceptions (distinguished by an asterisk), made in the deepest levels or accessible parts of the respective mines.

Number.	Date of Observation.	Mines and Localities. (C. stands for Copper Mine. T. ditto Tin ditto. G. for Granite. K. for Killas.)	Depth from Surface in Fathoms.	Temperature of Rock, Fahrenheit.	Depth in Fathoms.	Temperature of Water by Fahrenheit.	Depth in Fathoms.	Temperature of Air, Fahrenheit.	Depth in Fathoms, repeated.	Mean Depth.	Temperature, repeated.	Mean Temperature and Increase.	Ratio of Depth to a given Increase of Temperature = 10°.
1	1822	South Huel Towan, St. Agnes, C.K. in cistern or reservoir at the bottom	... ...	... ...	45	60°	... ...	... ...	45	... ...	60°	... ...	Surface = 50°
2	1827	Huel Wellington, near Camborne, C.K. small stream from Western end of deepest level	... ...	... ...	50	57	... ...	... ...	50	... ...	57		
3	1827	Ditto ditto ditto Eastern ditto	... ...	... ...	50	58	50	58°·5	50	... ...	58		
4	1822	East Liscombe, near Tavistock, C.K. in deepest cistern	... ...	... ...	82	64	... ...	... ...	82	... ...	64		
5	1822	Huel Unity-Wood, Gwennap, Tin, K. in ditto	... ...	... ...	86	64	... ...	... ...	86	... ...	64		
										62·60		60°·60 − 50·00	
												10·60	fms. or 59·05 = 10°
6	1820	*Huel Unity, Gwennap, T. and C.K.	90	66°	... ...	... ...	... ...	... ...	90	... ...	66		
7	1820	Ting-Tang, ditto C.K., in the lode	110	68	... ...	... ...	... ...	... ...	110	... ...	68		
8	1820	Huel Gorland, ditto C.G.	... ...	... ...	... ...	... ...	110	68	110	... ...	68		
		Means of depths and temperatures	100·00	67°·00	62·60	60°·60	80·00	63°·25					
9	1822	Beer-Alston, Beer-Ferris, Devon, Lead and Silver K., in deepest cistern	... ...	... ...	120	66·5	... ...	... ...	120	... ...	66·5		
10	1820	Huel Squire, Gwennap, C.K., in deepest cistern	... ...	... ...	120	68	... ...	... ...	120	... ...	68		
11	1830	Huel Rose, Newlyn, Lead, K., deepest level	... ...	... ...	... ...	... ...	120	71	120	... ...	71		
12	1824	Chasewater, Chasewater, T.K., in E. end of deepest level	... ...	... ...	128	75	128	76					
13	1824	Ditto ditto, W. end of deepest level	... ...	... ...	128	68	128	74	128	... ...	71		
14	1824	Huel Trumpet, Wendron, T.G., in deepest level	... ...	... ...	128	65	128	63	128	... ...	64		
15	1827	*Huel Jewel, Gwennap, C. and T.G., 40 fathoms to the E. of shaft, and 4 feet from end of level	128	70	... ...	... ...	... ...	... ...	128	... ...	70		
16	1827	Ditto ditto, 18 ditto, W. ditto, and 4 feet ditto	138	71	138	71	... ...	... ...	138	... ...	71		
17	1827	Huel Vor, near Helston, T.K. in deepest level, 4 fathoms from shaft	... ...	... ...	139	69	136	69	137·5	... ...	69		
18	1822	*Consolidated Mines, Gwennap, C.K. in lode, 16 fathoms E. of Job's shaft	140	72	... ...	... ...	... ...	... ...	140	... ...	72		
19	1819	Treskerby, near Redruth, C.G. in lode	... ...	... ...	140	76	... ...	... ...	140	... ...	76		
20	1822	Poldice, Gwennap, T. and C.K., bottom of Trussel's shaft	... ...	... ...	144	78							
21	1822	Ditto ditto ditto of Oppy's shaft	... ...	... ...	144	80							
22	1822	Consolidated Mines, C.K., bottom of Job's ditto	... ...	... ...	150	76	... ...	... ...	150	... ...	76		
23	1822	Ditto ditto ditto of Taylor's ditto	150	80	150	80							
24	1820	Huel Damsel, Gwennap, C.G.	150	70	... ...	... ...	... ...	... ...	150	... ...	70		
25	1820	*United Mines, ditto, C.K., end of level N. of Sampson's shaft, and supposed 6 feet from lode	152	74	... ...	... ...	... ...	... ...	152	... ...	74		
26	1820	*Ditto ditto ditto, W. of ditto in lode	152	75	... ...	... ...	... ...	... ...	152	... ...	75		
27	1827	Huel Alfred, near Hayle, C.K., E. end of level	... ...	... ...	155	70	... ...	... ...	155	... ...	70		
28	1827	Ditto ditto ditto, W. ditto	... ...	... ...	155	67	... ...	... ...	155	... ...	67		
29	1820	*United Mines, K., in the lode E. of shaft	160	82									
		Means of depths and temperatures	146·25	74·25	138·50	72·11	128·00	70·60		132·81 − 59·05	...	70·13 − 60·00	
		Means of temperature	... ...	70·62	... ...	66·35	... ...	66·92					
		Deduct mean temperature of climate	... ...	50·00	... ...	50·00	... ...	50·00		73·76		10·13	or 72·81 = 10°
		Means of depths in fathoms, and of excess of temperature above climate	123·12	20·62	100·55	16·35	104·00	16·92					131·86 = 70°
		Which being estimated at the depth of 100 fathoms, give	100·00	16·75	100·00	16·26	100·00	16·27					
				50·00	... ...	50·00	... ...	50·00					
		Computed temperature at	100·00	66°·75	100·00	66°·26	100·00	66°·27					

Figure 2.4 The results of Fox's BAAS-funded report on the measure of heat in mines (Author's image, 2020).

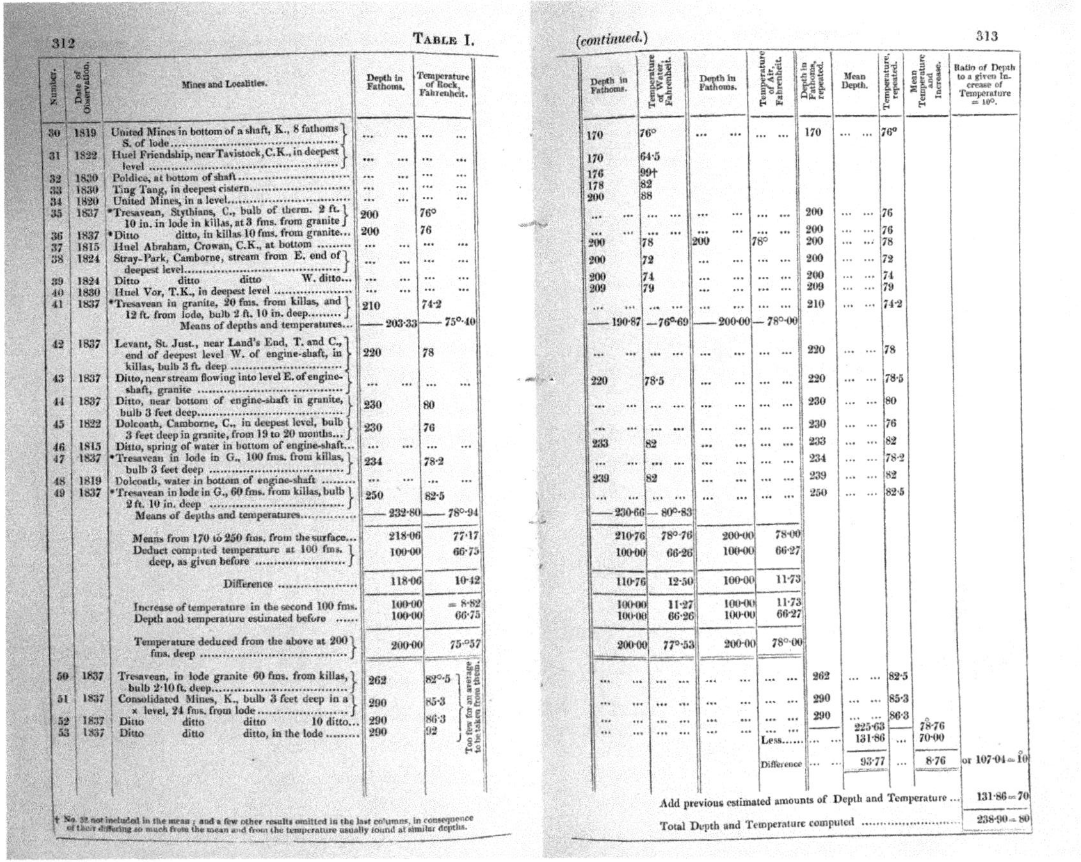

Number.	Date of Observation.	Mines and Localities.	Depth in Fathoms.	Temperature of Rock, Fahrenheit.	Depth in Fathoms.	Temperature of Water, Fahrenheit.	Depth in Fathoms.	Temperature of Air, Fahrenheit.	Depth in Fathoms, repeated.	Mean Depth.	Temperature, repeated.	Mean Temperature and Increase.	Ratio of Depth to a given Increase of Temperature = 10°.
30	1819	United Mines in bottom of a shaft, K., 8 fathoms S. of lode	... ...	... ...	170	76°	... ...	... ...	170	... ...	76°		
31	1822	Huel Friendship, near Tavistock, C.K., in deepest level	... ...	... ...	170	64·5							
32	1830	Poldice, at bottom of shaft	... ...	... ...	176	99†							
33	1830	Ting Tang, in deepest cistern	... ...	... ...	178	82							
34	1820	United Mines, in a level	... ...	... ...	200	88							
35	1837	*Tresavean, Stythians, C., bulb of therm. 2 ft. 10 in. in lode in killas, at 3 fms. from granite	200	76°	... ...	... ...	... ...	... ...	200	... ...	76		
36	1837	*Ditto ditto, in killas 10 fms. from granite	200	76	... ...	... ...	... ...	... ...	200	... ...	76		
37	1815	Huel Abraham, Crowan, C.K., at bottom	... ...	... ...	200	78	200	78°	200	... ...	78		
38	1824	Stray-Park, Camborne, stream from E. end of deepest level	... ...	... ...	200	72	... ...	... ...	200	... ...	72		
39	1824	Ditto ditto ditto W. ditto	... ...	... ...	200	74	... ...	... ...	200	... ...	74		
40	1830	Huel Vor, T.K., in deepest level	... ...	... ...	209	79	... ...	... ...	209	... ...	79		
41	1837	*Tresavean in granite, 20 fms. from killas, and 12 ft. from lode, bulb 2 ft. 10 in. deep	210	74·2	... ...	... ...	... ...	... ...	210	... ...	74·2		
		Means of depths and temperatures	203·33	75°·40	190·87	76°·69	200·00	78°·00					
42	1837	Levant, St. Just., near Land's End, T. and C., end of deepest level W. of engine-shaft, in killas, bulb 3 ft. deep	220	78	... ...	... ...	... ...	... ...	220	... ...	78		
43	1837	Ditto, near stream flowing into level E. of engine-shaft, granite	... ...	... ...	220	78·5	... ...	... ...	220	... ...	78·5		
44	1837	Ditto, near bottom of engine-shaft in granite, bulb 3 feet deep	230	80	... ...	... ...	... ...	... ...	230	... ...	80		
45	1822	Dolcoath, Camborne, C., in deepest level, bulb 3 feet deep in granite, from 19 to 20 months	230	76	... ...	... ...	... ...	... ...	230	... ...	76		
46	1815	Ditto, spring of water in bottom of engine-shaft	... ...	... ...	233	82	... ...	... ...	233	... ...	82		
47	1837	*Tresavean in lode in G., 100 fms. from killas, bulb 3 feet deep	234	78·2	... ...	... ...	... ...	... ...	234	... ...	78·2		
48	1819	Dolcoath, water in bottom of engine-shaft	... ...	... ...	239	82	... ...	... ...	239	... ...	82		
49	1837	*Tresavean in lode in G., 60 fms. from killas, bulb 2 ft. 10 in. deep	250	82·5	... ...	... ...	... ...	... ...	250	... ...	82·5		
		Means of depths and temperatures	232·80	78°·94	230·66	80°·83							
		Means from 170 to 250 fms. from the surface	218·06	77·17	210·76	78°·76	200·00	78·00					
		Deduct computed temperature at 100 fms. deep, as given before	100·00	66·75	100·00	66·26	100·00	66·27					
		Difference	118·06	10·42	110·76	12·50	100·00	11·73					
		Increase of temperature in the second 100 fms.	100·00	= 8·82	100·00	11·27	100·00	11·73					
		Depth and temperature estimated before	100·00	66·75	100·00	66·26	100·00	66·27					
		Temperature deduced from the above at 200 fms. deep	200·00	75·°57	200·00	77°·53	200·00	78°·00					
50	1837	Tresavean, in lode granite 60 fms. from killas, bulb 2·10 ft. deep	262	82°·5 (Too few for an average to be taken from them.)	... ...	... ...	... ...	... ...	262	... ...	82·5		
51	1837	Consolidated Mines, K., bulb 3 feet deep in a × level, 24 fms. from lode	290	85·3	... ...	... ...	... ...	... ...	290	... ...	85·3		
52	1837	Ditto ditto ditto 10 ditto	290	86·3	... ...	... ...	... ...	... ...	290	... ...	86·3		
53	1837	Ditto ditto ditto, in the lode	290	92	... ...	... ...	... ...	... ...	... ...	225·63		78·76	
								Less	... ...	131·86	...	70·00	
								Difference	... ...	93·77	...	8·76	or 107·04 = 10°
								Add previous estimated amounts of Depth and Temperature					131·86 = 70
								Total Depth and Temperature computed					238·90 = 80

† No. 32 not included in the mean; and a few other results omitted in the last columns, in consequence of their differing so much from the mean and from the temperature usually found at similar depths.

Figure 2.4 Continued

hospitality. Returning from Bristol, the Foxes travelled with Dr William Buckland (1784–1856), the University of Oxford's eminent professor of geology, and arrived in Falmouth in time for the opening of the RCPS's permanent new premises on Church Street. The design of architect George Wightwick, this handsome neoclassical building was a fitting temple to the town's rising scientific enterprise (Figure 2.5). Throughout the building's inaugural exhibition, held between 5 and 9 September, the Fox family hosted an elite scientific gathering who had ventured to Falmouth for the auspicious occasion. At Rosehill on the fifth, Fox held a dinner party, with guests including Buckland, Gilbert, and De la Beche. The following day, Charles Wheatstone arrived, along with news that Princess Victoria had consented to be the polytechnic's patron; a year later she became queen, ensuring the learned society numbered the monarch as a supporter. Along with dinner parties and musical evenings at Rosehill and Grove Hill, and trips to Kynance Cove, the RCPS's 1836 festival included a rich exhibition of local inventions and scientific apparatus, including several of Fox's design.[86] While Fox provided hospitality at Rosehill, Falmouth's new RCPS offered an incentive for prominent

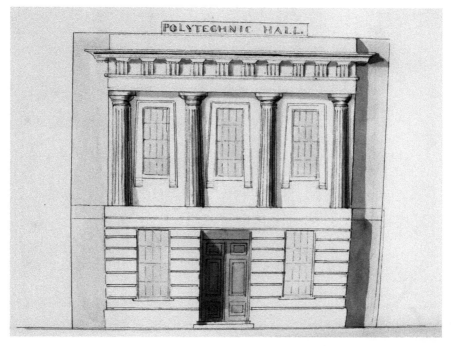

Figure 2.5 George Wightwick's architectural drawing for the RCPS's neoclassical building in the centre of Falmouth (Image by Michael Carver, 2020).

---

[86] Brett (ed.), *Barclay Fox's Journal*, 96–7.

British scientific figures to travel to Cornwall following the BAAS gathering at Bristol (Figure 2.6).

Undoubtedly Fox's most valuable acquaintance, however, was Henry De la Beche, who was a regular guest at Rosehill between 1836 and 1837. De la Beche had recently acquired government funding to support his colouring of Ordnance

Figure 2.6 The RCPS's completed building, located in Falmouth's main street (Author's image, 2020).

Survey maps, relative to rock type, which was widely believed to be of economic value for coal and mineral prospecting. After geologically mapping Devon, he formally established the British Geological Survey (BGS) in 1835, keen to extend this work to Cornwall, where Fox provided comfortable accommodation.[87] It helped that they shared a respect for each other's work. Their mutual associate, John Enys, wrote to De la Beche in early 1836 with news of Fox's investigations into mineral vein formation. From Truro, De la Beche responded in March, agreeing that he had, for some time, 'no doubt that the electro-magnetic theory of the formation of the non-mechanical part of mineral veins was the true one or at least the most probable'. Fox's investigation appeared to De la Beche 'to have been exceedingly well directed and of that true philosophic character which we should expect from him'. Declaring this a triumph for Cornish science, De la Beche thought it would be appropriate that when Fox eventually finalized his theory, he should deliver it first to a 'Cornish meeting, such as the Anniversary of the Polytechnic Society', before then presenting it 'to the scientific world at large' in the *Philosophical Transactions*. Above all, De la Beche was 'highly and particularly gratified that a Cornish man should be the one to put that matter of the filling of mineral veins straight'.[88]

De La Beche admired Fox's work, but this did not prevent him making the Cornishman's experiments a source of mirth. Drawn while overnighting in St Austell at 'The White Hart' in February 1837, De la Beche's parody newspaper, the *Mining Chronicle*, portrayed Fox's mine experiments on heat.[89] De la Beche was famous for his satirical sketches of august scientific institutions and individuals. Fox must have enjoyed the gentle mischief of a newspaper misreporting on his experimental inquiries in the hand of a friend and prestigious scientific collaborator. 'Experiment Extraordinary' gave an account of how Fox had discovered electrical action within a mine (Figure 2.7). Accompanied with a sketch of miners, disturbed by Fox's electrical experiments in the shaft, De la Beche reported how on

> Wednesday last as a gentleman well known to the scientific public was experimenting on the Electro-magnetic action of Copper Bottom North Lode, this action became so strongly developed in a shaft through which one of the experimenting wires passed, and down which three miners were descending at the time, as to draw the latter suddenly, by polarity, from the ladders to the north wall of the shaft.[90]

---

[87] David G. Bate, 'Sir Henry Thomas De la Beche and the Founding of the British Geological Survey', *Mercian Geologist*, 17/3 (2010), 149–65, at 156–62.

[88] RS, Robert Were Fox Papers, MS/710/31: Henry De la Beche to Robert Were Fox (27 Mar. 1836).

[89] 'Letter 377: Henry De la Beche (2 Feb. 1837), in T. Sharpe and J. McCartney (eds.), *The Papers of H. T. De la Beche (1796–1855) in the National Museum of Wales* (Cardiff: National Museums & Galleries of Wales, 1998), 38; Jenny Bulstrode, 'Men, Mines, and Machines: Robert Were Fox, the Dip-Circle and the Cornish System', Part III diss., History and Philosophy of Science Department, University of Cambridge, 2013, 13.

[90] RS, Robert Were Fox Papers, MS/710/124: 'Henry De la Beche, "The Mining Chronicle", 2nd Feb., 1837'.

Figure 2.7 The front cover of De la Beche's *Mining Chronicle*, with Fox's 'Experiment Extraordinary' appearing in the top-right corner (Reproduced by permission of the Royal Society, 2023).

Such open mockery of the Cornishman's mine experiments revealed that De la Beche was sure that Fox would not mistake such humour as derisive. The real highlight of De la Beche's newspaper, however, was an account of another of Fox's subterranean experiments. Complete with a drawing of a series of miners descending a shaft, losing their many layers of clothing as the mine's temperature rose with depth, 'Heat in Mines' was an imaginative portrayal of Fox's investigations into the Earth's interior heat. Here, De la Beche saw no need for thermometers. The *Mining Chronicle*'s editor reported that a

> very satisfactory mode of ascertaining the measure of heat as we descend in mines has lately been invented by a distinguished philosopher[.]...It consists...of a rope, suspended from a winze, to which rope is attached cross bars of wood at regular intervals. Upon each of these cross-bars a miner is placed, each miner being clothed in the same manner. The miners are then let down the shaft, and they are told that when they feel they have too much clothing on, they are to take as much off as will render them comfortable. By these means it is found that the heat in mines gradually increases with the depth, as the lower miners gradually despense [sic] themselves of all clothing while those at the top of the line retain their clothes.[91] [Figure 2.8]

When De la Beche arrived at Falmouth soon after his satirical musings, this marked a real coup for the Fox family. Though often frivolous, these networks were to prove crucial to the promotion of Fox's work on the formation of mineral veins.

In 1839, De la Beche delivered a very public endorsement of Fox's work in his *Report on the Geology of Cornwall, Devon, and West Somerset*. Providing a geological account of the survey of south-west England, this extensive document described the formation of rock and strata. It also offered a study of the rich seams of ore that ran through Devon and Cornwall. This was exactly the sort of economic question that had helped De la Beche win government funding and he accordingly provided an inventory of the region's mineral resources. There were few geological subjects, De la Beche contended, so important, yet so poorly understood, as the formation of mineral veins. Nevertheless, Fox's theoretical work, so weightily supported by the experiences of Cornish miners, offered the best understanding to date of such phenomena. That rock fissures were filled by 'electro-chemical agency' appeared to be supported by 'the present impression among the Cornish miners...that lodes are contemporaneous with the rocks in which they are found'.[92] De la Beche agreed with Fox's theory that heated water

---

[91] Ibid.
[92] Henry T. De la Beche, *Report on the Geology of Cornwall, Devon, and West Somerset* (London: Longman, Orme, Brown, Green, and Longmans, 1839), 352.

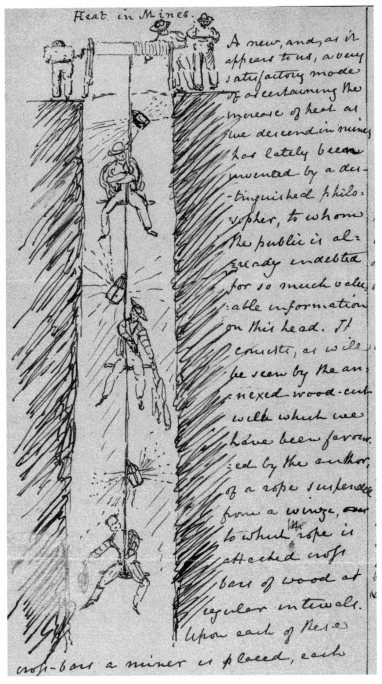

Figure 2.8 Inside the *Mining Chronicle*, a depiction of Fox's heat experiments, with miners discarding clothing as they descend deeper into a mine (Reproduced by permission of the Royal Society, 2023).

circulated through fissures, vaporizing, and depositing salt and minerals as the water transformed into steam: within the fissure, the circle of rapid currents of ascending hot steam and descending cold water caused the steam to condense, as the fissure cooled over time, leaving behind rich veins of ore.[93] Cornwall's mineral veins appeared to take a general direction parallel to the lay of the land. Employing Fox as his guide on this subject, De la Beche concurred that 'the general coincidence between the lines of tin and copper veins and those of the elven courses' was undeniable.[94] Cornwall's beds of tin and copper ore displayed a clear east-to-west direction. In total, almost all of chapter 12 of De la Beche's *Report* built on Fox's research. De la Beche concluded that Fox's experiments had proven 'that currents of electricity passing around the globe from east to west, have influenced the deposits from substances held in solution in the fissures'.[95] By 1839, with his social networking having paid off, Fox's theory over the formation of mineral veins and its implications for the study of terrestrial magnetism had considerable support.

2.5. Conclusion

Since the Royal Society's rejection of his paper on terrestrial magnetism in 1832, Fox's reputation and credibility had grown substantially. Even Fox's brutal Royal Society reviewer, Christie, conceded that Fox had met the demands of his 1832 report on the Cornishman's experimental work, writing to Lemon in December 1837 to praise Fox's RCPS report, *Observations on Mineral Veins*. 'His views,' Christie asserted, 'if well followed out, might explain many circumstances which are to be observed even in hard specimens of minerals...for I suspect that electricity is the great agent in the production of these forms.'[96] Fox's ongoing mine experiments, survey of miners, and growing support among the nation's scientific elites had persuaded even his sternest critic of his philosophical claims. His conclusions appeared to scale up the laboratory observation of electromagnetic phenomena within the Earth itself. As Faraday and Ørsted investigated the relationship between magnetism and electricity through carefully orchestrated arrangements of apparatus, so Fox employed similar techniques and instruments in mines across Cornwall to examine directly how these forces operated in Nature. Likewise, Fox's detailed collection of mining knowledge of veins drew on very different evidence to that usually employed in geological enquiry. He was providing not the observations of a gentlemanly specialist, but tacit know-how which took miners years of hard work to accrue. He was mobilizing the authority

---

[93] Ibid. 375–7 and 322.   [94] Ibid. 309.   [95] Ibid. 394.
[96] RS, Robert Were Fox Papers, MS/710/23: Samuel Hunter Christie to Charles Lemon (10 Dec. 1837).

of Cornwall's mining community in a way that was quite different to the elite products of laboratory science in the academic centres of London, Oxford, Cambridge, and Edinburgh. Yet knowledge fashioned in Cornwall was of uncertain worth. On the one hand, Fox had immensely privileged access to mines and miners, as well as a superb location from which to observe the aurora borealis. He could claim that, in Cornwall, he observed Nature without the need for contrived laboratory arrangements: Fox's credibility rested on assertions that he experimented directly on natural phenomena. This made his work persuasive, but also constrained the degree to which leading scientific audiences were prepared to accept it. Both Faraday's continuing scepticism over the existence of subterranean electrical currents and Christie's devastating review of Fox's submission to *Philosophical Transactions* in 1832 demonstrated how hard it was for scientific authorities to validate Fox's philosophical claims. Over the next few years, however, Fox would develop new strategies to build a consensus around his work, including careful networking, public displays, and ingenious experimental exhibitions.

As much as Fox's early magnetic experiments were conducted on a very local scale, it is evident not only that he conceived of his work as having global implications for the Earth's natural phenomena, but also that his growing instrumental experience offered the chance to extend this research beyond Cornwall. By the 1840s, Fox's attention had largely shifted towards the study of terrestrial magnetism. Fox's development of instruments with which to conduct experiments in mines shaped his future design and construction of new devices for measuring the Earth's magnetic field around the world. His use of magnetized needles to measure magnetic phenomena in mines and in the atmosphere, such as the aurora borealis, involved the careful management of temperature and exclusion of disturbing influences. In developing new instrumental arrangements, isolating needles, and producing more accurate data, Fox was hoping to contribute to what he believed should be a worldwide investigation. In calling for experiments, similar to his own in Cornwall, to be replicated in foreign mines, mountains, and both lower and higher latitudes, Fox revealed the global aspirations of his scientific enterprise. Within the context of increasing philosophical interest in terrestrial magnetism, both in Britain and in continental Europe, Fox's experiences of the mine and study of subterranean phenomena were to become increasingly significant.

3
Survey and Science

Polar Expeditions, Terrestrial Magnetism, and the Instruments of Empire, 1815–1839

> With a blow from the top-maul Ahab knocked off the steel head of the lance, and then handing to the mate the long iron rod remaining, bade him hold it upright... Then, with the maul, after repeatedly smiting the upper end of this iron rod, he placed the blunted needle endwise on the top of it, and less strongly hammered that, several times, the mate still holding the rod as before. Then going through some small strange motions with it—whether indispensable to the magnetising of the steel, or merely intended to augment the awe of the crew, is uncertain—he called for linen thread; and moving to the binnacle, slipped out the two reversed needles there, and horizontally suspended the sail-needle by its middle, over one of the compass cards. At first, the steel went round and round, quivering and vibrating at either end; but at last it settled to its place, when Ahab, who had been intently watching for this result, stepped frankly back from the binnacle, and pointing his stretched arm towards it, exclaimed, 'Look ye, for yourselves, if Ahab be not lord of the level lodestone! The sun is East, and that compass swears it!' One after another they peered in, for nothing but their own eyes could persuade such ignorance as theirs, and one after another they slunk away. In his fiery eyes of scorn and triumph, you then saw Ahab in all his fatal pride.
>
> Ahab fashions a magnetic compass, Herman Melville, *Moby-Dick* (1851)

Herman Melville's Ahab has long had a bad press. As captain of the infamous *Pequod*, he is remembered primarily for his monomaniac hunt for Moby-Dick, the cursed white whale. What is forgotten is that Ahab was, in fact, a highly skilled navigator. The morning after a thunder and lightning storm hit the *Pequod*, compromising its two magnetic compasses, fear seized the crew. Both compasses were pointing east, but the *Pequod* was unmistakably sailing west. Ahab remained calm: 'Thunder turned our compasses—that's all', he casually remarked to first mate Starbuck.[1] No one else on board had experienced this phenomenon before, but the old captain knew what to do. Standing before the binnacle, he took a bearing of the Sun, ordered the ship's course to be altered according to the deviation of the two corrupted compasses, and then fashioned his own compass. Ahab knew that it was practically possible to steer with an incorrect compass once the amount of deviation was known, but he also grasped that this 'was not a thing to be passed over by superstitious sailors, without some shudderings and evil portents'. On receiving the lance, needle, and maul, Ahab turned to his frightened sailors, boasting that 'the thunder turned old Ahab's needles; but out of this bit of steel Ahab can make one of his own'.[2] In full view of his astounded crew, he removed the lance's steel head, leaving a long iron rod. A crewman held this firmly while Ahab used the maul to beat the upper end of the rod before placing the needle on it and then hammering this too. He then suspended the newly magnetized sail-needle by a thread over a compass card taken from one of the defected instruments. After some quivering, the needle settled pointing north, giving an accurate bearing, confirming that Ahab truly was 'lord of the level lodestone'.[3] With his immense skill, Ahab demonstrated a godlike mastery of magnetic phenomena that restored the crew's faith in their obsessive captain and, at once, sealed their inevitable fate.

Melville's use of magnetic compass error to confirm both Ahab's navigational experience and his mesmeric control over his crew would have been immediately familiar to mid-nineteenth-century readers. Published in 1851, *Moby-Dick* appeared at a moment when questions of magnetism and unreliable compasses were increasingly urgent. The increasing use of iron in the construction of ships raised concerns over safe navigation, with the disturbing influence of the industrial metal on ships' compasses a growing problem in mid-nineteenth-century Britain. No one was more concerned with this threat than the celebrated whaling captain William Scoresby (1789–1857), who had personal experience of the dangers of unreliable compass readings. Indeed, his account of fashioning a new magnetic compass needle from the metal of a ship, published in 1823 in his *Journal of a Voyage to the Northern Whale-Fishery*, informed Melville's later narrative of Ahab's performance before his crew.[4] Scoresby attributed large British shipping losses to erroneous compass readings, including thirty ships wrecked on Portugal's coast in 1804, two men-of-war lost in Danish waters, and several warships and merchantmen sunk off the Isle of Texel between 1811 and 1812. These navigational disasters had, he argued, resulted from the increasing amount of

---

[1] Herman Melville, *Moby-Dick, or The Whale* (1851; London: Macmillan, 2004), 684.
[2] Ibid. 685.   [3] Ibid. 686.
[4] Jenny Bulstrode, 'Cetacean Citations and the Covenant of Iron', *Notes and Records*, 73/2 (2019), 167–85, at 173.

iron that naval ships carried.[5] It seemed that this problem would get even worse, as ships included ever more iron in their construction throughout the 1820s and 1830s, following the launch of Britain's first ocean-going iron-hulled vessel, the *Aaron Manby*, in 1821. By 1845, with Isambard Kingdom Brunel's iron-hulled SS *Great Britain* finally launched, the rise of the material in naval architecture appeared to be risking the safety of oceanic navigation and, with it, Britain's global dominance of trade. Increasingly, ships started to go missing, fuelling public fears over the devastating effects of terrestrial magnetism on iron ships and their compasses. Shipwreck figures escalated until, between 1852 and 1860, one in every 210 British ships was lost at sea, including 10,336 vessels wrecked around the British Isles alone.[6]

As Alison Winter has argued, debates over how to resolve this problem involved complex questions over who could be trusted as a scientific authority. While Cambridge mathematician George Biddell Airy put forward his own method of averting magnetic disruption by positioning magnets around a ship's compass, in reference to mathematical theory, Scoresby recommended placing compasses on wooden masts, high above a ship's iron. Airy's mathematical solution depended on calculations of the magnetic error a ship's hull induced, but it appeared a fragile remedy; the magnetic character of a ship was liable to change over time and space, as well as from the force of waves encountered at sea. As captains observed incorrect readings from compasses corrected to Airy's specifications, so their faith in his mathematical credentials and skill wavered. By the 1850s, this had escalated into a crisis of confidence, as public belief in iron for ship construction collapsed. In 1854, the newly built *Tayleur* ran aground, killing 290 passengers, despite having three compasses corrected by Airy's methods. Scoresby was adamant that the mathematician was to blame for failing to understand the sea's percussive impact to the ship. In contrast, Scoresby appeared a trustworthy alternative, with his evangelical faith and heroic endeavours as a whaling captain in the 1810s securing him a reputation as a reliable spiritual and scientific authority.[7]

In the context of this ongoing crisis in navigation, the question of the Earth's magnetism was transformed from a philosophical problem into an urgent challenge that reached the highest echelons of government. The public panic surrounding magnetic deviation ensured that this was a national concern for natural philosophers and statesmen alike. Combined with growing international interest

---

[5] John Cawood, 'Terrestrial Magnetism and the Development of International Collaboration in the Early Nineteenth Century', *Annals of Science*, 34/6 (1977), 551–87, at 566.

[6] Alison Winter, '"Compasses All Awry": The Iron Ship and the Ambiguities of Cultural Authority in Victorian Britain', *Victorian Studies*, 38/1 (Autumn, 1994), 69–98, at 71–2 and 76–7; see also Crosbie Smith, *Coal, Steam, and Ships: Engineering, Enterprise, and Empire on the Nineteenth-Century Seas* (Cambridge: Cambridge University Press, 2018), 157–8.

[7] Winter, '"Compasses All Awry"', 73–9 and 82–3.

in terrestrial magnetism and the continuing quest for the North-West Passage, Robert Were Fox's work on the relationship between subterranean heat, electricity, and magnetism would take on increased significance. Throughout the 1830s, it was to become part of a much broader scientific investigation into the Earth's physical properties. Fox increasingly focused his research on the nature of the Earth's magnetic force, fashioning new practices, instruments, and experimental arrangements to examine this phenomenon. In doing so, as this chapter demonstrates, he established Falmouth as a centre of magnetic science that would attract the nation's leading authorities on questions of terrestrial magnetism. These included naval officers, like John Franklin and Robert FitzRoy, as well as promoters of what would eventually become the government's magnetic survey, notably Edward Sabine and James Clark Ross. Above all, it was the promise of Fox's dipping needles, crafted through years of mine experiments, that drew these audiences to Cornwall.

3.1. The Science of Terrestrial Magnetism

Over time and space, terrestrial magnetism changes both in strength and in the wandering of the north and south magnetic poles. Caused by convection currents in the Earth's spinning molten core, which rotates at a different speed to the Earth's crust, the fluctuation of the world's magnetic field had been a well-known phenomenon since the late fifteenth century.[8] As early as 1188, Chinese navigator Shen Kuo recorded the northerly attraction of a magnetized needle. From 1418, Portuguese ships began setting their courses by compasses and, by 1492, the difficulties of calculating the difference between magnetic north and geographic north as depicted on navigation charts and determined by the Pole Star, known as magnetic 'variation', were familiar to transatlantic navigators. In 1514, João de Lisboa published the first treatise on a magnetic needle, *Tratado da agulha de marear*, including observations on the phenomena of magnetic variation. It took longer for the second characteristic of the Earth's magnetic field to be observed. In 1581, the London compass maker Robert Norman (*fl.* 1560–1605) noticed how one of his very long magnetic needles not only pointed north, but also appeared to dip down towards the pole. He subsequently mounted a needle horizontally to measure the incline of this force: this was effectively the world's first dipping needle. Soon after Norman's observation of dip, an Oxford mathematician, William Gilbert (1544–1603), published his *De Magnete, Magneticisque Corporibus, et de Magno Magnete Tellure* in 1600, where he claimed that the

---

[8] Andrew Lambert, *Franklin: Tragic Hero of Polar Navigation* (London: Faber and Faber, 2009), 73.

Earth's interior consisted of iron and operated like a huge magnet.[9] By the mid-seventeenth century, it was also apparent that the Earth's magnetism changed over time, with measurements of London's magnetic variation having declined by 7° over the course of fifty-four years. Edmond Halley's (1646–1742) work in the 1680s and 1690s extended this body of magnetic knowledge further, as he developed his theoretical understanding of the Earth having four magnetic poles, two fixed on the surface of the globe and two revolving within the planet's inner nucleus. Suitably impressed, King William III gifted Halley the *Paramour* with which to go to sea and test his hypothesis. He subsequently began mapping lines of equal magnetic variation over charts for the Atlantic, demonstrating that magnetic meridians did not form regular arcs. In 1722, the instrument maker George Graham (1673–1753) presented his findings that magnetic needles moved throughout the day to the Royal Society. He found that a dipping needle took time to come to rest, arguing that this revealed the magnetic intensity at work on the needle and could be used to calculate the varying strength of the Earth's magnetic force in different locations.[10]

By the nineteenth century, then, there were assumed to be three characteristics to the Earth's magnetism, each of which would require surveying on a global scale to be fully understood. First, there was magnetic variation, or 'declination', being the difference between magnetic north as shown by a compass and geographic north as shown by the Pole Star. This could be calculated using an accurate compass, known as an azimuth. The second property was inclination, or 'dip', which was the horizontal angle given by a magnetized needle pointed towards the magnetic north pole. Finally, there was intensity, this being the strength of the Earth's magnetic force. The result of this survey work would be the production of charts, with isogonic maps showing lines of equal deviation, isoclinic maps illustrating lines of equal dip, and isodynamic maps displaying lines of equal intensity. For measuring the latter two of these qualities, specific instruments were required: dipping needles for inclination and magnetometers, consisting of a suspended magnetic needle put in motion, for intensity.

During the late eighteenth century, France had taken a lead in this science, with the Paris Observatory the centre of the study of terrestrial magnetism. But it was the arrival of Alexander von Humboldt (1769–1859) that would elevate Paris to the forefront of magnetic investigation. Having arrived in the city in 1798 to purchase instruments, Humboldt worked at the observatory, gaining experience with its dipping needles, before commencing a six-year expedition to South America in 1799. Humboldt's magnetic investigations showed that the Earth's

---

[9] Anita McConnell, 'Surveying Terrestrial Magnetism in Time and Space', *Archives of Natural History*, 32/2 (2005), 346–60, at 347–8.
[10] Granville Allen Mawer, *South by Northwest: The Magnetic Crusade and the Contest for Antarctica* (Edinburgh: Birlinn Ltd, 2006), 5–7 and 10; McConnell, 'Surveying Terrestrial Magnetism in Time and Space', 348–9.

magnetic force increased as an observer moved away from the equator and travelled closer to the magnetic poles.[11] Humboldt returned to Paris in 1804, where he spent the next two decades developing his 'cosmical' vision of Nature in which terrestrial magnetism was but one force among many connected natural phenomena. These views gained influence following Ørsted's discovery that electricity and magnetism were related in 1820.[12] At the core of Humboldt's cosmical tradition was a commitment to mapping global patterns of natural phenomena, combining local details to reveal larger worldwide systems.[13] Determined to collect further magnetic data, Humboldt and his colleague, François Arago (1786–1853), organized a loose network of observatories throughout Europe to investigate magnetic phenomena.[14] Of particular interest was the correlation between magnetic needles and atmospheric auroras, these being electrical discharges which manifest themselves as light. Irish astronomer Henry Usher had reported on these in 1798, claiming that auroral discharge occurred in the plane of the magnetic meridian. Believing that this phenomenon was related to sunspots, with such solar activity producing massive electric currents that affected the Earth's atmosphere, Arago observed the aurora borealis over Paris in February 1817, which caused his magnetic needle to fluctuate in what Humboldt had termed 'magnetic storms' in 1806. The spectacular auroras of 1817 and 1818 that were witnessed across Europe, following the sunspot maximum of 1816 to 1817, stimulated new interest in terrestrial magnetism throughout France, Germany, and Britain. Along with establishing correspondence and cooperation with Carl Gauss (1777–1855) and Wilhelm Weber (1804–91) at Göttingen, in 1822 Humboldt and Arago travelled to England to make magnetic observations, complementing several French expeditions to gather magnetic information. In addition to a series of land surveys, the voyages of the *Coquille* in 1822, the *Thétis* between 1824 and 1826, and the *Astrolabe* from 1826 until 1829 all collected magnetic data for the French-driven project. Humboldt and Arago orchestrated an expansion of magnetic measurements with which to substantiate their conviction that the Earth's heat, including the Sun's influence through aurora, was related to its magnetic phenomena.[15]

Despite Humboldt's international eminence, Paris's dominance of the science of terrestrial magnetism was waning by the 1830s. Adopting Humboldt's cosmical framework, Jean-Baptiste Biot (1774–1862), a French mathematician, and

---

[11] Mawer, *South by Northwest*, 10; Christopher Carter, *Magnetic Fever: Global Imperialism and Empiricism in the Nineteenth Century* (Philadelphia: American Philosophical Society, 2009), 9.

[12] John Cawood, 'The Magnetic Crusade: Science and Politics in Early Victorian Britain', *Isis*, 70/4 (Dec. 1979), 493–518, at 495–7; Cawood, 'Terrestrial Magnetism', 561–5.

[13] John Tresch, 'Even the Tools Will Be Free: Humboldt's Romantic Technologies', in David Aubin, Charlotte Brigg, and H. Otto Sibum (eds.), *The Heavens on Earth: Observatories and Astronomy in Nineteenth-Century Science and Culture* (Durham, NC: Duke University Press, 2010), 253–84, at 272.

[14] Cawood, 'The Magnetic Crusade', 496; Cawood, 'Terrestrial Magnetism', 566.

[15] Cawood, 'Terrestrial Magnetism', 569–70, 574–6, and 582.

Christopher Hansteen (1784–1873), Norway's leading magnetic authority, developed a mathematical theory of the Earth's magnetic phenomena, revitalizing Halley's theory of four magnetic poles. In Germany, this was to come under increasing attack. After conducting survey work in Hanover to determine the shape of the earth between 1818 and 1832, Gauss rejected the four-pole interpretation of terrestrial magnetism in his influential *Allgemeine Theorie der Erdmagnetismus*. He analysed the earth as a collection of magnets, limiting terrestrial magnetism to the surface and interior of the planet. Although this would not be published until 1838, Gauss's 1832 paper on the intensity of terrestrial magnetic force provided the mathematical foundations for what he hoped would be a global network of magnetic observatories.[16] At Göttingen, Gauss and Weber envisioned a far more extensive international system of magnetic inquiry than Humboldt's scheme, establishing the Göttingen Magnetische Verein in 1834. Gauss envisaged this scheme as a worldwide investigation, promoting his instruments, daily routine of observations, and synchronization of magnetic readings as international standards.[17] By 1840, Gauss and Weber had persuaded over twenty international observatories to join their Magnetische Verein, eclipsing Humboldt and Arago's earlier Paris-led scheme. But it was not just Göttingen and the German lands that had taken the initiative in the investigation of terrestrial magnetism. Within Gauss and Weber's network of magnetic observers was an increasingly influential British element.

In Britain, the significance of terrestrial magnetism had always possessed a quite different character to Continental investigations. As well as questions of navigation, Gauss and Humboldt were equally concerned with the philosophical implications of magnetic observations and collected data mostly from observatories and land-based surveys. However, for an island with an immense navy and unrivalled mercantile commerce, understanding terrestrial magnetic phenomena carried far more practical urgency for the nation's shipping and, as a result, was focused on maritime magnetic observations. Yet for all these anxieties over navigation and iron, in the 1830s the Admiralty was not overly concerned with the risks of compass deviation: Captain Matthew Flinders (1774–1814) had, it seemed, solved this problem in 1805. The protégé of Captain William Bligh (1754–1817), himself a student of Captain James Cook (1728–79), Flinders had been extremely interested in questions of navigational error and believed that iron was frequently to blame. As a remedy, he prescribed the swinging of a ship through thirty-two points of the compass while at anchor, with the magnetic variation at each heading being recorded. Captains could, in this way, know the

---

[16] Cawood, 'The Magnetic Crusade', 497–8; James Gabriel O'Hara, 'Gauss and the Royal Society: The Reception of His Ideas on Magnetism in Britain (1832–1842)', *Notes and Records*, 38/1 (Aug. 1983), 17–78, at 25–8; W. K. Bühler, *Gauss: A Biographical Study* (Berlin: Springer-Verlag, 1981), 95–109.
[17] O'Hara, 'Gauss and the Royal Society', 28.

magnetic disturbance induced from the iron on their ships.[18] So while the practical challenge of compass deviation was of increasing importance for British philosophical investigations into terrestrial magnetism, throughout the early nineteenth century it was Britain's Continental rivals who led research into this subject.

Things changed in 1818. Following the final defeat of Napoleon Bonaparte at Waterloo in 1815, the Royal Navy found itself with little to do. With the boredom of peace, a generation of naval officers were desperate for new challenges and opportunities for advancement. In 1817, news arrived in England of just such a chance. Scoresby reported that the ice in Baffin Bay had cleared, opening the region up to navigation for the first time since the seventeenth century and raising the possibility that a North-West Passage from the Atlantic to the Pacific might be found. Emphasizing the scientific and commercial potential of discovering the fabled trade route, the Second Secretary to the Admiralty, John Barrow, organized an expedition to the Arctic under the command of the experienced captain John Ross (1777–1856). It was the first of a series of Arctic voyages in the hunt for the passage that would obsess Barrow until his death in 1848.[19] These expeditions would, increasingly, characterize British magnetic science.

As well as searching for a navigable channel, the expedition's officers were to collect scientific data through magnetic, astronomical, and meteorological observations. With Ross commanding the *Isabella* and Edward Parry in charge of the *Alexander*, the expedition sailed in January 1818 for the Davis Strait and Baffin Bay. Along with his dashing young nephew, James Clark Ross (1800–62), Ross's officers included a host of future celebrities of Arctic exploration and magnetic science, including Edward Sabine (1788–1883), John Franklin (1786–1847), and George Back (1796–1878). The expedition confirmed that Baffin Bay was open, but when John Ross mistook low-lying cloud at Lancaster Sound for a mountain range, the so-called 'Croker Mountains', he returned to England without closer inspection. Furious that no sea route had been found, Barrow blamed Ross entirely and despatched a new expedition under Parry's command to find out if the Croker Mountains could be passed. By 1820, Parry had confirmed Ross's error and sailed through Lancaster Sound on the *Hecla*, but this new endeavour likewise failed to find a passage. Barrow's efforts were not, however, just aimed at charting a North-West Passage. Under Parry's command, Barrow appointed James Clark Ross to conduct scientific work on the *Hecla*. The rising star of the Royal Navy, young Ross won a reputation for his zeal in magnetic observations, which secured him posts on Parry's two subsequent voyages to the Arctic on the *Fury* between 1821 and 1823 and from 1824 until 1825. On Parry's 1827 expedition, back on the *Hecla*, Ross served as second in command and managed to trek

[18] Mawer, *South by Northwest*, 12; Lambert, *Franklin*, 24–5.   [19] Lambert, *Franklin*, 10–14.

over the ice to 82° 45′ N, just 500 miles south of the geographical north pole: a record that would stand until 1875 (Figure 3.1).[20]

For all that James Clark Ross was the undoubted pin-up of Arctic exploration, it was Sabine, his trusted colleague, who garnered the most scientific reputation for the study of terrestrial magnetism. Born in Dublin, Sabine grew up in a family steeped in military tradition, with his great-grandfather a general in William III's army. Aged 14, Sabine entered the Royal Military Academy at Woolwich, where he was commissioned a second lieutenant in the Royal Artillery in 1803. Extremely well connected socially, he turned his attention to science and exploration following the peace of 1815. Through his brother-in-law, Henry Browne, Sabine became acquainted with Captain Henry Kater, a fellow of the Royal Society with a specific interest in terrestrial magnetism, who provided the aspiring artillery officer with his first magnetic instrument. With the Royal Society's support, Sabine accepted an appointment as astronomer to John Ross's 1818 Arctic Expedition, with instructions to determine latitude and longitude, to measure the direction and intensity of the Earth's magnetism, and to observe a host of natural phenomena including tides, gravitation force, currents, aurora, and the salinity of seawater.[21] On returning, Sabine's magnetic work, delivered as a two-part account to the Royal Society in February 1819, secured great attention with Britain's science elites. Employing Browne's dipping needle, built by Nairne & Blunt, Sabine performed measurements by drawing the needle to a horizontal position, aimed in the direction of the magnetic meridian, and then releasing it 'at an observed moment of time'. He then watched and timed how long the needle oscillated until the arcs became 'too small to be readily distinguished', as it returned to the angle of dip. This furnished Sabine with readings of magnetic intensity. After an initial dip measurement of 70° 7′ 39″ in Regent's Park in London, observations continued throughout the expedition, recording a dip of 85° in the Arctic. Experiments 'on ice' provided especially good opportunities for magnetic results away from the disturbance of the ship's iron. On 9 June 1818, for instance, Sabine made observations on a huge iceberg to which the ships were anchored; he later made a series of intensity readings on the ice of Baffin Bay.[22]

---

[20] Ibid. 19; Elizabeth Baigent, 'Ross, Sir James Clark (1800–1862)', *Oxford Dictionary of National Biography*, https://www.oxforddnb.com/display/10.1093/ref:odnb/9780198614128.001.0001/odnb-9780198614128-e-24123, 23 Sept. 2004, accessed 24 Mar. 2020.

[21] Gregory A. Good, 'Sabine, Sir Edward (1788–1883)', *Oxford Dictionary of National Biography*, https://www.oxforddnb.com/view/10.1093/ref:odnb/9780198614128.001.0001/odnb-9780198614128-e-24436, 23 Sept. 2004, accessed 24 Mar. 2020; on naval survey sciences, see Richard Dunn and Rebekah Higgitt (eds.), *Navigational Enterprises in Europe and Its Empires, 1730–1850* (Basingstoke: Palgrave Macmillan, 2016).

[22] Edward Sabine, 'Observations on the Dip and Variation of the Magnetic Needle, and on the Intensity of the Magnetic Force; Made During the Late Voyage in Search of a North West Passage', *Philosophical Transactions of the Royal Society of London*, 109 (1819), 132–44, at 132–4.

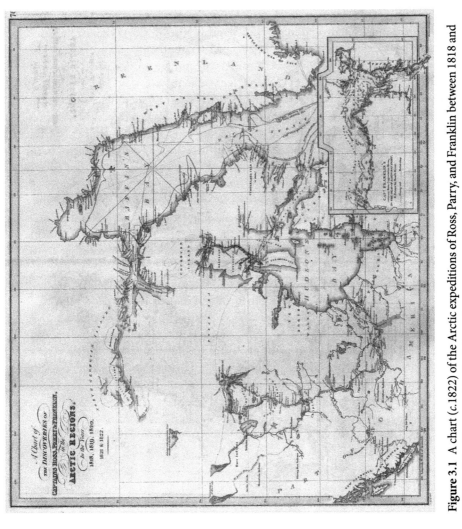

Figure 3.1 A chart (c.1822) of the Arctic expeditions of Ross, Parry, and Franklin between 1818 and 1822, by John Thomson (Author's image).

In sharp contrast to John Ross's humiliating failure to find the North-West Passage, Sabine's magnetic observations were triumphantly published in the Royal Society's *Philosophical Transactions*. Importantly, however, he also claimed to have advanced knowledge of the relationship between terrestrial magnetism and the influence of a ship's iron on its compass. With an azimuth compass, Sabine compared readings of magnetic variation for the North Pole made onboard the *Isabella* with identical measurements on nearby ice. At one point, the ship gave a compass bearing of 90° 20' in contrast to 75° 30' obtained on ice.[23] This was just one of many such disruptions. Equally crucial, Sabine asserted, was the influence of magnetic dip and intensity on a ship, in terms of how its iron disrupted the compass. 'The influence of the ship's iron on their compasses increasing, as the directive power of magnetism diminished', Sabine explained, produced 'irregularities that rendered observations on board ship of little or no value towards a knowledge of the true variation'. Sabine's azimuth comparisons emphasized 'how essential it is to navigation in high latitudes, that the nature of errors which the ship's attraction produces in her compasses should be understood'.[24] This was a firm declaration of the urgency not just of measuring magnetic deviation, but dip and intensity as well. Without an accurate understanding of terrestrial magnetism, Barrow's search for the passage was in grave danger of navigational error.

Indeed, Sabine connected his claim that dip affected compasses to those of Flinders. From his magnetic observations on the coast of New Holland and subsequent experiments on Royal Navy vessels back in Britain, Flinders had established rules for how to correct iron-induced compass error, but this was still a poorly understood phenomenon. On crossing the Atlantic, Sabine reported that the binnacle compasses of the *Isabella* and *Alexander* were differing by a worrying 11¼°.[25] While Flinders had found that a compass, once fixed in place, showed no error 'when the ship's head was on the magnetic north or south points', the disruption from iron on a compass appeared to vary at different locations around the globe. The navigational problem of magnetism was not, therefore, confined to deviation between the magnetic and geographic poles, but was inseparable from the Earth's fluctuations in magnetic dip and intensity.[26] Picking up where Flinders had left off, Sabine was adamant that investigating and understanding terrestrial magnetism was critical to securing reliable navigation. He sailed again for the Arctic on Parry's 1819 expedition, continuing his magnetic observations in collaboration with the captain and James Ross. Along with winning him the Royal

[23] Ibid. 144.   [24] Ibid. 142.
[25] Edward Sabine, 'On Irregularities Observed in the Direction of the Compass Needles of H.M.S. *Isabella* and *Alexander*, in Their Late Voyage of Discovery, and Caused by the Attraction of the Iron Contained in the Ships', *Philosophical Transactions of the Royal Society of London*, 109 (1819), 112–22, at 114.
[26] Ibid. 116 and 119.

Society's 1821 Copley Medal, this work impressed Barrow, securing Sabine a powerful ally at the Admiralty. Within the Royal Society, Sabine's influence was also growing rapidly, having been a fellow since 1818 and subsequently serving as secretary and a council member. Opposed to reformers who targeted the Royal Society as a bastion of aristocratic patronage and elitism, Sabine built an invaluable network of supporters in the eminent institution, as well as within the army and Admiralty. Eventually ascending to the Royal Society's presidency in 1861, throughout the 1820s and 1830s Sabine was seen by many as the archetypal establishment figure, symbolic of the society's culture of aristocratic patronage.[27]

Sabine's growing scientific reputation was confirmed by the Royal Society's invitation to deliver its prestigious Bakerian Lecture in November 1821. He used this opportunity to discuss his experiments on magnetic dip in London the preceding August, as well as recent developments in the instruments of magnetic observation. Lamentably, Sabine observed that there had been no improvement in dipping needles for over fifty years and those that existed produced results 'which can only be considered as approximate, even when the observer has made himself well acquainted with the various sources of inaccuracy in the instrument, and has adopted precautions to guard against, or remedy them'. The big problem, he surmised, was that the needles usually oscillated on an imperfectly made axis, due to the 'difficulty which even the most skilful artists experience, in the endeavour to make the axis of motion pass through the centre of gravity'.[28] As well as inaccuracies due to friction, it was difficult to adjust the balance of a needle once it had been magnetized: Sabine claimed to have taken steps to remedy these instrumental deficiencies, commissioning a new 11½-inch needle from the London instrument-making family firm Dollond, built to the design of the German astronomer Tobias Meyer (1723–62). To his Royal Society audience, Sabine concluded that this device determined dip with 'a much smaller hint of uncertainty, than has hitherto been the case by needles of the usual construction'.[29] Nevertheless, he conceded that Dollond's instrument was far from perfect: for magnetic science to advance, improved dipping needles were desperately required.

Throughout the 1820s, Sabine continued his study of the Earth's physical properties, including expeditions to the Caribbean, South Atlantic, and Spitsbergen. Magnetic research took centre stage, though often with great difficulty. Cooperation with a ship's captain was crucial to undertaking scientific measurements, as

[27] Gregory A. Good, 'Sabine, Sir Edward (1788–1883)', *Oxford Dictionary of National Biography*, https://www.oxforddnb.com/view/10.1093/ref:odnb/9780198614128.001.0001/odnb-9780198614128-e-24436, 23 Sept. 2004, accessed 24 Mar. 2020.
[28] Edward Sabine, 'The Bakerian Lectures: An Account of Experiments to Determine the Amount of the Dip of the Magnetic Needle in London, in August 1821; With Remarks on the Instruments Which Are Usually Employed in Such Determinations', *Philosophical Transactions of the Royal Society of London*, 112 (1822), 1–21, at 1–2.
[29] Ibid. 4 and 21.

Sabine found off the Rio Grande on the *Iphigenia* in 1822. He was unhappy with the expedition as accurate magnetic observations had been impossible due to the ship's head being inconsistent and its course unclearly known. As the vessel's compasses were 'subject to different systems of attraction', the magnetic difference varied, making trustworthy dip readings hard to obtain.[30] Yet it was not just magnetism that occupied Sabine's attention. At the request of the Royal Society's president, Humphry Davy, Sabine undertook measurements of temperature at considerable depths of freshwater lakes in the tropics and 'of basins of salt water out of reach of the currents from the polar regions', with a view to obtaining 'such knowledge in reference to the important question of the high temperature of the interior of the globe'.[31] Sabine undertook these readings for Davy on HMS *Pheasant* while in the Caribbean and Mexican seas in 1822. Using a weighted thermometer, his experiments produced mixed results, but enhanced his standing with the Royal Society as a travelling natural philosopher who could acquire reliable scientific measurements from around the world.[32] Sabine also measured geographical variations in gravitational force, using a seconds pendulum to determine the ellipticity of the globe, and secured evidence of a Gulf Stream on European coasts through his continuing temperature readings. His physical investigation into the Earth's physical properties earned him the Lalande gold medal from the Institut de France in 1826, which encouraged his efforts to cooperate with Continental natural philosophers.[33]

Having been commissioned alongside John Herschel and two French natural philosophers in 1825 to measure the difference in longitude between Paris and Greenwich observatories, Sabine worked with Arago to compare gravity pendulums and magnetometers at Paris and Altona observatories. In December 1826, using four needles sent from Hansteen and two that Dollond had built to the same specifications, Sabine made four experiments in the garden of the Horticultural Society at Chiswick, repeating these near Tunbridge Wells and then performing them again in January 1827 in the garden of Paris Observatory. With Arago's assistance, Sabine observed how these needles oscillated when suspended by a silk fibre in a wooden box to measure magnetic intensity. With London's dip of 69° 45' taken as an index of 1.000, Paris's mean dip, 67° 58', gave a relative measure of 1.0714. Sabine was delighted at the consistent results given by his six needles. To the Royal Society the following summer, he concluded that 'the absolute intensity of terrestrial magnetism was greater at London than at Paris at

---

[30] Scott, Ms862: Letter from Edward Sabine to James Rennell, 11 July 1822, in Edward Sabine, 'Journal 30 April to 2 Nov 1818, and Letters, 1807–1836', 165–73, at 165.
[31] Edward Sabine, 'On the Temperature at Considerable Depths of the Caribbean Sea', *Philosophical Transactions of the Royal Society of London*, 113 (1823), 206–10, at 206.
[32] Ibid. 207 and 209.
[33] Good, 'Sabine, Sir Edward (1788–1883)', *Oxford Dictionary of National Biography*.

the period of these experiments by about eleven parts in a thousand'.[34] He followed this up with further experiments in August 1828, employing a small dipping needle in Regent's Park that Franklin had acquired for him from the government's Colonial Office and that had been used on his recent overland expedition in North America. Comparing his new observations with those of August 1821, Sabine found that London's dip had diminished by 17′ 5″ in seven years. Connecting this to the growing enthusiasm for terrestrial magnetism investigation in Europe, Sabine observed that this was consistent with Humboldt and Arago's measurements on the Continent.[35] By the end of the 1820s, Sabine was the undoubted British authority of terrestrial magnetic phenomena and would, in the 1830s, lead Britain's contribution to Gauss and Weber's Magnetische Verein. Yet for all Sabine's magnetic research, it was to be James Ross who seized the greatest prize in expeditionary science and propelled terrestrial magnetism to the forefront of Britain science.

In 1829, John Ross led another expedition to the Arctic—his first since his Croker Mountains debacle. However, his misfortunes continued, with his ship, the *Victory*, becoming trapped in ice at Felix Harbour. After getting a little further south the following year, the expedition was frozen in a second time. It was James Ross, however, who won acclaim on this venture, far outshining his uncle. With a small exploration party, young Ross set out on sledges to Peel Sound in mid-May 1831. There they took a dip of 89° 41′. Seeing that the route north followed the coast, they trekked 35 miles over four days to a limestone beach where, on 1 June, they became the first European explorers to reach the magnetic north pole. Despite this achievement, in May 1832 the *Victory* had to be abandoned. At Fury Beach, the desperate crew endured a fourth winter surviving on an Inuit diet and sheltering in a hut built out of the wreckage of the *Fury*, on which James Ross had been wrecked seven years earlier. In the summer of 1833 John Ross's old ship, the *Isabella*, spotted the beleaguered crew and sailed them back to England. The expedition had been a grim failure. However, once back in London, James Ross met with Peter Barlow (1776–1862), the Woolwich-based mathematician, and the two magnetic observers compared each other's readings from June 1831. These measurements seemed to confirm that the north magnetic pole had indeed been attained. Provoking the jealousy of his uncle, James Ross was now Britain's foremost hero of Arctic exploration and magnetic science as he claimed the prized objective of the pole for Britain (Figure 3.2).[36]

---

[34] Edward Sabine, 'Experiments to Ascertain the Ratio of the Magnetic Forces Acting on a Needle Suspended Horizontally, in Paris and in London', *Philosophical Transactions of the Royal Society of London*, 118 (1828), 1–14, at 5.

[35] Edward Sabine, 'On the Dip of the Magnetic Needle in London, in August 1828', *Philosophical Transactions of the Royal Society of London*, 119 (1829), 47–53, at 51–3.

[36] Mawer, *South by Northwest*, 15–17.

SURVEY AND SCIENCE 109

Figure 3.2 The romantic image of polar exploration. James Clark Ross, which John Robert Wildman (1788–1843) originally painted in 1834, with the Pole Star shining in the top-right corner and his magnetic dipping needle in the bottom-right corner (Reproduced by permission of the National Portrait Gallery, 2023).

3.2. Fox's Dipping Needle

Combined with Barrow's quest for the North-West Passage and concerns over navigation, James Clark Ross's locating of the north magnetic pole stimulated a boom in the study of terrestrial magnetism. Yet there remained a serious challenge to the accurate investigation of the Earth's magnetic phenomena: the construction of reliable instruments. And it was this question of instrumentation that was to bring Robert Were Fox to prominence with the nation's leading scientific and naval authorities. The problem with measuring magnetic dip and intensity was that it was hard to suspend a magnetic needle so that it rolled freely on an axis of agate planes. In other words, to perform accurate observations required a delicate, perfect balancing of a magnetic needle, so that it could oscillate with ease, responding to the Earth's magnetic force. At sea, such suspension was even more troublesome. Sabine had long complained of Admiralty dipping needles, while Samuel Christie claimed in 1833 that British instruments were poor, not allowing for the free movement of needles. He extended this criticism to French instrument maker Henri-Prudence Gambey (1787–1847) of Paris, whose sensitive instruments were celebrated on the Continent, but performed poorly at sea.[37]

By the early 1830s the most prominent magnetic needles were those of Hansteen's design. After travelling to London in 1819 with a small portable instrument with a horizontally suspended needle for taking intensity observations, known famously as 'Hansteen's magnetometer', the Norwegian natural philosopher built up a network of British supporters, keen to adopt his instruments to standardize international magnetic research. In Britain, Hansteen's assertions that the Earth had four magnetic poles found sympathetic audiences, given that this was a revitalization of Halley's own account of terrestrial magnetism. Appointed professor at Christiania University in 1811, Hansteen's *Untersuchungen über den Magnetismus der erde*, published in 1819, secured much attention in Britain, especially in the context of Parry and Ross's recent Arctic expeditions. His subsequent Russian expedition to find a second north pole in Siberia between 1828 and 1830 proved inconclusive, but it was not until Gauss published his *Allgemeine Theorie des Erdmagnetismus* in 1839 that these differences over the number of poles appeared to have been finally resolved.[38]

With British audiences enamoured with the Norwegian's magnetic theories, Hansteen's instruments and practices found favour in London. His early magnetic apparatus were the works of the Christiania instrument maker Henrik Clausen. The magnetometer consisted of a small wooden box with glass panels for viewing,

---

[37] Trevor H. Levere, 'Magnetic Instruments in the Canadian Arctic Expeditions of Franklin, Lefroy, and Nares', *Annals of Science*, 43/1 (1986), 57–76, at 60.

[38] Vidar Enebakk, 'Hansteen's Magnetometer and the Origin of the Magnetic Crusade', *British Journal for the History of Science*, 47/4 (Dec. 2014), 587–608, at 588–9 and 593.

in which a horizontal magnetical cylinder suspended from a silk fibre vibrated freely. This could be positioned in the magnetic meridian, pointing north, with the needle oscillating over a paper circle showing degrees. To calculate the Earth's magnetic intensity, an observer had to put the needle into motion and record the time taken for it to complete 300 oscillations.[39] With this apparatus and a very hard magnetic needle of Dollond's construction, which appeared to retain a constant magnetic force, Hansteen left London and toured Europe, performing magnetic observations. In 1825 Sabine contacted the Norwegian magnetic specialist and the two collaborated over the next few years. After criticizing Sabine's magnetic results from his expeditions to Brazil and Spitzbergen, Hansteen recommended the use of his own instruments and needles. Along with supporting the Norwegian's four-pole theory, Sabine became a firm adherent to Hansteen's apparatus and methods. Throughout the late 1820s and 1830s, the standard magnetometer and magnetic needles in use in British magnetic survey work were to be built largely to Hansteen's specifications, including those taken on James Clark Ross's expedition to the North Pole in 1831.[40] Although London instrument makers provided an increasing number of these delicate works, the design was very much Hansteen's in principle. However, after 1833, with the news in of Ross's North Pole triumph, there was a growing market for more robust dipping needles, also known as 'dip circles' on account of the circular brass box in which the magnetized needle oscillated.

For mine experiments, Fox had developed a very different sort of apparatus for measuring dip and intensity to those of Hansteen and Gambey. Away from London's instrument trade, amid the dirt of Cornwall's mines, Fox had cultivated his own arrangements for measuring the behaviour of a magnetic needle under challenging conditions. Fox's claims that subterranean metalliferous veins formed in relation to consistent, governing laws of electricity and magnetism relied on his ability to map the spread and direction of such mineral deposits. To show the order of this phenomenon, he demonstrated how there was uniformity in these veins, traversing from east to west across Cornwall. From this, he argued that the direction of the Earth's electric currents, to which he attributed the formation of veins, was itself determined by terrestrial magnetism, which moved from east to west. As with his observations on the aurora borealis, Fox employed magnetized needles to examine this movement. Deep below ground, he relied on robust instruments, miners' compasses, that could give magnetic bearings in subterranean regions. Not only was Fox, as a result, deeply invested in understanding terrestrial magnetic phenomena, but he also oversaw the construction of magnetic needles that were both sensitive enough to measure the magnetic force of the Earth and robust enough to be used in the physically challenging conditions of

[39] Ibid. 594.   [40] Ibid. 596–600 and 604.

the mine. Both these qualities put Fox in a privileged position during the 1830s, as natural philosophers grappled with the problem of understanding the Earth's magnetic character.

In 1832, fresh from his observations of the aurora borealis and his subterranean experiments on magnetic dip and intensity, Fox placed an order for a dipping needle of his own design with the London instrument makers Watkins & Hill of Charing Cross. The big challenge for Fox was to balance a magnetic needle within the device's brass circular box, so that it could oscillate freely to measure the Earth's dip and intensity. While British instrument makers traditionally suspended needles by resting them on an axis which sat on supports, Fox used jewelled holes, in which the axis of a needle could be held by friction alone. The friction of these two jewelled holes, or 'pallets', kept the needle in place while allowing it to rotate with little resistance. To remove any friction that might prevent the needle oscillating freely, Fox included a pin on the reverse of the instrument's dial, connected to the pivots, which could be rubbed with an ivory disc to manage any static electricity acting on the needle. While needles resting on agate planes were unstable, both at sea on maritime expeditions and in mines, where the finely balanced instruments would have to be carried up and down thousands of fathoms of ladders and through small, hot, damp tunnels, Fox's method of suspension promised robustness. Fox believed his solution highly original, but he was not the first to trial the arrangement, with Continental instrument makers having earlier proposed a similar construction. However, with its potential to offer reliable measurements at sea, Fox's design was well timed and prudently located: more than their Continental contemporaries, British magnetic authorities in the 1830s were eager for instruments that would be of value to maritime expeditions.[41]

At his foundry at Perranarworthal, Fox, his son Barclay, and William Henwood made their first trials with the newly built dipping needle on 9 January 1832. Barclay described how he 'went with Papa to Perran to try the intensity of the magnetic with William Henwood, first in the valley and then on top of the hill. We varied ½ a degree.'[42] His assistant, Henwood, made intensity observations with Fox's first device 'in a few of our Cornish deep mines' which suggested that there was no increase of magnetic intensity 'at the depth of 1000 or 1200 feet below the surface of the sea.'[43] These results were included in Fox's paper which

---

[41] Trevor H. Levere, *Science and the Canadian Arctic: A Century of Exploration, 1818–1918* (Cambridge: Cambridge University Press, 1993), 152; Jenny Bulstrode, 'Men, Mines, and Machines: Robert Were Fox, the Dip-Circle and the Cornish System', Part III diss., History and Philosophy of Science Department, University of Cambridge, 2013, 32; Levere, 'Magnetic Instruments in the Canadian Arctic Expeditions of Franklin, Lefroy, and Nares', 60–1.

[42] Barclay, quoted in R. L. Brett (ed.), *Barclay Fox's Journal* (London: Bell & Hyman, 1979), 33.

[43] RS AP/15/20: Robert Were Fox, 'On Certain Irregularities in the Magnetic Needle Produced by Partial Warmth, and the Relations Which Appear to Subsist between Terrestrial Magnetism and the Geological Structure, and Thermo:electric Currents of the Earth' (1832), 5.

the Royal Society rejected for publication later in 1832. Henwood's measurements were consistent with earlier experiments Fox had 'made in several of our deep mines with a vibrating needle, exhibit[ing] also such discrepancies in their results as to prevent my making any very satisfactory comparisons with the intensity at the surface; they proved, however, that there is no very decided change in it at the greatest accessible depths in Cornwall'.[44] Though these observations formed but a small part of Fox's work in 1832, within the context of increasing philosophical interest in terrestrial magnetism, the dipping needle was to take on new significance.

Just a year before Ross's return from the Arctic and declaration of having reached the north magnetic pole, the timing of Fox's trials was fortuitous. The first published account of Fox's new device appeared in the *Philosophical Magazine* in February 1834. Fox here alleged that scientific knowledge of terrestrial magnetic phenomena was poor due to 'the imperfection of our instruments, and the time and minute attention which the best of them require, to enable us to obtain results which can be considered even as approximations to accuracy'. The answer, Fox declared, was at hand; he boasted that he had designed an instrument that would be crucial 'in the prosecution of magnetic researches in any part of the world'.[45] He proceeded to describe how his circle suspended its magnetized needle in a system of pivots and jewelled holes: the needle's axis had extremely delicate pivots which sat in a pair of aligned jewelled holes, to allow for the free oscillation of the magnetic needle. The needle could be vibrated with a gentle tap, then the angle of dip read off against the graduated circle on the instrument's face, on which a magnifying glass could be adjusted to give precise measures in degrees, minutes, and seconds.[46]

The real advantage, however, of Fox's dipping needle, was that in addition to dip, it could also measure magnetic intensity. The 'most valuable property of the instrument', Fox explained,

> is the facility with which it will indicate the intensity of the earth's magnetism in every latitude. To accomplish this object, steel magnets are employed to deflect the needle from its natural dip, the greater or less intensity existing at the place of observation being determined by the extent of the deflection. I have employed two magnets, each three inches in length...and after having been exposed to the heat of 150° at least, have inclosed them in two brass tubes. These are made to

[44] Ibid. 6.
[45] Robert Were Fox, 'Notice of an Instrument for Ascertaining Various Properties of Terrestrial Magnetism, and Affording a Permanent Standard Measure of Its Intensity in Every Latitude', *Philosophical Magazine*, 3rd ser., 4/20 (Feb. 1834), 81–8, at 81; for Fox's draft, see CUL, Add.9942/18(i): Robert Were Fox, 'Notice of an Instrument for Ascertaining Various Properties of Terrestrial Magnetism, and Affording a Permanent Standard Measure of Its Intensity in Every Latitude' (c.1833).
[46] Fox, 'Notice of an Instrument for Ascertaining Various Properties of Terrestrial Magnetism', *Philosophical Magazine*, 3rd ser., 4/20 (Feb. 1834), 81–2.

slide in a larger tube, six inches in length, on the principle of the spy-glass, one being fixed at each end, so that the magnetic poles alternate with each other.[47]

To take a measurement of intensity, Fox instructed users to turn the vernier on the circle's reverse side so as to match the dip indicated by the needle. If the force of the Earth's magnetic field acting on the needle increased, this would reduce the impact of the deflectors so, in theory, these magnets would have a greater influence over the needle as the circle was moved away from the magnetic poles, towards the magnetic equator, where the Earth's magnetic force was weakest. With the deflectors fixed behind the needle in line with this correlation, the needle would be deflected. The angle of this disturbance revealed the effect of the deflecting magnet, which would vary in relation to the influence of the Earth's magnetic force acting on the needle. This process was to be repeated as many times as possible, with the experiments made both with the needle facing east and then turned 180° to face west. A mean figure could therefore be calculated for intensity.[48]

Fox continued to place new orders for instruments from Watkins and Hill throughout the 1830s, gradually modifying his design. Faraday, for one, was impressed, writing to Fox in early 1834 to congratulate him on the completion of his new instrument at Mr Watkins's, as described in the *Philosophical Magazine* in the same year. He wondered if he might display it at a Royal Institution Friday evening lecture before it was sent down to Falmouth.[49] Several months later, in July 1834, Fox published a second description of his dipping circle in the *Philosophical Magazine*. Here Fox detailed experiments that could be made with his dipping needle through the use of tiny weights. Fox had tried balancing a magnetic bar on a knife-edge point using small metal hooks suspended from one end of the bar and a magnet at the other. He had then suspended a magnet at the weighted end, observing the relationship between weight and magnetic force. Using a paper dial, he recorded the ratios of distance each weight induced when working against the impulse of the magnet. As these little hooks were applied, the needle came into balance. A tenth of a grain weight gave a distance of 1.33 inches, but as the weights increased, the distance diminished. A weight of 3.5 tenths of a grain reduced the distance to 0.67 inch, and eventually he applied 42.5 tenths of a grain to give a distance of just 0.17 inch.[50] Fox incorporated this principle into his dipping needle as an alternative means of measuring intensity, adding a grooved wheel to which weights could be attached to act on the needle itself. This was a

---

[47] Ibid. 83.   [48] Ibid. 84.
[49] Letter 697: Michael Faraday to Robert Were Fox, 23 Jan. 1834, in Frank A. J. L. James (ed.), *The Correspondence of Michael Faraday*, ii: *1832–December 1840: Letters 525–1333* (London: Institution of Electrical Engineers, 1993), 163–4, at 164.
[50] Robert Were Fox, 'On Magnetic Attraction and Repulsion and on Electrical Action', *Philosophical Magazine*, 3rd ser., 5/25 (July 1834), 1–11, at 1–2.

brilliant arrangement that Fox was eager to show off. At the inaugural exhibition of the Royal Cornwall Polytechnic Society (RCPS) on 23 December 1833, Fox's circle took pride of place.[51] In the RCPS's first annual report, published in 1834, the dipping needle was accorded a substantial description, which was subsequently reproduced as a stand-alone booklet (Figure 3.3). In this, more substantial details of the instrument's system of weights were given than in Fox's earlier contributions to the *Philosophical Magazine*.

This time, however, Fox combined the use of weights with the instrument's deflectors to directly measure the pull of the Earth's magnetic field. After attaining the dip by directing the needle towards the North Pole and recording the angles given, the dial on the circle's rear could be lined up to correspond with the angle of dip, before the deflectors were to be screwed in. The needle would then be repelled by this application of magnetic impulse. Fox observed that it was sufficient to simply record the degree of deflection this gave to get an intensity reading, relative to results from the same process at different locations.[52] However, it was at this point that the weights could be used to literally weigh the Earth's magnetic intensity. By removing the circle's glass cover and exposing the needle itself, a silk thread could be looped around the needle's grooved wheel, and hooks attached to the thread. Weights could then be added to coerce the needle from its angle of deflection back to its angle of dip. With the deflecting magnets giving a constant magnetic force, the amount of weight required to induce the needle back to its measured dip therefore indicated the strength of the Earth's magnetic force. A greater intensity would undermine the relative impact of the deflectors and require more weights to return the needle to its indicated dip. This way of making weights equal to the intensity of the Earth was a unique character for Fox's apparatus.[53] By using the weights at different observation stations, relative intensity could be known in reference to a single fixed site that was initially measured and used as an index. Confidently, Fox claimed that his needle, in this way, could show the 'difference of intensity at places situated at less than half a degree of latitude from each other'.[54]

After exhibiting his first device to 'several scientific individuals in London' and at the Oxford meeting of the BAAS in 1832, he resolved to prepare a new instrument on an improved plan.[55] Importantly, this was to be built in Falmouth by a local instrument maker, over whom Fox could have direct influence, and who could also share in experimental work aimed at enhancing the design. By late

---

[51] Brett (ed.), *Barclay Fox's Journal*, 59.
[52] Robert Were Fox, *Description of R. W. Fox's Dipping Needle Deflector* (Falmouth: Jane Trathan, c.1835), 6.
[53] Ibid. 7–8. [54] Ibid. 8.
[55] CUL, Add.9942/18(ii): Robert Were Fox, 'Notice of an Instrument for Ascertaining Various Properties of Terrestrial Magnetism, and Affording a Permanent Standard Measure of Its Intensity in Every Latitude' (c.1833).

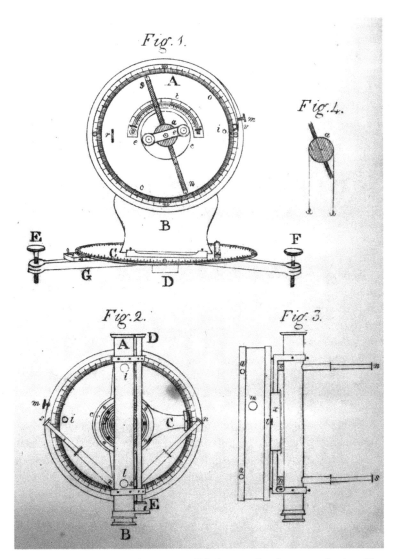

Figure 3.3 Figure 1 shows the dipping needle's front with A being a cylindrical brass box, fixed in a vertical position on the stem (B) and horizontal plate (C). These sit on an axis, D, which is grounded in a tripod, with two legs showing at E and F. G represents a nonius and tangential screw fixed to the leg of the tripod. The needle, *n s*, oscillates in the box on a wheel, *a*, on a concentric disc, *c*, and a bracket, *e*. A thermometer is at *t*.

Figure 2 represents the back of the device, with AB being a telescope. The arm, C, is provided with a nonius to subdivide the vernier on the box's back. DE is a small tube for solar observations.

Figure 3 shows the side of the instrument.

Figure 4 represents the grooved wheel at the centre of the instrument, with the axis of the needle fixed at *a*.

1832, Fox believed he had found his man. The son of a Quaker engineer and an artist by training, Thomas Brown Jordan (1807–90) was originally from Bristol, but moved to Falmouth in his early twenties, taking employment as drawing master for Barclay Fox in 1832. Fox encouraged his mechanical interests and he gradually became a professional instrument maker. Given their shared Quaker connections and the timing of Jordan's move to Falmouth, coinciding with the completion of Watkins & Hill's first dipping needle, it is likely that his appointment as Barclay's teacher was part of Fox's broader plan to have future instruments built locally. Although his marriage to Sarah Dunn in 1837 might have taken him out of his Quaker community, his shared religious values helped make Jordan a trusted associate for the Fox family. After his first drawing lesson with Barclay in September 1832, he became a regular guest at Rosehill throughout 1833 and 1834, building a solid working relationship with Fox.[56] On 21 January 1835, for example, Jordan took tea with Fox, staying until after eleven to perform experimental investigations with his fellow Quaker.[57] By late 1833, Jordan was at work on his first dipping needle, marking the beginning of what was to be a productive partnership. Initially specializing in miners' dials for determining magnetic north while underground, Jordan worked with Fox to develop a portable magnetometer during 1834 and became a frequent exhibitor at the RCPS.[58] In 1839, Jordan published his own account of Fox's new dipping needle in the *Annals of Electricity* (Figure 3.4). With its two verniers, one on the front and one on the back, Jordan's instrument drew heavily on the construction of theodolites for survey work. The circle itself was built of brass and included a thermometer. As with Watkins &

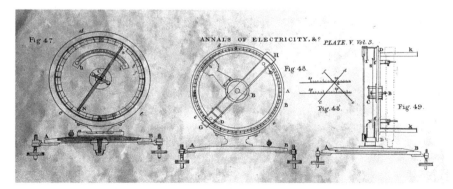

Figure 3.4 Fox's Falmouth-built dipping needle as portrayed in Jordan's article in *Annals of Electricity* in 1839 (Author's image, 2020).

[56] Brett (ed.), *Barclay Fox's Journal*, 41–2, 45, and 59.   [57] Ibid. 74.
[58] Anita McConnell, 'Jordan, Thomas Brown (1807–1890)', *Oxford Dictionary of National Biography*, https://www.oxforddnb.com/view/10.1093/ref:odnb/9780198614128.001.0001/odnb-9780198614128-e-15123, 23 Sept. 2004, accessed 24 Mar. 2020.

Hill's 1832 model, Jordan used jewelled holes, a small grooved wheel for the weights, and two small deflectors in tubes that could be applied at the rear. A pin connected to the back of the jewelled pallet could be rubbed with an ivory disc to produce friction and steady the needle's vibration.[59] Dip and intensity could be measured with the needle alone, or by the application of deflecting magnets. As with Watkin & Hill's instrument, Jordan's circle employed weights to determine the Earth's magnetic intensity.

Fox's initial magnetic investigations had been focused on the mine, but from 1832 he became increasingly aware of the broader maritime and exploratory implications of understanding magnetic phenomena and, concomitantly, the value his dipping might have. Indeed, with his family's firm, G. C. Fox, Falmouth's leading shipbrokers, Fox would have been all too aware of the dangers of compass deviation. Along with news of wrecked ships along the perilous Cornish coast, any rumours of navigational error from magnetic interference would have been prominent in Falmouth. In particular, the Fox family kept a close eye on the rising use of iron and steam for ocean navigation. In May 1838, for instance, Barclay Fox attended the launch of the 1,836-ton *British Queen* in London. The largest ship yet built, Barclay thought her 'a superb sight. Her debut from the dock was as if a mountain removed into the midst of the sea'.[60] The following year, he had tea in Falmouth with Joseph Robinson of the Coalbrookdale Iron Company. Boasting a monthly output of 1,378 tons of iron, Robinson's company had just received a huge order from the Great Western Steamship Company, which had only recently launched the wooden SS *Great Western*. With Brunel as their engineer, Barclay learnt that the company was now 'building another steamer, larger than the former & all of iron, which seems to be a material likely to come into general use for ship building'.[61] This would be the SS *Great Britain*, that would run aground on the Irish coast seven years later, fuelling rumours of iron-induced compass error. In 1839, however, it seemed clear to Barclay that the future of ocean navigation and shipbuilding would be iron. On 7 November 1843, Barclay travelled to see the *Great Britain* for himself in its Bristol dock. Nearing completion, he was impressed with its propeller, 'a circular fan with 6 arms', and its engines, resembling 'a huge watch chain'. Awestruck, Barclay concluded that Brunel's iron monster was surely 'the greatest experiment since the creation'.[62] For a devout Quaker, this was some declaration. Likewise, at the 1841 BAAS meeting in Plymouth, Barclay heard shipbuilder John Scott Russell's paper on '*forms* of ships' and his investigation into the water resistance of various shaped hulls. Three days later, Barclay joined the BAAS excursion to Devonport to witness the dramatic launch of the Royal Navy's latest warship, HMS *Hindustan*.[63] New developments in the

---

[59] T. B. Jordan, 'Description and Use of a Dipping Needle Deflector Invented by Robert Were Fox, esq. by Mr. T. B. Jordan, Philosophical Instrument Maker, Falmouth', *Annals of Electricity*, 3 (July 1838–Apr. 1839), 288–97, at 288–90.
[60] Barclay, quoted in Brett (ed.), *Barclay Fox's Journal*, 126.   [61] Ibid. 143.
[62] Ibid. 361–2.   [63] Ibid. 239–40.

materials and theory of ship architecture were of paramount concern to the Foxes. As much as their associations with the mines, their maritime connections at Falmouth informed the cultivation of Fox's magnetic instruments. And in 1834 it was to be this maritime context that would transform Fox's dipping needle from an object of Cornish natural philosophy into a crucial device of British naval exploration and global scientific inquiry.

3.3. Sir John Franklin and the Magnetism of the British Isles

In late 1834, Fox's dipping circle attracted the interest of an especially influential patron. Fox's published accounts of his needle had piqued the curiosity of the celebrated Arctic explorer Sir John Franklin, who was excited at the prospect of an improved dipping circle for naval expeditions. Few rivalled his experience of magnetic inquiry at sea, having been a midshipman on HMS *Investigator* under Flinders during the early 1800s, where the captain's fascination with magnetic phenomena and compass error rubbed off on Franklin during their circumnavigation of Australia and voyage to Tahiti. After serving on HMS *Bellerophon* at the Battle of Trafalgar in 1805, Franklin was commissioned lieutenant in 1808, seeing further action at New Orleans in 1814.[64] In the peace that followed 1815, Franklin found himself on half-pay back in Britain. In Barrow's plans to mobilize the Royal Navy to find the North-West Passage, Franklin saw an opportunity for advancement. In 1818 he served alongside Sabine and Parry on John Ross's voyage to the northern seas, before commanding the brig *Trent* as part of an Arctic expedition to Spitzbergen. Bigger prizes were to come. In 1819, Barrow appointed Franklin to lead his overland expedition in Canada, tasked with following the Coppermine River to the Arctic coast, as part of the ongoing search for the passage.[65]

All went well at first, with Franklin displaying his considerable navigational skill as his expedition reached Cumberland House in Saskatchewan in 1819, before spending their second winter at Fort Chipewyan. However, driven by his determination to conduct survey work, including magnetic observations, Franklin delayed his expedition's return. With their food supplies exhausted, the journey back down the Coppermine River in June 1821 was a grim affair, with Franklin and his men scratching a living off rock lichen. Ten men died, with one being shot and eaten, having himself killed and attempted to eat one of his colleagues.[66] Despite this, the expedition reached the sea on 18 July and returned home to England the following year with Franklin, 'the man who ate his boots', confirmed as a hero of Arctic exploration. Almost a third of his bestselling

---

[64] Lambert, *Franklin*, 22–6.
[65] Ibid. 30–1; B. A. Riffenburgh, 'Franklin, Sir John (1786–1847)', *Oxford Dictionary of National Biography*, https://www.oxforddnb.com/view/10.1093/ref:odnb/9780198614128.001.0001/odnb-9780198614128-e-10090, 23 Sept. 2004, accessed 24 Mar. 2020.
[66] Lambert, *Franklin*, 32.

Narrative of a Journey to the Shores of the Polar Sea discussed magnetic phenomena, earning him election as a fellow of the Royal Society in 1823.[67] As a founding member of the Athenaeum Club in 1824, Franklin became increasingly well known to eminent natural philosophers and statesmen, including the club's secretary, Faraday, as well as John Herschel and future prime ministers Robert Peel and Viscount Palmerston.[68] Franklin led another expedition in search of the North-West Passage between 1825 and 1826, conducting magnetic and atmospheric readings throughout the long dark winter, observing correlations between the two phenomena: magnetic needles were influenced by changes in the atmosphere. After another near disaster in which local Inuit saved Franklin's men from a repeat of the Coppermine fiasco, he went back to London in 1827. At home, he married the ambitious Jane Griffen in 1828 and, a year later, collected an honorary doctorate from the University of Oxford and a knighthood.

So when Franklin showed an interest in Fox's dipping needle, this was a hugely significant moment for the Cornish natural philosopher. Few naval officers possessed such experience and knowledge of terrestrial magnetic phenomena and the challenges of survey work. Though presently unemployed, Franklin was eager to do all he could to advance the science of magnetism. He had written to James Clark Ross in March 1834, supporting the lobbying of Parliament for further financial investment in discovery and science. Franklin was sure the navy was close to locating the North-West Passage and that a recently appointed House of Commons committee would recommend the 'revival of further research towards its attainment'. At the same time, Franklin was anxious to see Ross's new observations 'on the deviations of the Needle caused by the Aurora' made during his recent expedition to the Arctic.[69] This final remark reflected Franklin's own interest in the relationship between atmospheric and magnetic phenomena. It cannot have escaped his notice that, just three years earlier, Fox had published an account of the aurora's influence on a magnetic needle in Falmouth.

Having read of Fox's new dipping needle, Franklin went to Falmouth in October 1834 to examine the instrument for himself. He was not disappointed with what he found, staying late into the evening and enjoying dinner with the Fox family. Barclay noted in his diary that Franklin appeared as 'a stout man with a splendid forehead and piercing eye, but very blue about the beard. He gave us some very interesting accounts of his own adventures & sufferings', but more importantly, the Arctic explorer seemed highly 'delighted with Papa's instrument'.[70] For both Franklin and Fox, the evening had been a triumph and the start of a crucial relationship for promoting the instrument. Franklin's chance to

---

[67] Adriana Craciun, *Writing Arctic Disaster: Authorship and Exploration* (Cambridge: Cambridge University Press, 2016), 82–123.
[68] Lambert, *Franklin*, 33–5.
[69] Scott, Ms1524: James Clark Ross to John Franklin (25 Mar. 1834).
[70] Barclay, quoted in Brett (ed.), *Barclay Fox's Journal*, 71.

advance Fox's needle came soon after, with an expedition to the Euphrates River launched in the hope of establishing 'a communication by steam with India, through that Channel & the Persian Gulf'.[71] On 22 November, Franklin breakfasted with Captain Francis Chesney (1789–1872) of the Royal Artillery, who was to lead the expedition and take responsibility for its survey instruments. As these would include devices for measuring magnetic phenomena, Franklin encouraged Chesney to take a Fox dipping needle. After a productive breakfast, Franklin was pleased to inform Fox that Chesney had expressed enthusiasm at 'the opportunity of making experiments with the instrument, and would gladly take it with him for that purpose if you approved', but warned that as

> the instrument has not yet been in general use I fear the Government in these economical times would not sanction it...[being] purchased at once for the use of the Expedition though they might perhaps sanction its purchase after it had been repeatedly tried—if the report were favourable.[72]

It only remained for Fox to entrust Chesney with an instrument and draw up instructions for the captain's use. Although Franklin regretted not being able to provide Chesney with a direct demonstration of a Fox type, he promised to raise the subject with Captain Francis Beaufort (1774–1857), the Royal Navy's hydrographer with responsibility for the production of naval charts, at the next opportunity. Recognizing the expedition's potential for securing government patronage, Fox wrote back immediately, promising to send his original Watkins & Hill-built instrument. Chesney was grateful and proposed delegating special responsibility for the device to the expedition's second-in-command, James Bucknall Estcourt (1803–55), a captain in the 43rd Regiment of Foot. Franklin and Chesney agreed that the best way to prepare Estcourt for this work would be to send him to Falmouth to stay with Fox. Franklin assured Chesney that Estcourt 'would be certain of receiving from [Fox]...every kind attention as well as full instruction about the Instrument and also much valuable information relative to magnetical observations in general'.[73]

There was more good news for Fox. Franklin had shared details of the new dipping circle with Beaufort who, it turned out, had already heard about the instrument from Davies Gilbert. According to Franklin, Beaufort 'thought so highly of what he had heard & read of it—as to make him desire much to see the Instrument. He thinks the Astronomical Society a very proper place for its being

---

[71] CAN, R3888-0-7-E, J. Franklin Correspondence, 1834–6: John Franklin to Robert Were Fox, 22 Nov. 1834, 1–4, at 1.
[72] Ibid. 1.
[73] CAN, R3888-0-7-E, J. Franklin Correspondence, 1834–6: John Franklin to Robert Were Fox (6 Dec. 1834), 5–9.

sent'.[74] Fox's reputation was, evidently, growing rapidly within the most influential realms of the Admiralty. The following August, this partnership between Franklin and Fox secured further success at the 1835 BAAS meeting. The Fox family travelled with Sir John and Lady Franklin by steamer from Falmouth to Dublin, where Fox exhibited his instrument to the BAAS's Physical Section. Although Barclay regretted that the section's 'committee were puzzled & bothered him with questions instead of letting him get on', he was confident that the minority who understood Fox's description of the device recognized its merits.[75] Importantly, Fox subsequently met with Sabine and a small group of influential magnetic specialists for a private examination of the instrument. The meeting included James Clark Ross and Humphrey Lloyd (1800–81) who were Sabine's collaborators in the embryonic BAAS scheme to survey the magnetism of the British Isles.[76] This was to prove an auspicious moment. While Franklin had secured Fox's dipping needle attention with the Admiralty and leading scientific authorities, the Dublin BAAS proved to be the beginning of Fox's own personal recruitment into Sabine's ambitious magnetic enterprise.

In displaying his instrument to this network of magnetic specialists, Fox found himself at the very heart of Britain's investigation into terrestrial magnetism. Calls for a domestic survey of Britain's magnetic dip and intensity, made at the BAAS's founding meeting in 1831, had remained largely unfulfilled. Scoresby had been the initial choice to undertake the task, but the commission had fallen to his friend, Thomas Stewart Traill, who managed just eight intensity observations in 1832 before abandoning the project.[77] At its 1833 meeting in Cambridge, the BAAS had passed a new resolution for a magnetic survey of the British Isles. This time, however, Sabine seized control of the project. Recently posted to Limerick with the army, he commenced the survey in Ireland, where he formed a productive partnership with Lloyd, Trinity College Dublin's recently appointed professor of natural and experimental philosophy. A mathematician by training, Lloyd became increasingly interested in terrestrial magnetism and contacted Gauss in 1835, eager to join his network of observatories. Between 1837 and 1838, he followed Gauss's instructions in constructing a magnetic observatory at Dublin, where he designed his own magnetic instruments according to Gaussian principles.[78]

[74] Ibid.; Levere, 'Magnetic Instruments in the Canadian Arctic Expeditions of Franklin, Lefroy, and Nares', 61–2.
[75] Barclay, quoted in Brett (ed.), *Barclay Fox's Journal*, 78 and 81.
[76] Matthew Goodman, 'From "Magnetic Fever" to "Magnetical Insanity": Historical Geographies of British Terrestrial Magnetic Research, 1833–1857', PhD thesis, University of Glasgow, 2018, 91.
[77] Ibid. 253.
[78] James G. O'Hara, 'Lloyd, Humphrey (1800–1881)', *Oxford Dictionary of National Biography*, https://www.oxforddnb.com/view/10.1093/ref:odnb/9780198614128.001.0001/odnb-9780198614128-e-16840, 23 Sept. 2004, accessed 24 Mar. 2020; James Gabriel O'Hara, 'Gauss and the Royal Society: The Reception of His Ideas on Magnetism in Britain (1832–1842)', *Notes and Records*, 38/1 (Aug. 1983), 17–78, at 45–50.

From 1834, Lloyd provided Sabine with an industrious partner in the magnetic surveying of Ireland, with James Clark Ross joining their exploits in the summer of 1835. At the BAAS Dublin meeting, Lloyd reported on the progress of this work, with Sabine, Ross, and Lloyd having produced a magnetic chart for Ireland, showing lines of magnetic inclination and intensity.[79] However, throughout their work in Ireland, they had found the unreliability of their Hansteen needles a constant problem, with the instruments losing magnetic force over time.[80] As Matthew Goodman has shown, these early encounters with needles in surveying the British Isles were to have an important impact on the subsequent study of terrestrial magnetism around the globe, as well as the rapid adoption of Fox's dipping circle. At Trinity College Dublin, Lloyd found that his instruments performed well but became unpredictable as they travelled. In 1834, Sabine sent Lloyd the 11-inch Dollond circle that Franklin had taken to the Arctic between 1825 and 1827, but this was found to be riddled with error. Lloyd himself favoured his own smaller circle, probably from London instrument maker Thomas Charles Robinson (1792–1841), whose business was located at 38 Devonshire Street in Westminster.[81]

Despite the growing experience of the magnetic surveyors, dipping needles remained perplexing devices throughout the investigation. When Sabine began surveying Scotland in 1836, disaster struck. On taking a steamer from Helensburgh to the island of Great Cumbrae on the Scottish west coast, a rough sea caused his dipping needle to fall from the table. His subsequent results differed greatly from earlier readings, with the needle having lost magnetic force. Even gentle jarring and vibrations from travel compromised the integrity of needles, as it became increasingly apparent to Sabine and Lloyd that instrumental improvement would be essential to completing the BAAS's magnetic survey. Their solution was twofold: to establish fixed testing sites and to acquire superior instruments. The first of these was relatively straightforward. Sabine and Lloyd identified several locations at which the magnetic character was known and where needles could be tested for error. In London, Sabine chose Westbourne Green and Regent's Park for this function, these being as free as possible of magnetic interference. Lloyd employed the Provost's Garden at Trinity College Dublin for the same function. These were standardized places of magnetic experiment, against which observations made further afield could be tested and needles trialled for error.[82]

---

[79] Jack Morrell and Arnold Thackray, *Gentlemen of Science: Early Years of the British Association for the Advancement of Science* (Oxford: Clarendon Press, 1981), 526.
[80] Goodman, 'From "Magnetic Fever" to "Magnetical Insanity"', 89.   [81] Ibid. 93–4.
[82] Matthew Goodman, 'Proving Instruments Credible in the Early Nineteenth Century: The British Magnetic Survey and Site-Specific Experimentation', *Notes and Records*, 70 (2016), 251–68, at 255–62; Goodman, 'From "Magnetic Fever" to "Magnetical Insanity"', 98–9.

The second solution, of finding reliable dipping needles and trustworthy craftsmen to build them, was far more challenging. Fox seemed to offer an answer. Initially favouring French instrument makers like Gambey, Lloyd and Sabine avoided English manufacturers, believing the workmanship of Continental mechanics to be superior. The leading English dipping-needle maker, Robinson, took time to earn the confidence of the magnetic investigators. In 1837, for instance, Ross found the axles of a Robinson type so poor that the needle would not return to rest correctly. Robinson made four new needles, modelled on European examples, and by July 1838 was providing far more credible instruments.[83] But in 1835, it was Fox that seemed best placed to resolve the question of trustworthy magnetic needles and it was his dipping circle that would continue to promise greater reliability for magnetic observations made while travelling. His devices were to earn a reputation for being more transportable than those of rival British instrument builders, making them ideal both for naval expeditions and the BAAS's British Isles survey. While France and continental Europe offered one source of instrumental solution, Cornish science provided an alternative far closer to home.

In Dublin, Fox effectively joined Sabine's magnetic survey of the British Isles, before the project recruited its fifth and final member, John Phillips (1800–74). Together, Sabine's five-man team zealously surveyed Britain and Ireland's magnetic properties. After his Irish exploits, Sabine examined twenty-seven stations in Scotland during the summer of 1836, while Lloyd surveyed fourteen in England. From August 1837 until late 1838, Ross covered fifty-eight stations across England, Ireland, and Scotland, with Sabine completing twenty-two coastal stations around the British Isles during the same period.[84] Fox's contribution was considerable, providing Sabine with a regular stream of magnetic observations, including some made previous to their Dublin acquaintance. During late 1832 and early 1833, he had observed the dip through Darlington, Melrose, Edinburgh, Glasgow, and Liverpool with his original 7-inch needle from Watkins & Hill.[85] Following the Dublin meeting, Fox went on to survey much of Ireland, including Galway and the Giant's Causeway in August 1835, before taking measurements throughout September on his journey home to Falmouth via Holyhead, Banger, the top of Snowdon, and Malvern.[86] Back in Falmouth, Fox delivered a report of his tour to the RCPS at the end of the year, publishing this as a report in 1836. Along with a new 6-inch needle, between August and September 1837 he employed a 7-inch needle 'of very early date;—1832, or 33', to take readings across

[83] Goodman, 'Proving Instruments Credible in the Early Nineteenth Century', 259–60.
[84] Edward Sabine, *Report on the Magnetic Isoclinal and Isodynamic Lines in the British Islands: From Observations by Professors Humphrey Lloyd, and John Phillips; Robert Were Fox, Esq; Capt. James Clark Ross; and Major Edward Sabine* (London: Richard and John E. Taylor, 1839), 50.
[85] TNA, Sabine Papers, BJ3/19/5: 'Dip Results 1839', 5.
[86] TNA, Sabine Papers, BJ3/19/9–10: 'Dates of Robert Were Fox's Magnetic Observations' [c.1838–9], 9.

Britain, including at Darlington, Edinburgh, Glasgow, Keswick, and Liverpool. This instrument had 'pivots about four times larger' than he was presently having them made, making his results a little 'less uniform' than could be expected from newer devices.[87] In total, during the summer of 1837, Fox determined the dip at twenty stations in northern England and Scotland, adding another eight in southern England the following summer, from London to the Scilly Isles. By 1838, he was contributing data obtained with a revised instrument with a 4-inch needle: throughout the survey, he was constantly looking to improve and refine his magnetic apparatus. Overall, for the English part of the survey, Sabine covered nineteen stations, Ross thirty-six, Lloyd fourteen, Phillips twenty-four, and Fox twenty-nine.[88] Fox did not cover Cornwall alone. In November 1837, Ross visited the county, stopping for magnetic observations at Padstow before reaching Falmouth on the eighteenth, where he stopped for dinner with the Foxes, regaling them with stories of his Arctic voyages.[89] After a few experiments in Rosehill garden, Ross performed measurements at Pendennis Castle on the twenty-first and twenty-second, before taking readings in a field just east of the 'First and Last Inn' at Sennen, near Land's End, England's most westerly tip.[90]

Fox's contribution, however, did not just consist of providing magnetic observations and instruments; he also delivered invaluable experience, acting as a guide in the study of terrestrial magnetism to Phillips, the survey's most inexperienced magnetic observer. Indeed, while Fox was, in many respects, the ideal participant in the BAAS scheme, Phillips struggled to perform accurate magnetic experiments and was never quite at ease with the survey's instruments. He had been interested in magnetic phenomena since 1830, working on a modified dipping needle of his own and taking pains to improve its balance in 1831 and, like Fox, was familiar with the dip of mineral veins and strata through his geological exploits, applying such knowledge in his failed attempts to measure the effects of auroras on a Hansteen needle in 1832.[91] At the 1834 BAAS meeting in Edinburgh, Phillips had discussed his interests in magnetic needles with Arago and, after the 1835 Dublin gathering, demonstrated a growing enthusiasm for the science. As Fox united mining expertise with philosophical pursuits, so too did Phillips at York, helping to found the industrially focused Geological and Polytechnic Society of the West Riding of Yorkshire in 1837, which echoed the Fox family's formation of the RCPS four years earlier. So when, in February 1836, Lloyd invited Phillips to join the magnetic survey of the British Isles, Fox gained a colleague with whom he had much in common: they soon became frequent correspondents.[92]

---

[87] Ibid. 10; BJ3/19/11–12: Robert Were Fox to Edward Sabine (14 Aug. 1838), 12.
[88] Sabine, *Report on the Magnetic Isoclinal and Isodynamic Lines in the British Islands*, 50 and 85.
[89] Horace N. Pym (ed.), *Memoirs of Old Friends: Being Extracts from the Journals and Letters of Caroline Fox of Penjerrick, Cornwall from 1835 to 1871* (London: Smith, Elder, & Co., 1882), 24.
[90] Sabine, *Report on the Magnetic Isoclinal and Isodynamic Lines in the British Islands*, 77.
[91] Jack Morrell, *John Phillips and the Business of Victorian Science* (Aldershot: Ashgate, 2005), 77.
[92] Ibid. 116 and 121.

Phillips recognized his own inexperience of magnetic observations and was grateful to have Fox as a correspondent, especially concerning questions of comparing local variations in magnetic intensity resulting from geological causes. In April 1836, Fox sent Phillips his essay on terrestrial magnetism. Phillips was sure Fox's observation of a line of dip of 71° at York would excite 'the whole of the Empire', but could not account for why this measurement varied so much from his own measurements in the city. 'Was the difference of dip indicated by the large needle always in excess? Why was this? More friction on pivots might give errors both ways. Do you find the dip <u>always</u> in excess in houses?', asked Phillips, bombarding Fox with questions over the practices of magnetic observation.[93] Fox replied, expressing his interest in Phillips's 'information relative to the dip of the needle at York', but wondered if the magnetic novice had tried his experiments 'out of the City, & at a distance from iron pipes'. In terms of geology, Fox surmised that some

> rocks clearly affect the needle...I think that their influence is very much modified, like iron, by their position with respect to the magnetic meridian, & it is possible that the break in the lines between England & Ireland, may be owing to this circumstance: thus the SW of Ireland may act the part of a N. Pole, & N. Wales of a South Pole, at least, iron placed in the same positions would do so, & the dip would thus be increased in N. Wales, & diminished on the opposite coast.[94] [Figure 3.5]

With Fox a trusted magnetic guide, Phillips charted isoclinal curves throughout 1836, linking places of equal dip across Yorkshire, to try to show the effects of hills on magnetic measurements. He presented this hypothesis at the 1835 BAAS's Dublin meeting, arguing that in flat areas these magnetic curves bent to the south, while hills caused northerly deflections. This later turned out to be due to Phillips's own experimental errors. Aware of his shortcomings, he subsequently agreed to work under Lloyd's instruction in taking magnetic observations across Yorkshire in 1837, but warned that Lloyd would 'inwardly bemoan the folly' which induced Phillips 'to meddle with the cold iron of a magnetic needle'.[95] Lloyd remained circumspect over Phillips's contributions to the survey, but he nonetheless proved an industrious magnetic observer, taking measurements at twenty-two stations across Yorkshire, the Lake District, Newcastle, Birmingham, and on the isles of Wight and Man.

Together, Fox and Phillips's experiences were shaping new practices and understandings of magnetic science. However, through their correspondence, Fox revealed early ambitions to extend his work beyond Britain. Fox was pleased to inform Phillips that their efforts were, along with Gauss and Weber's Magnetischer Verein, contributing to a subject which had been 'very deficient in information'. In particular, he was impressed with the efforts of Adolphe Quetelet, the director of Brussels Observatory, who was providing a productive link

---

[93] RS, Robert Were Fox Papers, MS/710/83: John Phillips to Robert Were Fox (2 Apr. 1836).
[94] OUM, Papers of John Phillips, Phillips 1836/16: Robert Were Fox to John Phillips (9 Apr. 1836).
[95] Phillips, quoted in Morrell, *John Phillips*, 122.

SURVEY AND SCIENCE 127

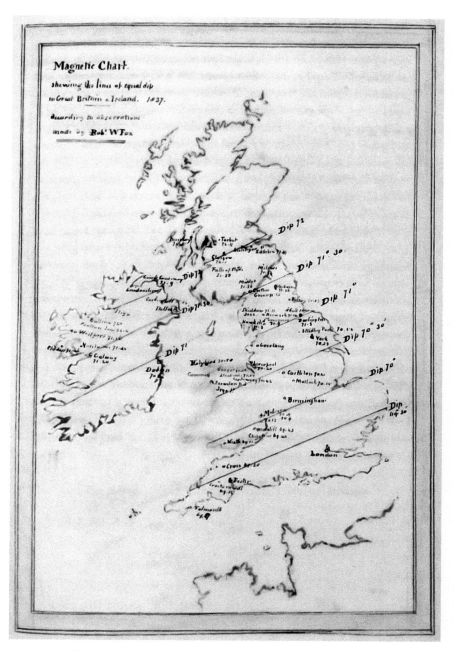

Figure 3.5 Fox's magnetic chart for dip across the British Isles in 1837 (Image by Michael Carver, 2019).

between British and Continental magnetic programmes.[96] Within this expanding European survey, Fox believed his dipping needles would play an important role, confidently reporting to Phillips that the 'Russian Govt have ordered two of my dipping needles, but they will not be so complete as they ought to have been'. The cause of this incompleteness, he explained, was that Watkins & Hill had received the request but not consulted Fox early enough for the Cornishman to provide accurate instruction. Nevertheless, Fox hoped one of these two devices would 'be sent to the Siberian pole, & I have just sent a small one for R. King, to take with him to the neighbourhood of the American magnetic pole. It will be an interesting object to ascertain the relative intensity of these two poles'.[97] Not only was Fox now securing international orders for his instruments, but he also revealed here his own views of terrestrial magnetism, subscribing to Hansteen's theoretical account of the Earth having four magnetic poles—a theory that both Sabine and James Clark Ross agreed with throughout the 1830s. Thanks to his association with Franklin and his assured performances on the magnetic survey of the British Isles, Fox had become one of Britain's leading specialists in terrestrial magnetism. Both the Cornish natural philosopher and his dipping needle were about to secure international acclaim.

3.4. Europe and the Atlantic

As important as the BAAS construed the surveying of the British Isles' magnetic phenomena to be, this domestic enterprise was but a part of much grander, global aspirations. What Sabine envisioned was an investigation of the magnetic character of the world. With its extensive navy and network of overseas trading posts and colonies, the British state had unrivalled resources for advancing philosophical understandings of terrestrial magnetism. To investigate Halley's theory that there were four magnetic poles, slowly rotating within the globe on two axes, there would have to be an expedition to the southern hemisphere to locate both southern poles. As early as 1826, Hansteen had written to Sabine observing that 'if the English Government will not give a generous and powerful hand to researchers...there is little hope that the theory shall advance'.[98] A government-sponsored voyage of magnetic discovery to the Antarctic was, therefore a long-held ambition for natural philosophers. At the BAAS Oxford meeting of 1832, Lloyd, Samuel Christie, James Forbes, George Peacock, and John Lubbock promoted such a scheme, arousing considerable interest with the meeting's delegates.[99] In the BAAS's 1833 report, Christie published his 'State of Our Knowledge

---

[96] OUM, Papers of John Phillips, Phillips 1836/16: Robert Were Fox to John Phillips (9 Apr. 1836); see Fabien Locher, 'The Observatory, the Land-Based Ship and the Crusade: Earth Sciences in European Context, 1830–1850', *British Journal for the History of Science*, 40/4 (Dec. 2007), 491–504.

[97] OUM, Papers of John Phillips, Phillips 1836/16: Robert Were Fox to John Phillips (9 Apr. 1836).

[98] Hansteen, quoted in Carter, *Magnetic Fever*, 11.

[99] Cawood, 'The Magnetic Crusade', 500; on the BAAS lobby, see Morrell and Thackray, *Gentlemen of Science*, 353–70.

Respecting Geomagnetism', emphasizing the problem that variation between magnetic and geographic north presented. He claimed that, as this varied over time, it was still unknown how to correct the resulting compass deviation with any great certainty. Christie therefore called for a global network of magnetic observatories. At stake here was a wider theoretical question. Adherents to Hansteen's four-pole theory, such as Ross, Sabine, and Fox, favoured exploratory voyages sent out to locate the supposed two southern poles, while those who endorsed Gauss's two-pole account of terrestrial magnetism, such as John Herschel, preferred establishing fixed observatories to register changes in the Earth's magnetic field over time.[100] In 1834, Arago joined the BAAS's calls for the British government to establish magnetic observatories throughout its colonial territories. However, these early efforts were an utter failure. With Sabine otherwise engaged in military service and survey work, and Herschel at the Cape of Good Hope, the promoters of the science lacked both political and scientific clout. Yet despite these setbacks, Sabine was building an influential network of magnetic promoters through his connections at the Royal Society and BAAS. Sabine was convinced that Hansteen had found a north pole in Siberia to complement the North Pole Ross had reached in the Arctic in 1831, appearing to substantiate the four-pole theory. This placed increased weight behind his calls to launch an expedition to find the two magnetic poles in the southern hemisphere.[101]

At its 1835 Dublin meeting, the BAAS resolved to renew its campaign for state sponsorship of an expedition to the Antarctic. In this context, Sabine, Lloyd, and Ross's magnetic survey of Ireland provided important evidence of the practicality of such an international operation. As the BAAS became more established and its magnetic survey of the British Isles collected and published data, Sabine took a leading role in uniting the Royal Society and BAAS to encourage state sponsorship of a global magnetic investigation. He went to Germany in 1835 to persuade Humboldt to write a letter requesting that the Royal Society lobby the government to establish colonial magnetic observatories. The Royal Society responded with a resolution in June 1836, supporting Humboldt's request and establishing an eminent committee, consisting of Sabine, Christie, George Airy, William Whewell, and Lubbock, to coordinate its efforts with those of the BAAS. This achieved little, other than a Treasury grant of £500 towards the costs of magnetic instruments.[102] Nevertheless, the campaign was gaining momentum. Among this escalating enthusiasm for a global magnetic investigation, the BAAS's surveying of the British Isles played an important part. To be able to fulfil the promise of a worldwide magnetic inquiry, Sabine had to demonstrate that he had the practices, know-how, and instruments with which to conduct such an enterprise. He had to show that British scientific practitioners could attain sufficiently accurate

---

[100] Carter, *Magnetic Fever*, 12–15.
[101] Cawood, 'The Magnetic Crusade', 501–2; Carter, *Magnetic Fever*, 15.
[102] Cawood, 'The Magnetic Crusade', 505.

observations of magnetic phenomena to justify government investment. Within this, Fox and his dipping needle played a crucial role, especially as his instruments were construed as being superior to rival devices for measurements made at sea.

Published in 1839, Sabine's report presented what he asserted to be 'the actual state of the phaenomena of the magnetic dip and intensity of the British islands'.[103] Importantly, Sabine stressed the constant scrutiny that had been paid to the survey's instruments, supporting his claims that dipping needles of sufficient accuracy were attainable for an expanded examination of the Earth's terrestrial magnetism. Throughout 1837 and 1838, the survey's participants undertook regular instrumental examinations at Westbourne Green. Most were of Robinson's production, along with two French Gambey dipping needles, made originally for Captain Robert FitzRoy, who had lent them to Sabine. The final report explained how Ross had performed eight to ten readings with eight different needles, producing dips which varied by as much as forty-one minutes. These differences, Sabine contended, were 'very far beyond the limits of the errors of observation' and included an assessment of all the needles used which revealed several 'perplexing discrepancies'. Lloyd admitted that 'we are forced to conclude that there may exist, even in the best needles, some source of constant error which remains uncorrected by the various reversals usually made' and, as a result, that no repetitions 'with a needle so circumstanced can furnish even an approximation to the absolute dip'.[104] Much error was due to the axle, on which the needle rested on the agate planes, not being perfectly cylindrical. For this reason, Continental instrument-makers preferred needles which were secured in place with friction alone. Ross requested that Robinson make four needles on this principle, with the London-based chronometer maker Charles Frodsham (1810–71) fashioning the axle for one of them. Although Frodsham's axle appeared superior, it was clear that British-made needles were still poorly suspended. Were the axles perfect, the needle would show the same dip when trialled in any position, but differences of over forty minutes were found in one of Robinson's needles, and between seven and eleven minutes for the Frodsham axle.[105] These tests, performed between June and July 1838, confirmed that only perfect axles could be relied on to show true dip.

In contrast, Fox's needles had performed well. Sabine explained that Fox's 'needles do not rest on a cylindrical axle supported by planes, but the axle is terminated by exceedingly fine and short cylindrical pivots', resting in jewelled holes. The deflectors for these needles allowed dip to be deduced from various parts of the circle, revealing any errors due to the axle or magnetism of the circle. Confidently, Sabine concluded that the 'performance of these needles sufficiently indicates the great care bestowed on their workmanship'.[106] Under trial on Westbourne Green on 8 June 1838, Fox's instrument produced a dip of 69° 18', compared to the survey's mean of 69° 17' 2" for this location.[107] Fox's

---

[103] Sabine, *Report on the Magnetic Isoclinal and Isodynamic Lines in the British Islands*, 50–1.
[104] Ibid. 51–2.   [105] Ibid. 52–3.   [106] Ibid. 56–7.   [107] Ibid. 57.

instruments also excelled in intensity readings. Sabine reported that Fox's 'needle has a small grooved wheel on its axle, which received a thread of unspun silk, furnished with hooks, to which weights may be attached'. At Regent's Park and Westbourne Green on 22 May and 4 and 8 June 1838, this method had been trialled in front of the survey's investigators, before further trials at Eastbourne and Falmouth. Sabine then visited Fox at Falmouth to perform intensity experiments between 24 and 26 August.[108] 'The needles employed by Mr. Fox appear to give results extremely consistent with one another, and with those of other needles', declared Sabine.[109] Fox and Jordan had provided a credible alternative to the persistently troublesome wares of London mechanics. The magnetic survey of the British Isles was a coup for Fox's dipping needle, but this product of Cornish science was also enhancing Sabine's assertion that a global study of the Earth's magnetic field was feasible.

It was not just the BAAS's magnetic survey of the British Isles which enhanced the reputation of Fox's device. Beyond Sabine's scheme, there was growing demand for Fox's magnetic apparatus. In July 1837, Gilbert wrote to congratulate Fox over 'the extended use of your magnetic Apparatus', evidently pleased to see the contribution of Cornish natural philosophers and instrument makers to the leading scientific enterprise of the day.[110] That same summer, De la Beche took advantage of a visit to Rosehill to try out a Fox type for himself, conducting his own magnetic-dip measurements in West Cornwall throughout July and August. Back in Falmouth, he calculated the variation in the Earth's magnetic field, and discussed the instrument with both Fox and Jordan.[111] Suitably impressed with the device's performance, De la Beche placed an order with Jordan for a Fox dipping circle of his own. On the geologist's behalf, the Devonian lightning specialist William Snow Harris (1791–1867) wrote to Fox in October asking that he forward the instrument and some thermometers to the Royal Hotel where De la Beche would collect them.[112] Through Fox and John Enys, De la Beche remained in close contact with Jordan and, on 1 November 1838, recommended a Fox type from the instrument maker for the Ordnance Survey, to be used in the production of maps.[113] Not all shared such confidence. In 1837, James Forbes had warned Airy that the performance and regularity of Fox's instrument seemed 'superhuman', implying that its results were too good to be credible.[114] Nevertheless,

---

[108] Ibid. 143 and 147.   [109] Ibid. 111.
[110] RS, Robert Were Fox Papers, MS/710/44: Davies Gilbert to Robert Were Fox (10 July 1837).
[111] Letter 381: Henry De la Beche (25 July–2 Aug. 1837); Letter 382: Henry De la Beche (25 July–4 Aug. 1837); Letter 383: Henry De la Beche to A. W. Robe (9 Aug. 1837), all in T. Sharpe and J. McCartney (eds.), *The Papers of H. T. De la Beche (1796–1855) in the National Museum of Wales* (Cardiff: National Museums & Galleries of Wales, 1998), 38.
[112] RS, Robert Were Fox Papers, MS/710/49: William Snow Harris to Robert Were Fox (14 Oct. 1837).
[113] Letter 388: Henry De la Beche to T. Colby (1 Nov. 1838), in Sharpe and McCartney (eds.), *The Papers of H. T. De la Beche*, 38. Thomas Colby reported at the end of the month that the board had evaluated Fox's dipping needle; see Letter 288: Thomas Frederick Colby to Henry De la Beche (20 Nov. 1838), in Sharpe and McCartney (eds.), *The Papers of H. T. De la Beche*, 32.
[114] Bulstrode, 'Men, Mines, and Machines', 31.

Fox and Jordan's instruments were proving of value to BAAS's magnetic survey, as well as to the nation's leading geological and ordnance projects.

3.5. Conclusion

In 1837, William Sturgeon's *Annals of Electricity, Magnetism, & Chemistry* reviewed the state of knowledge surrounding terrestrial magnetism and ascribed to Fox a prominent position. After Ørsted's observations that an electric current could make a magnetic needle twitch and Faraday's discovery that passing an electric current through a freely suspended wire caused it to oscillate around a magnet, Ampère had extended the implications of these experiments to the Earth's magnetism. He asserted that Ørsted's principle, if applied to the question of terrestrial magnetism, suggested the existence of 'electric currents in planes perpendicular to the direction of the dipping needle, and which move from east to west'.[115] The existence of electric currents, 'continuously flowing in the earth', appeared obvious to the reviewer, who claimed that 'Nature's laboratory is well stored with apparatus of this kind, aptly fitted for incessant action, and the production of immense electric tides'. It was here that Sturgeon's journal placed Fox's philosophical contributions. That 'local currents do absolutely traverse the interior of the earth has been amply manifested by the experiments of Mr. Fox and Mr. Henwood, in the Cornish mines'.[116] Fox and Henwood had, it seemed, confirmed Ampère's theory that currents of electricity continually moved under the surface of the Earth, from east to west, just as 'similar currents in every magnet of steel' could be observed through the experiments of Ørsted and Faraday. As much as Fox had primarily intended to produce an account of how metalliferous deposits formed through heat and electromagnetic phenomena he had, in fact, advanced the philosophical investigation of terrestrial magnetism. From his earliest use of magnetized needles, suspended simply from thread to measure magnetic influences in mines and atmospheric aurora, Fox had developed a robust, complex apparatus, ideally suited to the survey work of expeditionary science. He would now take this work even further, as he and Sabine looked to fashion a global system of magnetic data collection. Fox's investigations into subterranean heat and magnetic phenomena, conducted during the 1820s and 1830s, put him a unique position from which to contribute to the rapidly escalating philosophical study of terrestrial magnetism. His instrumental and experimental skills and experience would secure Falmouth eminence as a centre of magnetic science, alongside London, Paris, and Göttingen. The circulation of Fox's dipping needles was about to take on a very global character as Sabine looked to instigate a worldwide survey of terrestrial magnetism.

[115] Anon., 'A General Outline of the Various Theories Which Have Been Advanced for the Explanation of Terrestrial Magnetism', *Annals of Electricity, Magnetism, & Chemistry; and guardian of experimental science*, 1 (Oct. 1836–Oct. 1837), 117–23, at 122.
[116] Ibid. 123.

4
The Antarctic Foxes
Dipping Needles on James Clark Ross's South Pole Expedition, 1838–1843

Just off the Lizard on 5 October 1839, HMS *Erebus* passed the Cornish coast. Under the command of the dashing Captain James Clark Ross and bound for the Antarctic, this would be her last sight of England for four long years. A little further up the English Channel, delayed in a gale, HMS *Terror* followed.[1] After rendezvousing at Madeira, the two ships progressed to Cape Town and Hobart, before sailing on to the Antarctic. Ross's expedition had been tasked with locating the south magnetic pole and measuring terrestrial magnetism in the southern hemisphere. To do this, both ships carried Robert Were Fox's dipping needles, designed and built in Falmouth. This expedition marked what has traditionally been understood as the start of the British Magnetic Scheme (BMS) which, with Edward Sabine's almost fanatical direction, would mobilize the Royal Navy, the British Army, and Britain's overseas territories in the worldwide surveying of the Earth's magnetism. There were two parts to the proposed magnetic survey: an expedition to locate the south magnetic pole, and a global system of observatories in Britain's colonies. This project represented the fruition of years of campaigning for state sponsorship, under the leadership of Britain's most eminent scientific authorities. This magnetic lobby, organized through the BAAS and Royal Society, had pressed Parliament and the Admiralty over the urgency of understanding the globe's magnetic phenomena. During the next decade, observatories would be established in Canada, South Africa, Van Diemen's Land (Tasmania), India, and Singapore. Surveying expeditions would circumnavigate the world, from the Arctic and Antarctic to Japan, Korea, Borneo, and West Africa, and on the Atlantic, Indian, and Pacific oceans. The BMS was a scientific venture of unprecedented scale in terms of its geographical extent and its government support. Its promoters were keen to portray this as the first truly global state-orchestrated experiment. And it was Fox's dipping needles, the products of experiments deep underground in Cornwall's mines, that took a leading role in this scrutinizing of the Earth's magnetic properties.

---

[1] James Clark Ross, *A Voyage of Discovery and Research in the Southern and Antarctic Regions, During the Years 1839–1843*, 2 vols. (London: John Murray, 1847), vol. i.

Through expeditions like Ross's voyage to the Antarctic, Fox's Cornish natural philosophy was transformed into a scientific enterprise spanning the world. Reflecting on the BMS in the second edition of his *History of the Inductive Sciences* (1845), William Whewell first declared that the 'manner in which the business of magnetic observation has been taken up by the government of our time makes this by far the greatest scientific undertaking which the world has ever seen'.[2] If Whewell was right, then Cornwall's scientific resources had been conscripted within the world's most prestigious scientific venture; in what historian John Cawood termed the 'Magnetic Crusade' in 1979. This term, in fact, originated in North America in the early 1840s and was not generally used to denominate the British magnetic campaign. There is no evidence that it was used in Britain until April 1842 and, even then, it was not a commonly used category.[3] Regardless of its title, the Magnetic Crusade or, rather, the BMS combined philosophical investigation and government support at a moment of dramatic imperial expansion. This was an unprecedented system of state-backed global science, central to the growth of empire.[4] The lobbying for the BMS involved a reorganization of British science, with the well-established Royal Society and its new rival, the BAAS, uniting to secure substantial government funding for a scientific enterprise. At over £100,000 for Ross's Antarctic Expedition alone, this was an experimental investigation of unparalleled scale, representing a crucial moment in the relationship between the British government and scientific inquiry.[5] And yet, for all that Ross's departure in 1839 marked an epoch in state–science relations, this was the result of a growing integration between Britain's scientific and governmental resources. While historians traditionally pinpoint the start of the BMS to the Antarctic Expedition, Sabine and Fox had been developing a British international survey long before this formal government commitment, in which the Royal Navy and Admiralty had featured prominently. Throughout the 1830s, Fox had secured his dipping needles use with leading scientific naval officers, providing instruments and advice from his home at Falmouth. By 1837 he was looking to undertake his own magnetic research on the Continent, and from 1838 he was receiving magnetic data from across the Atlantic, measured with his magnetic devices. Between 1837 and 1839, Fox's circles saw use in Canada, Brazil, the Falklands, France, Switzerland, Italy, Germany, and the Netherlands.

---

[2] Claim first made in 1845 edition and repeated in the 1857 edition, see William Whewell, *History of the Inductive Sciences, from the Earliest to the Present Time*, 3 vols. (3rd edn., London: John W. Parker and Son, 1857), iii. 55.

[3] Examined in Matthew Goodman, 'From "Magnetic Fever" to "Magnetical Insanity": Historical Geographies of British Terrestrial Magnetic Research, 1833–1857', PhD thesis, University of Glasgow, 2018, 16–17; see also John Cawood, 'The Magnetic Crusade: Science and Politics in Early Victorian Britain', *Isis*, 70/4 (Dec. 1979), 493–518; Christopher Carter, *Magnetic Fever: Global Imperialism and Empiricism in the Nineteenth Century* (Philadelphia: American Philosophical Society, 2009).

[4] Carter, *Magnetic Fever*, 151–2.   [5] Cawood, 'The Magnetic Crusade', 517–18.

This effectively marked the start of Sabine's international surveying of terrestrial magnetism, representing the escalation of Britain's magnetic enterprise from domestic inquiry to global investigation. It was also a complete triumph for Fox and a coup for Cornish natural philosophy and instrument making. Instead of seeing Ross's Antarctic Expedition as the commencement of the BMS, the global circulation of Fox types reveals that the government's commitment to Sabine's investigation in 1839 marked the expansion, rather than the beginning, of the state's patronage of British magnetic science. Yet, as ever, the success of this venture very much depended on the production and management of delicate dipping needles. In resolving these challenges, Fox and his mechanic, Jordan, took a central role. At Falmouth, they built, repaired, and examined the instruments with which Ross surveyed the Earth's magnetic properties. But they also helped to interpret and analyse the incoming data from the Antarctic Expedition, correcting errors and helping Sabine make sense of the enormous volume of measurements from around the world, taken with Fox-type dipping circles. In these endeavours, Fox's earlier experiences of large-scale magnetic survey work from 1837 to 1838 would prove invaluable.

4.1. International Fox: The Navy, Europe, and the Atlantic

Since Franklin's intervention in 1834, the stock of Fox's instrument had risen dramatically. Crucially, it was with the Royal Navy's discovery service that this growing reputation was to have increasingly global implications. Fox's home and gardens at Falmouth, as the final port of departure for naval vessels bound for the Atlantic and beyond, as well as their first port of call on returning, made the ideal venue for officers to witness the instrument first-hand. A trip to Rosehill, where lunch, tea, or dinner accompanied demonstrations of the new dipping circle, was easily accessible to the navy's most magnetically enthusiastic officers. Many of them brought first-hand experiences of magnetic observations acquired from all over the world. The most famous of these visits took place in 1836. On 2 October, after its five-year circumnavigation of the globe, 'a dirty vessel, with her old rigging worn to shreds', came into Falmouth harbour: it was HMS *Beagle*. The day before, the little ship had ploughed through a gale off Land's End, during which its celebrated naturalist, Charles Darwin, had been seasick one last time.[6] Desperate for land, Darwin left behind his extensive collection of botanical and animal specimens, taken from around the world, and hurried off to Shrewsbury, determined never to embark on another sea voyage ever again. The *Beagle*'s captain, FitzRoy, however, had different plans. Overnighting in Falmouth, he went to

---

[6] Christ College Library, Cambridge, STAN1/32: '32. Safe Home', 1.

see Fox at Rosehill to examine a Fox-type circle. 'He came to see Papa's dipping needle deflector, with which he was highly delighted', noted Caroline Fox, recording that 'He has one of Gambey's on board, but this beats it in accuracy. He stayed till after eleven'.[7] After exchanging stories of the *Beagle*'s exploits and 'fly-catcher' Darwin for details of the dipping needle over afternoon tea and then a lengthy dinner, FitzRoy retired to bed. But the very next morning he returned to Fox's home, clearly overstimulated at the revelation of having seen the Fox dipping needle. Barclay Fox was delighted that FitzRoy 'breakfasted here & reports that he was not able to sleep from the philosophical excitement of last evening. He and father were engaged in comparing observations with their respective instruments in the field all the morning'.[8] Despite five years' exploration at sea, FitzRoy could not resist a trialling of the *Beagle*'s Gambey against Fox's device in the tropical peace of Rosehill garden. Evidently, the instrument had an almost mystical power of exciting even the most well-travelled of naval officers. Darwin must have regretted missing out on this dramatic culmination of the *Beagle*'s voyage: the expedition's final discovery, perhaps, was not a natural specimen, but a Falmouth-built magnetic instrument.

Years later, in October 1843, Lieutenant Robert Hammond, who had been a midshipman on the *Beagle* with FitzRoy, dined at the Foxes' Falmouth residence, regaling them with further tales from the voyage, including how at Tierra del Fuego the crew had been approached by angry Fuegians armed with hatchets.[9] With its natural harbour and central position for naval shipping, Falmouth provided a regular flow of Royal Navy officers eager to exchange stories from their voyages for Fox's insights on his dipping needle experiments. Not only did this make it easy for Fox to procure tropical plants for his gardens, but it secured him audiences with visitors possessing first-hand knowledge of magnetic phenomena from across the globe. Fox's dipping needle was also gaining increased attention with elite naval audiences up in London. Earlier, in 1836, the Geographical Society sent a deputation to the Admiralty to press claims for equipping HMS *Terror* with instruments with which to complete the magnetic surveying of the northern coast of America on its forthcoming expedition to the North-West Passage. In March, Franklin recommended to Beaufort that a Fox type be taken and requested Fox prepare a device by May. Complementing Fox on his needle's success in the magnetic survey of the British Isles, Franklin was 'pleased with the results of your observations in Ireland—and with the improvement you have made in the needle' and was 'more convinced that the instrument must be adopted when its comprehensive merits & uses are known'.[10]

---

[7] Ibid. 2; Horace N. Pym (ed.), *Memoirs of Old Friends: Being Extracts from the Journals and Letters of Caroline Fox of Penjerrick, Cornwall from 1835 to 1871* (London: Smith, Elder, & Co., 1882), 7–8.
[8] Barclay, quoted in R. L. Brett (ed.), *Barclay Fox's Journal* (London: Bell & Hyman, 1979), 98.
[9] Pym (ed.), *Memoirs of Old Friends*, 184.
[10] CAN, R3888-0-7-E, J. Franklin Correspondence, 1834–6: John Franklin to Robert Were Fox (26 Mar. 1836), 9–12.

In late April, Beaufort, Ross, and Sabine examined Fox's dipping needle in the Admiralty Garden at Greenwich and were duly impressed.[11] When *Terror* sailed for the North American coast under Captain George Back's command, she carried the dipping needle that Fox had himself employed during his return to Falmouth from Dublin in 1835. Before *Terror*'s departure, Franklin requested a second magnetized needle be made for the circle, 'whose Poles might be reversed' so that the principal needle could be 'kept solely for taking the intensity'.[12] He also arranged for Lieutenant Edward Barnett (1799–1879), responsible for the *Terror*'s instruments, to visit Rosehill and learn from Fox how to perform magnetic experiments. Franklin was eager he 'go at once to see' Fox to 'receive full instruction as to the purposes and modes of observing' with Fox's circle.[13] Despite this, the responsibility for the expedition's magnetic observations eventually fell to *Terror*'s second lieutenant, Owen Stanley (1811–50). Over the next year, he took dip and magnetic measurements with the Fox type and, on returning to Chatham in November 1837, reported that the dipping circle had performed well in the summer. In the cold winter, however, Stanley had found Fox's delicate hooks and weights for intensity measurements difficult to handle with his frozen fingers.[14] Despite this, Fox was delighted to inform Sabine in 1838 that 'Lieut Stanley has given me as good a character of it as I could desire'.[15] Captain Back's North American expedition aboard *Terror* marked the start of a rapid escalation in naval orders for Fox types. Along with his continuing work for the BAAS's magnetic survey of the British Isles, Fox's collaboration with the Admiralty was putting his instrument at the very heart of British magnetic science.

Fox not only encouraged the use of his instruments on naval expeditions but, on his own initiative, extended his work for Sabine's BAAS magnetic survey of the British Isles to Continental Europe between late 1837 and 1838. Armed with a newly built 4-inch dipping needle, Fox travelled through France, Italy, Germany, and the Netherlands, making regular magnetic observations in a string of European cities, including Paris, Grenoble, Aix, Baden, Geneva, Cologne, Rotterdam, and Arnhem.[16] From this data, he produced charts showing both

---

[11] Trevor H. Levere, *Science and the Canadian Arctic: A Century of Exploration, 1818–1918* (Cambridge: Cambridge University Press, 1993), 155; see also Andrew Lambert, *Franklin: Tragic Hero of Polar Navigation* (London: Faber and Faber, 2009), 83.
[12] CAN, R3888-0-7-E, J. Franklin Correspondence, 1834–6: John Franklin to Robert Were Fox (5 July 1836), 13–14.
[13] CAN, R3888-0-7-E, J. Franklin Correspondence, 1834–6: John Franklin to Robert Were Fox (25 July 1836), 16–17.
[14] Trevor H. Levere, 'Magnetic Instruments in the Canadian Arctic Expeditions of Franklin, Lefroy, and Nares', *Annals of Science*, 43/1 (1986), 57–76, at 62.
[15] TNA, Sabine Papers, BJ3/19/11–12: Robert Were Fox to Edward Sabine (14 Aug. 1838), 11.
[16] CUL, Add.9942/23: 'Results of Experiments Made by R.W. Fox in the Spring of 1839 on the Magnetic Intensity and Dip in Italy & Some Other Parts of Europe; His Perran Apparatus Was Employed as in His Previous Experiments in the British Islands the Needle Only Having Been Since Altered' (1839).

isoclinal and isodynamic lines, which he forwarded to Sabine in early 1839.[17] Despite having little time at each point of observation and 'the annoyance of lookers on & questioners; & the smallness of my needle', Fox had managed a series of experiments.[18] Although finding that Jordan's needles lost magnetic force as they travelled, the results were encouraging. Intensity measures in Paris, for example, suggested considerable consistency. Using the weights, Fox recorded an intensity of 0.9794 relative to an index of 1.000 for London, as well as a dip of 67° 14' in the garden of the École des mines. This was favourable considering that the dipping needle gave 67° dip and 0.9761 intensity in the English garden at the Château de Fontainebleau.[19] By the end of May 1839, Fox had written up his Continental results, giving details of stations and intensity with weights and deflectors.[20] With Sabine eager to publish these findings, Fox sent further measurements of intensity, including with both weights and deflectors, in March 1840.[21] Fox observed that 'the distance between the lines of equal dip increases in proceeding towards the eastern & northern parts' of Britain, which was also the case on mainland Europe.[22] Fox also used his European tour to exhibit his dipping needle to leading Continental scientific audiences. At Paris's Académie des sciences on 2 April 1838, Caroline observed how 'Our fellow-traveller, the Magnetic Deflector, excited strong interest', as she boasted that her father's method of needle suspension appeared far superior to all rivals.[23]

Importantly, Fox's European experiences would inform the construction of new Fox types for naval expeditions. Fresh from his Continental tour, he promised Sabine that he would oversee Jordan's construction of three superior instruments, with 7-inch needles, each including two grooved wheels for suspending weights, custom-built for use aboard ships. On 22 January 1839, Fox predicted that these would take three months to complete, with Jordan already at work on the castings.[24] Not only were these available to Sabine for future expeditions, but they came with the assurances of a magnetic observer fresh from an overseas expedition of his own. With the magnetic survey of the British Isles complete, Fox and Sabine were, by the late 1830s, looking to extend their investigations beyond the British Isles. As valuable as Fox's European observations were, this was but one part of a global expansion, in which his dipping needles were to take a lead-

---

[17] TNA, Sabine Papers, BJ3/19/40: 'Results of Magnetic Observations Made in 1838 by Means of a 4 Inch "Dipping Needle Deflector"' [c.1838], 40.
[18] TNA, Sabine Papers, BJ3/19/34–7: Robert Were Fox to Edward Sabine (22 Jan. 1839), 35.
[19] Ibid. 36.
[20] TNA, Sabine Papers, BJ3/19/52–3: Robert Were Fox to Edward Sabine (6 July 1839), 52.
[21] TNA, Sabine Papers, BJ3/19/62–3: Robert Were Fox to Edward Sabine (22 Jan. 1839), 62–3; BJ3/19/64–5: 'Magnetic Intensities Indicated by the Deflectors Adjusted to the Dip at Each Station & Repelling the Needle' (1840), 64–5.
[22] TNA, Sabine Papers, BJ3/19/17–20: Robert Were Fox to Edward Sabine (16 Nov. 1838), 17.
[23] Pym (ed.), *Memoirs of Old Friends*, 26.
[24] TNA Sabine Papers, BJ3/19/34–7: Robert Were Fox to Edward Sabine (22 Jan. 1839), 37.

ing role: while he toured Europe, at least two other Fox types were in overseas service. Estcourt, promoted major since the Euphrates Expedition, took a Fox dipping needle to Canada in 1839, making observations at Halifax on 9 and 17 April and at Montreal between 28 and 31 May. At Montreal he found a mean dip of 76° 17' 8" and a mean intensity of 1.3137 relative to Falmouth as an index of 1.000. At Halifax the dip was 74° 57' 9". Fox had these results in hand by November.[25] Estcourt proceeded to Drummondville, repeating his experiments near Niagara Falls during August and September and observing a mean dip of 74° 33' 8" over five days of measurements.[26] These were valuable results, but it was the second overseas magnetic expedition that was to provide a decisive model of how Fox's expertise and instruments would be conscripted into Sabine's magnetic enterprise.

Born and raised in Tregew, just outside Falmouth, and having served under FitzRoy on the *Beagle* between 1831 and 1836, Lieutenant Bartholomew James Sulivan (1810–90) was well acquainted with Fox's now celebrated dipping needle. In what was the first oceanic trial of the circle, Sulivan travelled on HMS *Arrow* from Falmouth to the Falkland Islands in 1838, returning to Rio de Janeiro on HMS *Stag* the following year, before catching the packet ship *Opossum* back to Falmouth.[27] Equipped with a Fox-type circle and 4-inch needle, both of Jordan's construction, Sulivan observed dip without the deflectors and used weights applied to the needle's wheeled axle to measure intensity. His observations provided invaluable lessons for subsequent expeditionary surveys. It became apparent that measurements were more reliable under sail than at anchor, 'because the ship's head cannot be kept so steadily on one point; and by the frequent variation in the direction of her head relatively to the magnetic meridian, the observed dip is rendered greater than the true'.[28] Likewise, the data collected on *Opossum* were superior to those Sulivan had made on the Royal Navy's vessels because its head was placed on the point of the compass which corresponded with the course of the voyage.

For these intensity observations, Sulivan took Rosehill as his index against which to calculate how the Earth's magnetic field changed over space. Sabine explained how the values of the intensity experiments were 'given relatively to unity at Mr. Fox's garden at Falmouth, which was the force in London = 1.372,

---

[25] TNA, Sabine Papers, BJ3/19/17–20: Robert Were Fox to Edward Sabine (16 Nov. 1838), 18–19; BJ3/19/32–3: 'Results of Major Estcourts Previous Observations' and 'Results of Major J. Bucknall Estcourts Observations with the Dipping Needle Deflector on the Dip & Intensity at Drummondville, Near the Falls of Niagara' [c.1838–9], 32.

[26] TNA, Sabine Papers, BJ3/19/32–3: 'Results of Major Estcourts Previous Observations' and 'Results of Major J. Bucknall Estcourts Observations with the Dipping Need Deflector on the Dip & Intensity at Drummondville, Near the Falls of Niagara' [c.1838–9], 32–3.

[27] Edward Sabine, 'Contributions to Terrestrial Magnetism', *Philosophical Transactions of the Royal Society of London*, 130 (1840), 129–55, at 130.

[28] Ibid. 132.

and at Falmouth 1.374'. For this, observations were made in Fox's garden both before and after the voyage to provide an intensity index, as well as to check the needle's magnetic intensity and 'show that no notable alteration took place in this respect during its employment'.[29] Sabine was not just mobilizing Fox's expertise and instrument, but also employing his Falmouth home as a central site of magnetic investigation. With comparative observations taken on 6 July 1838 and 24 June 1839, Fox capitalized on Falmouth's position for Atlantic-going magnetic expeditions. Rosehill offered a point of reference for calculating the magnetic influence of the *Arrow*'s iron, with its on-board dip of 69° 27′ compared to that of Fox's garden at 69° 12′ 5″.[30]

On 7 January 1839, Fox informed Sabine that he was receiving a regular income of magnetic results from Estcourt and Sulivan. Falmouth was now a centre of magnetic knowledge production and intelligence. Fox forwarded these Atlantic and North American measurements on to Sabine in London, providing a pivotal communications link in the conveyance of data. From Rio, Sulivan had written to Fox, expressing how he hoped

> the line of dip & intensity carried on at sea will give you satisfaction; it, (the instrument) works so much better on board than I anticipated, that it has quite overcome my idea that it is an inferior instrument; tho of course it co^d be more satisfactory to work with one like that belonging to the Ordnance (the dipping n.d. which T. Jordan made for the Ordnance). The intensity I think can be relied on more than the dip, as I have found that in different positions when the dip varied as much as a degree, the same weight always deflected the needle at the same angle on each side of what the needle stood at.[31]

Fox regretted that Sulivan had been issued with a lighter needle than was ideal for intensity experiments, but was confident Sabine would be impressed with this report. 'I often wished you could have seen how nicely the intensity could be obtained even in a good swell', mused Sulivan, offering Fox assurance over his instrument's performance at sea.[32] Fox promised to check both Estcourt's and Sulivan's instruments for any change of force on their return and proposed that Jordan make heavier needles in future, especially for sea measurements.

With Falmouth Sulivan's first port of call on the completion of his survey, he went straight to Rosehill where Fox retested his dipping needle. He found that the instrument indicated 'the same intensity...as it did before he sailed for the Falkland I's'. Sulivan's needle was, Fox concluded, 'a good one' aside from a little rust due to 'accidental exposure to moisture at Bahia', which had made one pole

---

[29] Ibid. 130. [30] Ibid. 131.
[31] TNA, Sabine Papers, BJ3/19/27–30: Robert Were Fox to Edward Sabine (7 Jan. 1839), 27–9.
[32] Ibid. 28.

slightly heavier than the other.³³ Fox was more than able to make corrections for this error. Unfortunately, Sulivan's stop in Falmouth was so brief that Fox did not have the time to confirm the effect of temperature variations on his needle.³⁴ Over the following months, Fox played a central role in helping Sabine interpret and prepare Sulivan's observations for publication. After Sulivan reached the Falklands, Fox warned that the results obtained so far exhibited error resulting from temperature variations and required correction. Fox had possessed the lieutenant's 'needle so short a time that I could not determine this correction satisfactory', and had made subsequent experiments on a different needle to provide a solution. He also recommended Sulivan repeat his observations on 'a return voyage, as the <u>head</u> of the ship, containing much iron, will then become a N. Pole by induction, instead of a S one, as it must have been on his outward voyage'.³⁵ Once Sulivan's return voyage was complete, Fox wrote again in July 1839, deeming the lieutenant's observations of dip and intensity very satisfactory. Along with regulating and correcting Sulivan's results, Fox delivered further details of Estcourt's observations in the run-up to Sabine's announcement of the magnetic data.

When Sulivan returned to Falmouth in October 1839 for the annual RCPS meeting, he was an honoured guest of the Foxes, relaying impressive accounts of his Fox type's performance on the Atlantic to an eager Cornish audience, including Lemon, Enys, and Gilbert. This was the ideal moment and location for yet another celebration of Cornish science, this time through Sulivan's experiences with his Fox dipping needle.³⁶ Along with Fox's magnetic observations in Europe, Sulivan's results would provide the data for what was effectively the start of the global extension of the BAAS's survey of the British Isles. Sabine struggled to procure figures with which to complement Sulivan's and give a thorough view of the Atlantic's magnetic character. However, from the Royal Society's archives he acquired Scottish astronomer James Dunlop's (1793–1848) observations for dip and intensity, taken in 1831 on-board a merchant ship to Australia. As astronomer of Parramatta Observatory in New South Wales, Dunlop also had measurements from 1834 to 1839. Combining Sulivan's and Dunlop's figures, Sabine mapped the Atlantic's magnetic as far as the meridian of the Cape of Good Hope.³⁷

Sabine initially delivered his 'Contributions to Terrestrial Magnetism' paper to the Royal Society on 19 March 1840, with Fox despatching Estcourt's Canadian intensity and dip figures just a few days before. These were, Fox attested, 'tolerably consistent' results.³⁸ Significantly, Fox and his dipping needle took the leading

---

[33] TNA, Sabine Papers, BJ3/19/52–3: Robert Were Fox to Edward Sabine (6 July 1839), 53.
[34] Ibid. 52.
[35] TNA, Sabine Papers, BJ3/19/34–7: Robert Were Fox to Edward Sabine (22 Jan. 1839), 34.
[36] Brett (ed.), *Barclay Fox's Journal*, 164.
[37] Sabine, 'Contributions to Terrestrial Magnetism' (1840), 133–4.
[38] TNA, Sabine Papers, BJ3/19/66: Robert Were Fox to Edward Sabine (10 Mar. 1840), 66; BJ3/19/68–9: Robert Were Fox to Edward Sabine (14 Mar. 1840), 68.

roles in Sabine's debut 'Contributions', with Rosehill the intensity index and primary testing site for Sabine's magnetic data and Sulivan employing his dipping needle. Sabine combined Sulivan's and Dunlop's Atlantic results with those of Fox's Continental tour, including this European data as an appendix. Sabine's initial 'Contributions to Terrestrial Magnetism', appearing in the Royal Society's *Philosophical Transactions*, represented an epoch in British magnetic science. This was the first of some fifteen such papers, spanning the next thirty-seven years, which would represent the primary output of Britain's magnetic enterprise. Along with large volumes of data, Sabine's 'Contributions' would include increasingly extensive charts for magnetic dip, declination, and intensity, fulfilling the survey's promise to record magnetic phenomena 'over the whole surface of the globe'.[39] Through his skill and zeal as a magnetic observer, his cultivation of reliable dipping needles, the locations and isolation of his tropical gardens at Rosehill and Penjerrick, and his assistance in data collecting for Sabine's 'Contributions', Fox had delivered the resources with which the BAAS's magnetic survey of the British Isles could be transformed into the worldwide study of terrestrial magnetism.

4.2. The British Magnetic Scheme

With the magnetic survey of the British Isles nearing completion, Fox busily developing his instruments, and Sabine overseeing an increasingly experienced team of magnetic observers, the BAAS and Royal Society launched a fresh campaign for government support of a global magnetic survey; what would become the BMS. In 1838, John Herschel's return from the Cape of Good Hope injected this bid with fresh impetus. Few natural philosophers commanded so much authority in Victorian Britain, especially within the corridors of power at Westminster and Whitehall. Herschel had long argued that mathematical laws secured increased credibility as the number of demonstrative observations increased and advocated large-scale data collection as the best means of producing reliable knowledge of natural phenomena. Having put this into practice at the Cape through years of astronomical observation, he was the ideal figure to reinvigorate the magnetic campaign. In particular, Herschel was adamant that the scheme include proposals for new 'physical observatories' which would record both magnetic and meteorological phenomena. Sabine had never intended to lobby for the establishment of permanent observatories, but acquiesced to Herschel's insistence that these be included within the campaign's aims. In August, Herschel presented the magnetic lobby's case to the BAAS's 1838 meeting, held at Newcastle, arguing that an expedition to the South Pole was essential for British

---

[39] Sabine, 'Contributions to Terrestrial Magnetism' (1840), 129.

science to catch up with research on the Continent. Armed with a £400 grant, Herschel then turned to championing the magnetic scheme, comprised of an Antarctic expedition and a network of overseas observatories, with the government. He requested an interview with the Whig prime minister, Lord Melbourne, in late August. Through his eminent social connections Herschel built powerful political support for the project, raising the subject of a polar expedition with Queen Victoria and Melbourne over dinner in October. He was also well known to the Secretary of State for the Colonies, Lord Glenelg (1778–1866); the secretary of the East India Trading Company, James Melvill (1792–1861); the Earl of Minto (1782–1859), an admiral of the fleet; the Royal Navy's hydrographer, Francis Beaufort (1774–1857); and the Royal Society's president and sixth son of King George III, the Duke of Sussex (1773–1843).[40] This was an extremely influential network for advancing the cause of magnetic science. In early November, Herschel's socializing paid off, securing a meeting with Melbourne and the Chancellor of the Exchequer, Thomas Spring-Rice (1790–1866), to discuss the venture in detail. In these negotiations, claiming the south magnetic pole for Britain was a valuable selling point. As James Clark Ross put it to Sabine in November, 'the South Magnetic Pole is a *sounding* name in the ears of the *Cabinet council* that may be safely used... they don't care one farthing for the expense, but they want something wherein to meet the House of Commons'.[41] Certainly, a polar expedition, securing celebrity and fame for Britain and the Royal Navy, was more attractive than Herschel's network of observatories, costing between £1,500 and £2,000 for a three-year period of observations.

After a meeting between Herschel, Wheatstone, Sabine, and the prime minister on 6 January 1839, the government finally consented to finance the expedition to locate the south magnetic pole.[42] This commitment marked the official commencement of the BMS. Herschel, however, had changed the nature of the proposed scheme. The addition of permanent observatories was inseparable from his conviction that to establish truths, there had to be fixed stations, producing continuous observations of magnetic variation over time. However, the question of whether the government would support a global network of observatories remained intractable, even after the confirmation of the Antarctic Expedition. In December 1838, Ross, Sabine, and Beaufort had considered abandoning the observatory element of the campaign altogether, fearing the government would be spooked at the overall cost. For Ross and Sabine, this presented little concern, with both men sure that all necessary observations could be made at sea on naval vessels. At the same time, they worried that land-based observatories were

---

[40] Carter, *Magnetic Fever*, 29–33; Cawood, 'The Magnetic Crusade', 507.
[41] James Clark Ross, quoted in Jake Morrell and Arnold Thackray, *Gentlemen of Science: Early Years of the British Association for the Advancement of Science* (Oxford: Clarendon Press, 1981), 353.
[42] Carter, *Magnetic Fever*, 508–10.

susceptible to geological errors arising from the presence of localized volcanic rock and iron.[43] Given their credentials as hardened Arctic explorers, this maritime bias is unsurprising. Herschel's absolute commitment to fixed observatories, compared to Sabine and Ross's flexibility on this point, reflected the theoretical divisions among the magnetic lobbyists. Those favouring Gauss's single-axis view of the Earth's magnetism, like Herschel, generally emphasized the importance of fixed observatories, while subscribers to Hansteen's four-pole theory, like Sabine, tended to promote expeditionary investigation.[44] Herschel refused to back down. Mobilizing his political allies, notably the influential Marquess of Northampton (1790–1851), who had succeeded the Duke of Sussex as president of the Royal Society later in 1838, and Spring-Rice, Herschel secured the Royal Society's commitment to the establishment of overseas observatories, believing these to be crucial to the large-scale collection of data that would be required to advance knowledge of the laws of terrestrial magnetism. Despite Minto's persisting concerns over the costs, Ross eventually provided the solution by proposing to establish magnetic observatories en route to the Antarctic.

Four government-financed observatories would be established—at Toronto, Van Diemen's Land, the Cape of Good Hope, and St Helena—along with East India Trading Company observatories at Madras, Bombay, Simla, and Singapore. These would cooperate with the growing international network of observatories that by 1842 totalled thirty-three, including at Cairo, Cadiz, Brussels, Breslau, Prague, Milan, Philadelphia, Algiers, Munich, and Cambridge, Massachusetts; eleven under Russian control, one of which was in Peking; and two in the princely Indian states of Lucknow and Trivandrum. It was planned that the four government observatories would be modelled on Lloyd's 1838 magnetic observatory, built in Trinity College Dublin's garden and itself based on Gauss and Weber's magnetic observatory at Göttingen.[45] Toronto's selection as an observatory location was based on Canada's position as the point of maximum magnetic intensity in the northern hemisphere; the city also appeared to suffer less from localized magnetic disturbances than its rival, Montreal. There were plans for an observatory at Aden, but this was abandoned when the instruments despatched to the small outpost arrived broken. The three remaining government observatories, at St Helena, Cape Town, and Van Diemen's Land, would all be built from scratch, with troops and equipment transported out as part of the expedition. To synchronize observations between ships and observatories, Lloyd suggested that measurements be taken at a fixed time each day, proposing one universal field day of intensive magnetic observations per month, plus another weekly day of minor measurements. Herschel preferred simultaneous measurements, performed daily,

---

[43] Ibid. 53–4.   [44] Ibid. 46.
[45] Goodman, 'From "Magnetic Fever" to "Magnetical Insanity"', 153–6; Cawood, 'The Magnetic Crusade', 513.

to correspond to the Magnetische Verein's synchronized observations throughout Europe. Eventually, however, the BMS settled on monthly Term Days, with observations performed at five-minute intervals according to Göttingen mean time. This coordination was the one thing that united Britain's magnetic enterprise with Gauss and Weber's European magnetic investigation.[46] In these places of controlled magnetic measurement, it was the delicate instruments favoured by Gauss, made in Paris and London, rather than the robust devices of Fox's devising, that were favoured.

On 8 April 1839, Ross officially secured command of the expedition with his appointment to HMS *Erebus*. The Admiralty ordered that he sail as far south as he could, determine the position of the south magnetic pole, and—if possible—reach it. Yet this was not primarily a voyage of colonial acquisition or discovery: it was a magnetic investigation.[47] With Ross commanding *Erebus*, Francis Crozier (1796–1848) offered a reliable second as captain of the expedition's accompanying vessel, HMS *Terror*. Born in County Down, Ireland, Crozier was well experienced in naval exploration, having served on William Parry's Arctic voyages between 1821 and 1827 aboard HMS *Fury* and HMS *Hecla*. A lieutenant in 1826 and a commander from 1837, he became friends with Ross while under his command on HMS *Cove* in 1835. His selection to act as Ross's second for the Antarctic Expedition in 1839 owed much to these connections, but also to the skill he had cultivated as a magnetic observer. Indeed, Crozier had been elected a fellow of both the Royal Society and the Royal Astronomical Society in recognition of his contributions to magnetic science.[48] The expedition leaders therefore boasted impressive scientific experience and would perform the majority of the venture's magnetic measurements. They were each to be issued with a dipping needle, one for *Erebus* and one for *Terror*, with which to survey the terrestrial magnetism from Falmouth to Cape Town and Van Diemen's Land, and the seas surrounding the Antarctic. This expedition required the very latest instruments for measuring the Earth's magnetic character while at sea, and it was to this that Fox provided the solution.

4.3. Dipping Needles on the Atlantic

Amid all the preparations, the most pressing question was which magnetic instruments to take. While observatories required delicate dipping needles,

---

[46] Goodman, 'From "Magnetic Fever" to "Magnetical Insanity"', 161–2 and 181–4; Carter, *Magnetic Fever*, 116–17.

[47] Edward J. Larson, 'Public Science for a Global Empire: The British Quest for the South Magnetic Pole', *Isis*, 102/1 (Mar. 2011), 34–59, at 34–5.

[48] Elizabeth Baigent, 'Crozier, Francis Rawdon Moira (1796–1848)', *Oxford Dictionary of National Biography*, https://www.oxforddnb.com/view/10.1093/ref:odnb/9780198614128.001.0001/odnb-9780198614128-e-6840, 23 Sept. 2004, accessed 24 Mar. 2020.

survey work demanded robust instruments, capable of taking measurements at sea. In 1839, fresh from the success of Sulivan's magnetic expedition to Rio and the Falklands, Fox was the front-runner to deliver such devices. He had been a constant supporter of the BAAS–Royal Society magnetic campaign, favouring Sabine and Ross's vision of the project, agreeing on the urgency of locating Hansteen's southern hemisphere magnetic poles and prioritizing the Antarctic Expedition over fixed observatories. Writing to Sabine in August 1838, Fox expressed his hopes that 'the projected expedition to the Pacific *pole* will go on & prosper'.[49] 'I cannot suppose', Fox asserted, 'that they [the government] will refuse to send out an expedition for the attainment of an object so important in every point of view.' Nevertheless, in November he expressed reservations over the location of the proposed observatories, wondering if St Helena was not 'objectionable as a magnetic station' due to its volcanic character and fearing that the island's variations in temperature would 'modify the local influences on the needle, just as the action of iron on a needle is increased by heat or warmth'.[50] In January 1839, Fox offered further encouragement, informing Sabine that the Royal Society's appeal for government support 'in the cause of terrestrial magnetism, is too forcible...to be easily refused'. He assured Sabine of his local MP's support, confident that 'Charles Lemon is disposed to do what he can to second the object by speaking to his Brother-in-law Lord Lansdown'.[51] Fox was well aware of the opportunity the proposed Antarctic Expedition represented for his dipping circle. In the background of the ongoing campaign, he subtly marketed the value of his instruments for expeditionary survey work. In early 1839, Fox forwarded a report to Beaufort, reporting his observations made on the Continent in 1838 with his 4-inch dipping needle of Jordan's construction. Should Parliament support an Antarctic expedition, Fox was sure 'that this small instrument will answer Cap R's purpose well if he cannot get a larger one completed in time'.[52]

In March, Fox was delighted to read in the newspapers that the 'Expedition to the Southern Hemisphere has been decided upon' and determined to aid Sabine in equipping the venture with reliable dipping needles. In this task, they were increasingly confident, in no small part due to their encouraging experiences of Sulivan's recent Atlantic survey. Fox and Jordan were keen to demonstrate that they had learnt valuable lessons from the recent performance of their dipping circle and needles and would apply these to ensure that Ross's expedition had the very best instruments that Cornish science could supply. Reflecting on the problems Sulivan had faced due to the lightness of the needle for intensity measurements, Fox performed experiments on the relative ease with which needles of

---

[49] TNA, Sabine Papers, BJ3/19/11–12: Robert Were Fox to Edward Sabine (14 Aug. 1838), 12.
[50] TNA, Sabine Papers, BJ3/19/17–20: Robert Were Fox to Edward Sabine (16 Nov. 1838), 18.
[51] TNA, Sabine Papers, BJ 3/19/34–7: Robert Were Fox to Edward Sabine (22 Jan. 1839), 34.
[52] RS, Sabine Correspondence, Vol. 2, D–J, Ms/258/567: Robert Were Fox to Beaufort (15 Jan. 1839).

different sizes could be magnetized. He found with a 4-inch needle 'that <u>one</u> or <u>two</u> touches on a powerful magnetic will impart to it as great a force as any additional number of touches'. He had also

> <u>recovered</u> the poles many times by drawing the needle in the same direction, & over the <u>same</u> part of a magnetic, & its force was always constant or very nearly so, as may be seen by the enclosed notes. If this be so, might not the intensity <u>also</u> be taken by means of a second needle, which sh^d have its poles changed by an old magnet.[53]

In contrast, Jordan and Fox found by experiment that a heavier needle was more prone to error due to temperature variations. 'If light needles are capable of taking a much higher charge than heavier ones, it must be an argument in favour of the former for observations on intensity as well as dip', surmised Fox.[54] Given that Jordan's newest needles were heavier than those which Sulivan had used, Fox took steps to refine his method of correcting results for temperature changes. Enclosing the instrument in a double case of sheet copper, the space between which was filled with water of varying temperatures, and then taking readings, Fox measured how temperature effected different needles.[55]

On hearing of the magnetic campaign's success, Fox initially proposed two dipping circles with 7-inch needles for *Terror* and *Erebus*. Enthusiastically, he informed Sabine of the near completion of the first of these, which would 'do Jordan credit'.[56] The instrument maker reckoned he could manufacture a second to the same patterns in a much shorter period of time, should Sabine request it for the expedition.[57] Not only would two standardized instruments enable comparative magnetic measurements in order to reduce error, but both Crozier and Ross would be capable of performing uniformed magnetic experiments. However, by April 1839, Fox regretted that the first of these 7-inch devices had been delayed 'for want of jewels with holes of a proper size, & a supply has been expected from London for the last fortnight'.[58] Fox had, in fact, ordered three new instruments to these dimensions from Jordan following Sulivan's return from Rio, prior to the magnetic lobby's success in securing government support for an Antarctic expedition. Fox had, no doubt, Ross's voyage in mind for these devices, but they were originally intended for general naval use. Early in 1839, there were proposals that the first completed dipping needle completed be sent to George Gray (1812–98) for his ongoing exploration of north-western Australia, but the chance for Jordan's

---

[53] TNA, Sabine Papers, BJ3/19/27–30: Robert Were Fox to Edward Sabine (7 Jan. 1839), 28–9.
[54] TNA, Sabine Papers, BJ3/19/52–3: Robert Were Fox to Edward Sabine (6 July 1839), 53.
[55] TNA, Sabine Papers, BJ3/19/47–8: 'Influence of Changes of Temp^e on Needles B & S', 47–8.
[56] TNA, Sabine Papers, BJ 3/19/38–9: Robert Were Fox to Edward Sabine (19 Mar. 1839), 39.
[57] TNA, Sabine Papers, BJ 3/19/41–2: Robert Were Fox to Edward Sabine (23 Mar. 1839), 41.
[58] TNA, Sabine Papers, BJ3/19/43–4: Robert Were Fox to Edward Sabine (16 Apr. 1839), 43.

latest instruments to take the leading role in Britain's most eminent magnetic investigation was too good to miss. Fox swiftly offered Jordan's latest wares to Sabine and Ross, who were glad to acquire the latest Fox type. Ross wrote to Fox's brother, Alfred, expressing how he was

> most sorry that there should have been any mistake about the order for your Brother's needle deflection for the New Holland [Australian] Expedition. I quite understand that one would be required for that service and am very sorry [to] find that they are to go without it. The one which has been finished for them will do quite well for me and I am authorized by Capt. Beaufort to request that you will send it directed to him at the Admiralty as it is intended for the public service of the Navy.[59]

While the second dipping needle would also be wanted for the expedition, Ross thought any more would be surplus to requirement, but had no doubt that Alfred would 'find plenty of applications for the third instrument that you are requested get made by Mr Jordan'.[60] Fox and Jordan were now, by mid-1839, the Admiralty's premier providers of magnetic instruments for maritime experiments. Despite this commercial success, delays in completing Fox's two 7-inch needles called for a backup plan.

By May, the first new Fox type was complete and safely in Beaufort's hands at the Admiralty.[61] In June, however, Fox was sorry that the second 7-inch dipping needle, intended for *Terror*, had been delayed due to a spate of 'unexpected interruptions'. To begin with, the jewels Jordan had ordered for the pallets were found, on arrival, to be too large. When new ones were delivered, these were the wrong shape, not being enough like a cone, but flatter. This was such a crucial point, determining how freely the needle oscillated, that Fox insisted on waiting for improved jewels. While Jordan reworked the pallets, Fox tested the two magnetized needles for this circle. His initial trials suggested 'Needle S' to be of a hard temper and less liable to suffer magnetic change. 'Needle B', in contrast, was more highly magnetized and sensitive, but this meant that its force was likely to diminish before reaching London.[62] Were the 7-inch instrument not ready in time, Fox suggested to Ross that he take Lieutenant Sulivan's 4-inch dipping needle, though conceded that it was not so well mounted for intensity observations as newer models.[63]

[59] Letter in private ownership from James Clark Ross to Alfred Fox (*c*.1839), sold 12 June 2019 in New York by Invaluable (Lot 126: 'James C. Ross Orders a Needle Deflection Compass for Research & Expeditions'). Printed copy in author's possession.
[60] See n. 59.
[61] TNA, Sabine Papers, BJ 3/19/45–6: Robert Were Fox to Edward Sabine (27 May 1839), 45.
[62] TNA, Sabine Papers, BJ 3/19/49–51: Robert Were Fox to Edward Sabine (28 May 1839), 49–50.
[63] RS, Sabine Correspondence, Vol. 2, D–J, Ms/258/570: Robert Were Fox to James Clark Ross (15 July 1839).

Along with instruments, Falmouth was a source of instruction for the Antarctic Expedition. Fox wrote to Ross in June, to convey his 'very sincere pleasure to hear of thy appointment to this highly interesting & important service, & I heartedly congratulate thee upon it, as well as all who take an interest in the advancement of our knowledge of Terrestrial Magnetism'. Fox promised to send Ross any further information required regarding the use of his instruments. 'Sabine is in possession of all needful details,' wrote Fox; 'I have particularly mention'd to him the method of ascertaining intensities when the deflector, or the second needle is used instead of the weights & also of knowing, at all times & places, if both or either of the needles have lost any force, & how much.'[64] Eager to know how the new 7-inch performed on *Erebus*, Fox requested that Ross keep him updated on the device throughout the voyage. In particular, he stressed how the new instrument afforded Ross enhanced means for a careful settling of the balance of the magnetic needle itself within the jewelled pallets. Ensuring that this was 'properly balanced' was especially difficult around the equator in 'latitudes where the dip is small'. Supposing the dip to be under 45°, Fox advised Ross to turn the angle of the face of the instrument to azimuth, where it would be 'clear that if perfectly balanced, the same force should, independently of instrumental errors or defects, raise it to 0°, as if acting in a contrary direction, would coerce it to 90°; & any difference in these forces must be attributed to an imperfection in balancing'. There might be imperfections in the instrument's pivots or weights, but by performing lots of observations in different azimuths, facing both east and west, Fox was sure that 'a near approximation to the truth may be arrived at sufficient to insure great confidence in the results of experiments on intensity'.[65] Above all, Fox assured Ross that the combination of weights and deflectors provided the expedition with an unprecedented package of tools for controlling and coercing the needle in all latitudes. Ross would, Fox had no doubt, 'find the agreement between the methods of weights & deflectors entirely satisfactory, & that it will never be necessary for him to expose his needles to the influence of the weather'.[66] By the summer it was clear that the second 7-inch Fox type would not be ready in time. Instead, Ross and Sabine agreed that the older 4-inch dipping needle, so well tested on Fox's Continental survey and Sulivan's Atlantic voyage, would be issued to HMS *Terror* for Crozier's use.

In July 1839, Jordan despatched the small 4-inch Fox type by steamer to Beaufort in London. In turn, Fox gifted a third magnetized needle for this instrument 'as a small contribution to the cause of magnetism', this addition being 'much lighter than the other two, & will, for this reason, have less swing at sea'.

---

[64] RS, Sabine Correspondence, Vol. 2, D–J, Ms/258/569: Robert Were Fox to James Clark Ross (18 June 1839).
[65] Ibid.
[66] TNA, Sabine Papers, BJ 3/19/72–72a: Robert Were Fox to Edward Sabine (7 May 1840), 72.

Believing this more susceptible to changes in magnetism, he hoped Ross would be able to test the relative merits of lighter and heavier magnetized needles at sea. It was, however, the jewelled pallets that caused Fox the greatest angst. When making the axle for a new needle for the 4-inch Fox type, Jordan had found 'that one of the jewels was cracked a little, caused no doubt, by a fall the instrument had some time ago from its stand,—nearly four feet high. This incident did not injure the pivots & as the needle worked well it was not suspected that the jewels were injured'. Another jewel had been inserted and the bearings of the pivots altered to fit, but this was a warning of the delicacy of the jewels and axle of the needle. Nevertheless, Fox remained confident that the needle would maintain a constant magnetic force throughout Ross's expedition. As 'the needle did not suffer any appreciable change of force during my absence from home when I went to France', he expected 'that it will retain its magnetism without any material change during the voyage to the Southern Hemisphere'.[67] The failure to equip the Antarctic Expedition with two brand new 7-inch Fox types was undoubtedly disappointing for Fox and Jordan, but the 1839 expedition nevertheless represented a triumph for Cornish science and its scientific instrument trade.

When Ross departed England in September 1839, he carried Fox's new dipping circle with a 7-inch needle aboard *Erebus*, while Crozier took Fox's old 4-inch instrument that he taken to Europe, equipped with a new magnetized needle, for *Terror*'s measurements.[68] Having performed preliminary trials on 28 August 1839 at Westbourne Green, designated as the expedition's index of relative magnetic intensity measurements, on-board observations commenced from 1 October, as Ross and Crozier travelled south down the Atlantic. After stopping for onshore measurements at Funchal on Madeira between 21 and 26 October, and passing Tenerife on 4 November, they reached Porto Praya at Cape Verde fourteen days later. From here, Ross despatched an account of Fox's dipping needles which revealed how the two captains were growing accustomed to their devices. 'Our Fox is doing now most admirably in both ships', Ross reported, but confessed that he had found the dipping needle difficult to use in the expedition's earlier stages. He took responsibility for these troubles, admitting that 'much of the abuse I heaped upon him [the Fox type] the early part of the voyage properly belongs to myself for the stupidity of not doing him justice and getting a proper table made'.[69] Although tricky to manage between Falmouth and Funchal, Ross was gaining confidence with the instrument by Cape Verde. More importantly, in terms of accuracy, Ross and Crozier found their Fox types gave exceedingly similar measurements. Ross was sure that Sabine 'would be somewhat surprised to see the

---

[67] RS, Sabine Correspondence, Vol. 2, D–J, Ms/258/570: Robert Were Fox to James Clark Ross (15 July 1839).
[68] TNA, Sabine Papers, BJ3/19/99–100: Robert Were Fox to Edward Sabine (5 Oct. 1841), 100.
[69] CUL, Add.9942/26: 'Letter to Major Sabine from Capt Jms Clerk-Ross, HMS Erebus off C Verd' (10 Nov. 1839).

interchange of signals between the two ships (Erebus & Terror) when the dip comes to the same minute, and it is seldom more than a few minutes apart when corrected for the direction due to each vessel'.[70] The expedition appeared to have the instruments with which to map out the line of least magnetic intensity from Cape Verde to Cape Town, via Trinidad and St Helena. Although uncertain of the error arising from the ships' magnetism, Ross resolved to continue to take onboard measurements and then make comparative onshore experiments on St Helena and Cape Town with which to calculate the influence of the ferrous metal on *Terror* and *Erebus*.

The expedition reached the South Atlantic island of St Helena early in 1840, where a detachment of two bombardiers and a corporal under the command of Lieutenant Henry Lefroy disembarked *Terror* to set up a magnetic observatory. On arrival, Ross and Lefroy met Governor George Middlemore at Plantation House, performing magnetic observations in Jamestown, before repeating these at Longwood House, where the exiled Napoleon Bonaparte had resided until his death in 1821. Lefroy and Ross mutually agreed on an observatory location, selecting a point of minimal magnetic intensity.[71] While Lefroy busied himself organizing a place of permanent magnetic observation, Ross and Crozier's priority was to examine their Fox types beyond the confines of their ships. However, St Helena's magnetic rock made it difficult to make the onshore checks of the Fox circles that Ross had hoped for to calculate any error arising from the iron on *Terror* and *Erebus*. The igneous character of the island persuaded the captain 'that no observations on shore can be depended on under any, even the most favourable circumstances & that eventually all absolute magnetic observations must be made at sea'. In contrast to St Helena, the experience of *Erebus*'s Fox needle had convinced Ross that on 'the ship you can do away with most of the causes of disturbance & even determine the exact value of the rest. On shore all is uncertainty & beyond the reach of determination'.[72] Though they produced some magnetic interference, *Terror* and *Erebus* were experimental spaces over which Ross and Crozier could exert significant control.

Despite the island's challenging magnetic properties, the two Fox circles performed well. Recording dips varying between 18.2° and 21.3°, the instruments differed little more than a few minutes from each other. Ross reported to Sabine on 9 February 1840, explaining that because of this continued consistency between his 7-inch and the older 4-inch Fox type, Crozier was 'very anxious to have one of Mr Fox's dipping needles 7 inches length'. He requested Sabine press Jordan for the completion of this second new instrument and have it sent with all

---

[70] Sabine, 'Contributions to Terrestrial Magnetism' (1840), 155.
[71] Goodman, 'From "Magnetic Fever" to "Magnetical Insanity"', 156.
[72] CUL, Add.9942/26: 'Letter to Major Sabine from Capt Jms Clerk-Ross, HMS Erebus off C Verd' (10 Nov. 1839).

haste to Van Diemen's Land, where the expedition would collect it.[73] Hoping that Beaufort would make arrangements for this, Ross asserted that a second new Fox circle was 'most desirable in case of accident to that now on board as we find those of the ordinary kind do not act so well at sea & if any accident sh^d disable that now in use, there w^d be an end of all our sea observations'.[74] Ross also ordered a spare set of the delicate jewelled pallets on which Fox's needles oscillated.

Back in London, Sabine did not receive Ross's letters from Cape Verde and St Helena until April 1840, but was immensely gratified by these reports, which arrived just in time to be included as a postscript to his first issue of 'Contributions to Terrestrial Magnetism'. Sabine detailed how Ross had written from St Helena, having found 'the observations of the magnetic dip and intensity made at sea with Mr. Fox's instrument succeed far beyond his most sanguine expectation'.[75] In May, Sabine forwarded Ross's Cape Verde and St Helena despatches to Falmouth for Fox's perusal. On reading of Crozier's desire to have a 7-inch circle, Fox was delighted at the performance of Ross's new device, which he found 'highly gratifying'.[76] Fox wrote back to Ross with news that Jordan was hard at work on the second 7-inch needle, but warned that it would take twelve weeks to complete. Fox had recently overseen the completion of a similar instrument for John Nichol (1804–59) at Glasgow Observatory, for which he had persuaded Jordan to adopt the practice of 'grinding his pivots in a cylindrical hob, instead of having them turned by a tool', ensuring a smoother oscillation, with reduced friction.[77] Having adopted this 'improved method of making the pivots', and ensuring they formed true cylinders, Jordan reported enhanced magnetic readings, with 'the indication of the needles being much more perfect than they were before'.[78] In the hope of equipping Crozier with a 7-inch dipping needle ahead of schedule, Fox and Jordan appealed to Nichol for his instrument to be sent to Ross, hoping this might even reach *Terror* by the time it arrived at Cape Town. Fox thought this new instrument 'decidedly superior' to both Crozier's 4-inch and Ross's 7-inch devices.[79] To his surprise, Nichol agreed for his instrument to be sent out for the expedition.

As for the performance of the Fox types on *Terror* and *Erebus*, the Cornishman was pleased that the angles measured with the deflectors 'appear to be very

---

[73] Ibid.
[74] 'Note from James Clark Ross to Captain Beaufort from St Helena' (9 Feb. 1840), in CUL, Add.9942/26: 'Letter to Major Sabine from Capt Jms Clerk-Ross, HMS Erebus off C Verd' (10 Nov. 1839).
[75] Sabine, 'Contributions to Terrestrial Magnetism' (1840), 155.
[76] RS, Sabine Correspondence, Vol. 2, D–J, Ms/258/571: Robert Were Fox to James Clark Ross (10 Apr. 1840).
[77] TNA, Sabine Papers, BJ 3/19/56-7: Robert Were Fox to Edward Sabine (14 Dec. 1839), 57.
[78] RS, Sabine Correspondence, Vol. 2, D–J, Ms/258/571: Robert Were Fox to James Clark Ross (10 Apr. 1840).
[79] Ibid.

consistent, & I doubt not will give results which will harmonize well with those deduced from the weights'. The great advantage of measuring intensity with deflectors rather than weights, Fox explained, was 'the greater protection it affords to the needle in damp weather', as the observations would be made without opening the instrument's glass face. He stressed that employing the second needle for this was favourable in case the application of the deflectors altered the magnetic force of the principle dipping needle.[80] From Ross's Cape Verde and St Helena correspondence, Fox inferred 'that the small dipping needle was doing better, as he mentions the accordance of the results obtained simultaneously in the two ships'. He voiced concerns over St Helena's suitability for a magnetic observatory, fearing that the 'nature' of the island would 'interfere with magnetic observations' and asked Sabine if 'the nature of the rocks, or their depth under the surface of the sea, modify the results?'[81] Fox was not alone in these anxieties. By April 1840, Lloyd too regretted the choice of St Helena, believing the island's local irregularities made it an inappropriate place for a magnetic observatory.[82]

While Sabine forwarded Fox's recommendations on to Van Diemen's Land for Ross, Fox passed on Ross's request for new pallets and a 7-inch circle for Crozier to Jordan. He was less pleased to learn that Sabine was irked at the combined costs of the two 7-inch instruments and spare parts. When Jordan learnt of this, he too was unimpressed. Thanking Fox for his 'valuable service in endeavouring to support my interest at the Admiralty', Jordan warned that had Sabine 'carefully estimated the amount of labour requisite to produce the two instruments he would have arrived at a different conclusion'. Jordan wrote directly to Sabine, proposing modifications that would reduce the costs of Crozier's device, but he could not 'reduce the time by trimming my own profits, they are already as low as they can or ought to be'. Indeed, he was sorry to admit that 'experience dearly bought has shown me that I commenced my business on much too low a scale of profits'. In short, he had entered the instrument-making business with little conception of the true costs of mechanical construction. 'On careful examination of the four years now passed,' reflected Jordan, 'I really think that, not charging enough, has been the great error.'[83] He was confident that his own costs compared favourably to London instrument makers and promised any customer who sent him a pattern from a mechanic in the city would receive a quote for a device of at least equal quality at the same price. Jordan had 'not the slightest fear of losing the Admiralty orders by their getting the same thing equally well made for the same price elsewhere'.[84] In total, Jordan's new instrument for Crozier would require a

---

[80] TNA, Sabine Papers, BJ 3/19/70–1: Robert Were Fox to Edward Sabine (14 Apr. 1840), 70.
[81] TNA, Sabine Papers, BJ 3/19/73–4: Robert Were Fox to Edward Sabine (8 May 1840), 74.
[82] Goodman, 'From "Magnetic Fever" to "Magnetical Insanity"', 157.
[83] RS, Robert Were Fox Papers, MS/710/60: Thomas Brown Jordan to Robert Were Fox (25 May 1840).
[84] Ibid.

fee of £33. 1s. 0d.[85] With the Admiralty finally giving way over the question of Jordan's fee and Nichol agreeing to exchange the 7-inch dipping needle for Glasgow Observatory for a later instrument, Fox checked the new device in his garden at Falmouth, providing an effective quality control before sending the instrument to Beaufort. By the end of May 1840, Jordan's second 7-inch dip circle and replacement jewelled pallets were on their way to Van Diemen's Land for Crozier and the voyage south to the Antarctic.

4.4. Into the Southern Hemisphere

Ross stayed several days on St Helena to determine the dip and intensity measurable at Longwood House with the two Fox types, before sailing for the Cape of Good Hope.[86] Anchoring in Simon's Bay, the expedition took magnetic readings on 18 and 19 March 1840, before Ross and Crozier employed their Fox types on Admiralty Jetty in Simon's Town a day later.[87] A detachment under the command of Lieutenant Frederick Wilmot established the expedition's second magnetic observatory. For this, the local governor, Sir George Napier, selected a portion of ground next to the astronomical observatory where Herschel had worked between 1834 and 1838. With measurements taken, *Terror* and *Erebus* sailed east across the Indian Ocean, mooring in Christmas Harbour on Kerguelen Island in time for the May Term Day. The expedition remained on the deserted island until the following Term Day, not leaving until 20 July 1840. Throughout the voyage it had become increasingly evident that, in moderate weather, the Fox needles gave more reliable readings on ships than could be attained on land.[88] From Kerguelen, the expedition pressed on towards Australia and Van Diemen's Land. Within Herschel's vision of a global network of magnetic observatories, Hobart took on a significant role, selected for its correspondence to the point of maximum magnetic intensity in the southern hemisphere. Not only was this site of scientific importance but, at Hobart, Ross and Crozier could be assured an enthusiastic welcome from the colony's governor, Sir John Franklin.

After three years of unemployment, Franklin had been grateful to secure the post of lieutenant-governor of Van Diemen's Land, where he arrived in early 1837. His acceptance of this posting was on the understanding that he would oversee a civilizing renaissance of the colony, in which magnetic science would take centre

---

[85] RS, Robert Were Fox Papers, MS/710/42: 'Thomas Brown Jordan, "Dipping Needle Costs", [c.1840]'.
[86] CUL, Add.9942/26: 'Letter to Major Sabine from Capt Jms Clerk-Ross, HMS Erebus off C Verd' (10 Nov. 1839).
[87] For drafted table of results, see CUL, Add.9942/21: 'Extracts from Cap Ross' Report: Magnetic Dip & Intensity Observations On Shore & On Board the Erebus with Needle' (1839–40).
[88] Granville Allen Mawer, *South by Northwest: The Magnetic Crusade and the Contest for Antarctica* (Edinburgh: Birlinn Ltd, 2006), 91–2.

stage: as governor, he could assume a key role within Antarctic exploration.[89] At Hobart, Franklin hosted the French explorer Cyrille Laplace (1793–1875) in January 1839, followed by Dumont d'Urville's (1790–1842) French expedition to the Antarctic in December. It was Ross's visit, however, between August and November 1840, that was to be the highlight of Franklin's governorship.[90] On arriving at Hobart, Ross went straight to Government House to meet Sir John and Jane, Lady Franklin (1791–1875). Together, they immediately selected a location for the observatory, naming it 'Rossbank' (Figure 4.1). Within nine days, the observatory was complete, ready in time for the Term Days of 27 and 28 August, on which readings were taken every two and a half minutes for twenty-four consecutive hours.[91] Rossbank took shape rapidly thanks to its director, Lieutenant Joseph Henry Kay (1815–75), having prefabricated the materials of the observatory's construction, and Franklin's provision of two hundred convict labourers to erect the building.

Figure 4.1 Rossbank Observatory as depicted in Ross's published account of the Antarctic Expedition (Image in author's possession, 2021).

[89] Lambert, *Franklin*, 110–14.
[90] Ibid. 127–32; on d'Urville, see Mawer, *South by Northwest*, 57–66.
[91] Ann Savours and Anita McConnell, 'The History of the Rossbank Observatory, Tasmania', *Annals of Science*, 39 (1982), 527–64, at 532.

Along with assisting in establishing Rossbank, Franklin awaited receipt of the new Jordan-built Fox circle, which Ross had ordered from Cape Verde, and which had since been dispatched to Hobart. Franklin also had with him Fox's original dipping circle, built by Watkins & Hill in 1832. As Fox told Sabine, 'Franklin has there, my first dipping needle, which I had with me at the British Assoc^n meeting at Oxford in 1832'.[92] On the reception of Crozier's new 7-inch circle, the gathering of Franklin, Crozier, and Ross marked the assembling of an impressive array of no fewer than four Fox-type dipping needles, on the opposite side of the world from Falmouth. With Fox's original dipping needle, his 4-inch device, and the latest two 7-inch models, Ross, Franklin, and Crozier were able to perform an intensive examination of their respective magnetic instruments in preparation for the expedition's forthcoming journey south. Importantly, Hobart proved to be the ideal location to finally make the onshore measurements with which the magnetic influence of *Erebus* and *Terror* could be determined, with experiments conducted in October 1840 and repeated on the ships' return the following June. For this purpose, the new magnetic observatory at Rossbank proved invaluable, as had the use of Franklin's own Fox needle.[93]

Ross's expedition left Hobart in November, with Kay remaining behind as director of the observatory until 1842. With observatories established at St Helena, Cape Town, and Rossbank, *Terror* and *Erebus* made a course towards the south magnetic pole, breaking through the dense pack ice to the south of New Zealand in January 1841. Less than a week after this triumph, Ross observed the mountainous peaks of a huge expanse of land, covered in enormous glaciers flowing down to the sea. They had, at last, reached the Antarctic. Ross claimed this land for Britain, planting a British flag on an offshore islet which he named 'Possession Island'. After fighting off the local penguin population, who attacked the expedition's crew 'with their sharp beaks, disputing possession', Ross named the region Victoria Land.[94] The highest peaks sighted were given names of those most prominent in British science, including Sabine, Herschel, and Whewell, while the ferocious 12,000-foot active volcano was named Mt Erebus. Ross named Cape Crozier after his reliable second, located at the foot of the newly christened Mt Terror. *Terror* and *Erebus* followed the land mass but found it impossible to reach the pole. Nevertheless, along with charting the Ross Sea and Great Ice Barrier, Ross used his Fox dipping needle to locate the south magnetic pole, measuring a dip of 88°, which he reckoned put them just 160 nautical miles from the pole. Although only a circumnavigation of the point of 90°, Ross's expedition had gone further south than any before, and Fox's instruments proved crucial to calculating the

---

[92] TNA, Sabine Papers, BJ 3/19/73–4: Robert Were Fox to Edward Sabine (8 May 1840), 74.
[93] Edward Sabine, 'Contributions to Terrestrial Magnetism: No. V', *Philosophical Transactions of the Royal Society of London*, 133 (1843), 145–231, at 152 and 164.
[94] Ross, quoted in Larson, 'Public Science for a Global Empire', 41.

location of the pole.[95] After a 300-nautical-mile voyage along the Great Ice Barrier, the ships returned to Van Diemen's Land. *Terror* and *Erebus* reached Hobart in April 1841, which Ross fondly referred to as 'our southern home' in recognition of the Franklins' hospitality.[96] There Ross and Crozier helped prepare Rossbank for the following Term Day, on 22 April, equipping it with additional instruments from the ships. When it arrived, Ross, Crozier, Kay, and Franklin spent the morning taking magnetic measurements together.[97]

On 7 July, the expedition again left Van Diemen's Land, this time for good. Ross sailed for Australia, but kept in communication with Franklin so as to coordinate subsequent magnetic observations. By the time Ross arrived at Sydney, Franklin was in possession of a large box of papers and a collection of packages sent out from London, on which the governor 'recognized Sabine's writing' and had 'no doubt of their containing much important matter'.[98] While Ross made magnetic observations off the Australian coast on Garden Island for the July Term Day, Franklin took similar readings at Rossbank. Having received Ross's measurements in August, Franklin's reply arrived the following month, expressing his 'gratification in witnessing the accordance between the movements of your needles at Sydney and ours on the July Term Day'. The declinator and horizontal magnetometers gave almost identical readings. There were some local influences that caused a slight variation in the amount of magnetic deviation in terms of intensity, but Franklin was confident that a 'minute analysis of the different elements that come within the range of observations on these occasions' would resolve any anomalies.[99] Franklin continued to monitor and develop the observatory until the end of his governorship of Van Diemen's Land. In May 1843, he was disappointed to find that Lloyd and Sabine were struggling to reduce the figures taken at Rossbank due to the observatory's considerable temperature variations. He wrote to Beaufort in London, requesting funds to resolve this, warning that it was essential, 'having undertaken this great Magnetic Enquiry'. It was crucial, he asserted, to remedy 'any known causes of doubt and render the observation as free from error as possible'.[100] In July 1843, he oversaw a new lining of the observatory's interior to resolve these frequent temperature fluctuations.[101] While Franklin wrested with the challenges of producing data at Rossbank, Ross's expedition returned to the Antarctic to continue its survey of terrestrial magnetism.

---

[95] Larson, 'Public Science for a Global Empire', 41.
[96] Savours and McConnell, 'The History of the Rossbank Observatory', 538.
[97] Scott, Ms248/291: 'Journal of Sir John Franklin, Tasmania, 8 April to 18 May 1841', 1, 2, and 39.
[98] Scott, Ms248/316/6: J. Franklin to J. C. Ross (4 Aug. 1841).
[99] Scott, Ms248/316/7: J. Franklin to J. C. Ross (16 Sept. 1841).
[100] Scott, Ms248/295/3: J. Franklin to Beaufort (23 May 1843).
[101] Scott, Ms248/316/10: J. Franklin to J. C. Ross (21 July 1843).

4.5. Magnetic Contributions

Ross would complete two further journeys south during the following two Antarctic summers before returning to Britain by way of the Falklands and Rio, arriving home in September 1843.[102] The south magnetic pole would not be reached until 1909, when Ernest Shackleton's *Nimrod* expedition achieved the feat by sledge. Nevertheless, the locating of the pole was immediately held up as a celebrated justification for the expedition's expense. Ross had calculated the position of the south magnetic pole on 28 January 1841, but word of this did not reach Cornwall until May (Figure 4.2). On 12 August, the Foxes were breakfasting with Lloyd and Sabine while at the Plymouth meeting of the BAAS, when news arrived that Ross had reached a latitude of 78°, this being further south than any previous venture. Caroline recalled the subsequent morning scene as one of jubilation for Sabine and her father.[103] Yet Sabine and Fox were unaware of the impossibility of Ross actually reaching the pole under sail. A month after the Plymouth meeting, on 16 September 1841, Fox wrote to Sabine, optimistically enquiring as to whether there was any news of Ross having obtained the pole itself.[104] Nevertheless, British audiences celebrated this claiming of the polar prize and the role Fox's dipping needle had played in this achievement. Further magnetic data was to come. Throughout the 1839–43 Antarctic Expedition, Ross and Crozier provided a regular flow of measurements back to Britain for Sabine and Fox to examine. On 20 January 1842, Sabine delivered an account of the expedition's initial results to the Royal Society. It was a dramatic moment, with Ross and Crozier's observations from England to St Helena, Cape Town, and Kerguelen Island given in full. Sabine outlined how Ross had employed Fox's method of using weights for intensity measurements on *Erebus*, except when this would have exposed the needle to bad weather, in which case the captain had relied on Fox's deflecting magnets. With their data corrected for temperature, both Ross and Crozier had completed tables drawn up 'under Fox's own direction, of the equivalent, or as they are termed by him, the *coercing* weights, for each deflector on each of the needles at the different angles which are likely to occur in the course of the observations'.[105]

Fox had initially recommended intensity be calculated by fixing the deflectors, causing an amount of deviation from the point of dip, and then applying weights to bring the needle back to the angle of dip. Effectively this converted the measure of the Earth's magnetic force from degrees into grains of weight. Sabine, however, regretted that

---

[102] Larson, 'Public Science for a Global Empire', 43.
[103] Pym (ed.), *Memoirs of Old Friends*, 124 and 134–5.
[104] TNA, Sabine Papers, BJ 3/19/95–6: Robert Were Fox to Edward Sabine (16 Sept. 1841), 96.
[105] Edward Sabine, 'Contributions to Terrestrial Magnetism: No. III', *Philosophical Transactions of the Royal Society of London*, 132 (1842), 9–41, at 9.

THE ANTARCTIC FOXES 159

Figure 4.2 Stephen Pearce's 1871 portrait of James Clark Ross with his Fox-type dipping needle on the desk (Reproduced by permission of the National Portrait Gallery, 2023).

Owing to accidental circumstances, no table of this description was prepared for this instrument before the Expedition sailed; the pressure of other duties prevented it being done at St Helena, the Cape of Good Hope, or at Kerguelen Island; and at Van Diemen Island the end of the axle of the needle being accidentally broken, the needle was returned to England to be repaired, and was thus separated from the instrument and from the deflectors. Under these circumstances we have no other resource for reducing the observations made with the deflectors, than to form a table from the observations of the weights and deflectors,...which shall answer the same purpose as a table of coercing weights.[106]

Despite this setback, with intensity measurements indexed to London, Ross had made seventy-four observations between England and the Cape of Good Hope with weights. He had found the deflector caused a 22° 57′ deviation in London, increasing to 34° at the equator, where the Earth's magnetic intensity was weakest, before returning to 29° 53′ at the Cape.[107] With the reduced influence of terrestrial magnetism encountered in equatorial regions, Fox's deflectors had exerted an increased impact on the magnetic needles. With the 'S' deflector alone, Ross took some 109 measurements from London to the Cape. A temporary 'diminished force in the deflector' interrupted these continual observations, occurring 'between the forenoon and afternoon observations of the 12th February' and causing a degree's error in observations. It returned to full force soon after but subsequently required fewer grains of weight to correct a deflection back to the dip.[108] In total, Ross conducted 647 magnetic observations of all arrangements before reaching Cape Town. As for Crozier on *Terror*, he had made do with Fox's older 4-inch needle. Sabine jested that this might be 'regarded by many persons as scarcely more than a philosophical toy...[but] in Captain Crozier's hands, fully justified the expectations which Mr. Fox, from his own experiments with it, had ventured to entertain'. Testament to both Crozier's skill and the instrument's integrity, the results from *Terror* and *Erebus* were almost identical. At Longwood House on St Helena, for instance, Ross's needle from *Erebus* gave an intensity of 0.586 relative to London, compared to 0.587 measured with Crozier's. At Jamestown, both registered identical measures of 0.611.[109]

These results were subsequently published in the Royal Society's *Philosophical Transactions*, constituting Sabine's third 'Contributions to Terrestrial Magnetism'. Here he boasted immense precision, which was testament to the 'late improvement in this class of observations, for which we are mainly indebted to the method and instrument devised by Mr. Fox, and to the zeal and unwearied patience of our naval officers'.[110] The second series of Ross and Crozier's observations appeared in Sabine's fifth 'Contributions', published in 1843. From these results, Sabine produced charts of the magnetic declination and intensity of the southern polar regions. This represented an extensive amount of data, but it also included a very public recognition of Fox's instruments, intended to build trust into the observations attained. Sabine called attention

---

[106] Ibid. 10.   [107] Ibid. 11.   [108] Ibid. 11–12.
[109] Ibid. 25.   [110] Ibid. 35.

to the invaluable aid for which magnetic science is indebted to Mr. Fox. Without his instrument and method, which render observations of inclination and intensity made at sea nearly or altogether equal to those which could be made on land or on ice, such were the difficulties of the navigation, and such the inaccessible though magnificent character of the coast that was discovered, that two of the three charts herewith presented, and especially that of the intensity, must have offered an appearance very different from that which they now exhibit.[111]

Sabine read a further account of the endeavours of *Terror* and *Erebus* at the Royal Society in April 1844, detailing measurements from June 1841 to August 1842, taken between Van Diemen's Land, the Antarctic, and the Falklands, which Ross had first reached in April 1842 before his third and final voyage south. Published as the sixth series of 'Contributions', Sabine reported that the Fox needles, when aboard the two ships, continued to produce staggeringly similar measurements.[112]

Ross's 1839–43 Antarctic Expedition produced a wealth of magnetic data, but there remained a large portion of the southern hemisphere unsurveyed. To fill in this gap, between the meridians of 0° to 125° east and 20° to 70° south, the Admiralty hired a 360-ton barque, HMS *Pagoda*, to complete the inquiry. Sabine equipped Lieutenant H. Clerk of the Royal Artillery with a Fox dipping needle and instructions on how to use the needle. Sailing from Cape Town for the Antarctic Circle on 9 January 1845, the *Pagoda* continued Ross's magnetic survey. Along with Clerk's Fox type, his assistant, Lieutenant Moore, took a Fox of his own, allowing the two officers to conduct comparative observations. With Clerk making morning measurements, and Moore using his device for the afternoon, the two Fox types were in constant use throughout the day, producing a mass of data over the course of six months. When Sabine published these results in 1846 as the eighth series of 'Contributions to Terrestrial Magnetism', he reported that the two Foxes had again produced almost identical measures of intensity.[113] With these Cornish instruments, naval and artillery officers had successfully surveyed the magnetic field of the southern hemisphere's polar seas.

Along with the magnetic data obtained on *Terror*, *Erebus*, and *Pagoda*, Ross's expedition's observatories at St Helena, Cape Town, and Rossbank provided Sabine with a stream of magnetic observations from fixed locations, complementing those taken at Toronto. Unlike on the ships, the instruments for these magnetic observatories were not Fox's, but made to Gauss and Weber's specifications.

---

[111] Edward Sabine, 'Contributions to Terrestrial Magnetism: No. V', *Philosophical Transactions of the Royal Society of London*, 133 (1843), 145–231, at 173.
[112] Edward Sabine, 'Contributions to Terrestrial Magnetism: No. VI', *Philosophical Transactions of the Royal Society of London*, 134 (1844), 87–118, 115*–118*, 119–224, at 99.
[113] Edward Sabine, 'Contributions to Terrestrial Magnetism: No. VIII', *Philosophical Transactions of the Royal Society of London*, 136 (1846), 337–432, at 342.

Along with a description of a 'standard' observatory arrangement, Gauss set out this uniform instrumentation in his instructions for observatories participating in the Göttingen Magnetische Verein.[114] However, these instruments had arrived at their respective ports in varying states, with *Terror* and *Erebus* providing very different transportation conditions. Lefroy's magnetic apparatus arrived at St Helena in good order, having travelled under the care of Crozier on *Terror*, as did Kay's for Van Diemen's Land. Wilmot, however, was distraught to find his instruments for Cape Town in a terrible state. In *Erebus*'s 'hot and damp' holds, the heat had melted the glue holding the dipping-needle boxes together. Once the lids had fallen off, the iron needles had soon rusted in the humid ship.[115] Likewise, the observatories provided measurements of varying reliability. St Helena turned out to be a terrible place for magnetic observations. Along with the island's shortage of timber, it soon became clear that its rock was volcanic and interfered with magnetic instruments. The island's lava possessed particular polarity, causing large deflections to a needle. After examining St Helena, Lefroy had identified a relatively manageable location where deep soil and clay covered the igneous rock. As the observatory took several months to construct, he transformed Longwood House into a temporary magnetic observatory, in which he performed his magnetic observations for May to July 1840. Using wooden casks as makeshift pedestals for his instruments, Lefroy's detachment removed the grates and iron fittings from the two rooms selected for experiments. Constructed with stone quarried locally at Jamestown, and with copper nails and fastenings, the new observatory was ready by August, complete with Yorkshire paving-stone pedestals for the dipping needles, brought out specifically, having been determined free of magnetism. Working to Göttingen time, one hour ahead of the rest of St Helena, Lefroy oversaw measurements on the island until 1842, when Lieutenant James Smyth of the Royal Artillery replaced him. By 1847, three sergeants, a gunner, and a bombardier from the Royal Artillery manned the station under Captain Charles Younghusband's (1821–99) command.[116] At Cape Town, Colonel Lewis of the Royal Engineers had erected the new Magnetic and Meteorological Observatory, which was complete in February 1841 and ready for observations by April. Taking eleven months to build, constructed with 12-inch logs and insulated with plaster on its interior, the observatory had a pitched, felt-covered roof, with all fastenings of zinc or copper. The instrument room had Purbeck paving-stone and sandstone pedestals to insulate the magnetic apparatus. Between April and September 1841, Wilmot directed observations at two-hour intervals, beginning hourly

---

[114] Savours and McConnell, 'The History of the Rossbank Observatory', 546.
[115] Goodman, 'From "Magnetic Fever" to "Magnetical Insanity"', 176–8.
[116] Edward Sabine, *Observations Made at the Magnetic and Meteorological Observatory at St Helena, i: 1840, 1841, 1842, and 1843, with Abstracts of the 1847 Observations from 1840 to 1845 Inclusive* (London: Longman, Brown, Green, & Longmans, 1847), 9–13.

measurements from October. This continued until July 1846, when the Royal Artillery detachment was recalled to England.[117]

The expedition's three observatories continued to yield magnetic data well after Ross's return to Britain in 1843. Despite its geological challenges, St Helena proved to be a crucial site for examining the Earth's equatorial magnetism. With Toronto Observatory's earliest results announced in 1845, Sabine published his *Observations Made at the Magnetic and Meteorological Observatory at St Helena* in 1847, including results from the observatory made between 1840 and 1845.[118] In particular, St Helena offered rich insights into the phenomenon of diurnal variation or, rather, how a magnetic needle moved throughout the day. Writing in *Philosophical Transactions*, Sabine explained how, in the northern hemisphere, a needle swung east to west from night until early morning, before returning westwards throughout the day. In the south, however, this process was reversed. Given Humboldt and Arago's speculations that there would be a location on Earth between the two hemispheres in which this hourly deviation did not occur, Sabine argued that St Helena's position was vital to resolving this question. The island was close to the line of least magnetic force, near the magnetic equator, but also in the quarter of the globe where this line fluctuated most extremely from the terrestrial equator and line of no dip. In short, the lines of least magnetic intensity and dip could be examined on the island. Singapore, Sabine observed, could have been an observation location, but the Far Eastern trade post turned out to be in an area where the lines of least force and no dip were closest to each other. The choice of St Helena in 1839, then, had been fortuitous for understanding the nature of diurnal variation.[119]

St Helena did indeed appear to have an unusual diurnal variation. During half the year, a magnetic needle on the island oscillated throughout the day as though in the northern hemisphere, moving towards the east until early morning, before retreating west until nightfall. For the alternate six-month period, the needle acted as though in the southern hemisphere. This diurnal variation suggested that St Helena was, at different times, both in the northern and southern hemispheres. Nor was this a purely seasonal phenomenon: in the northern hemisphere the movement of a magnet maintained its daily fluctuation, although the swing was more exaggerated in the summer months. Sabine asserted that St Helena had 'characters of the phenomena of both hemispheres', coinciding with the equinoxes around March–April and September–October: between November and February a needle on St Helena behaved as though in one hemisphere, and then acted as though in the other between May and August.[120] From these observations, Sabine

---

[117] Ibid., pp. i–ii.   [118] Ibid. 12–14.
[119] Edward Sabine, 'On the Diurnal Variation of the Magnetic Declination at St Helena', *Philosophical Transactions of the Royal Society of London*, 137 (1847), 51–7, at 51–3.
[120] Ibid. 55.

concluded 'that the line which has been supposed to exist by the eminent authorities [Humboldt and Arago]...which should be characterized by the absence of a diurnal variation of the declination, will not be found upon the globe'.[121] In short, the experiments on St Helena demonstrated that there was no magnetic region between the two hemispheres but, instead, one that fluctuated from northern to southern.

Sabine's publishing of the St Helena results aroused much philosophical interest. On reading Sabine's 1847 *Philosophical Transactions* paper, the Swiss natural philosopher Auguste de La Rive (1801–73) wrote to Arago, suggesting an explanation for the island's diurnal magnetic variation. Published in *Annales de chimie*, La Rive argued that temperature inequalities between the higher and lower strata of the earth's atmosphere generated electric currents which descended on the Earth's surface, causing St Helena to experience the magnetic characteristics of the northern hemisphere. As the currents diminished, the island assumed the character of the southern hemisphere.[122] In April 1849, Sabine emphatically rejected this speculation and accused La Rive of misinterpreting the data. St Helena's fluctuation between northern and southern hemispheres could not, Sabine contended, be determined by the Sun in terms of temperature changes, as the Sun's influence on the island varied throughout the day. If this was a solar phenomenon, the change between northern and southern characteristics would occur daily, which they did not. An odd phenomenon to be sure, but one that Sabine was confident indicated 'a line of *least magnetic force*'.[123] It was a question of the magnetic equator's, rather than the Sun's, influence.

4.6. Conclusion

On 25 September 1841, magnetic instruments around the world, including at St Helena, Toronto, Rossbank, and Cape Town, recorded a violent magnetic occurrence. Compass needles swung north to east, as the globe's magnetic field was thrown into chaos. This worldwide simultaneous recording of what was in a fact a common event, a magnetic storm, was a triumph for Sabine's system of global observations. At the Royal Observatory of Greenwich, Airy saw his magnetic declination needle change position through 2½° in eight minutes of time, with a significant increase in magnetic force. Similar observations were made in observatories across Europe and Britain's colonies.[124] Although initially shocking, with such an extensive magnetic disturbance previously unrecorded, this was a

---

[121] Ibid. 57.
[122] Edward Sabine, 'Remarks on M. de La Rive's Theory for the Physical Explanation of the Causes Which Produce the Diurnal Variation of the Magnetic Declination', *Abstracts of the Papers Communicated to the Royal Society of London*, 5 (1843–50), 821–5, at 821–2.
[123] Ibid. 823–4.   [124] Carter, *Magnetic Fever*, xv–xvi and 118–19.

vindication of how the international surveying and observing of the Earth's magnetic field could produce new knowledge of Nature. Back in Britain, this provided invaluable evidence in support of continuing the BMS. Sabine successfully lobbied the government in 1842 and 1845 for renewed funding of what, in April 1842, Lloyd referred to as the 'Magnetic Crusade'.[125] The zeal of its promoters certainly resembled a 'crusade', with enterprise which can be likened to a religious mission. Having managed magnetic observatories both on St Helena and at Toronto, Lefroy diagnosed Kay, Rossbank's director, with what he termed 'magnetic insanity' in 1844.[126] The hours of observation were both obsessive and exhausting. As Matthew Goodman has shown, this term 'magnetic insanity' aptly encapsulated the almost madness-inducing crusading spirit of the BMS's promoters. For Kay at Rossbank, this commitment to magnetic measurements even conflicted with his religious keeping of the Sabbath as a day of rest. As Goodman put it, Fox's instruments were central to a more general 'magnetic fever that gripped Britain's scientific community'.[127] The voyage of *Terror* and *Erebus* initiated what its advocates were sure would develop into a truly global investigation.

By any measure, Ross's Antarctic Expedition had been a staggering success, exhibiting how state-funding and the mobilization of Britain's naval, military, and colonial assets could advance the science of terrestrial magnetism. Despite his aversion to nationalist rhetoric, Herschel put it succinctly in 1840 when he declared that 'no nation was ever so favourably situated for such a purpose, nor so strongly called on as a maritime and commercial country for cooperation in a cause directly connected with nautical objects'.[128] Yet for all that this global surveying of the Earth's magnetic properties demonstrated the enormous resources that the British state had at its disposal for philosophical inquiry, the production and use of reliable dipping needles had been crucial to the Antarctic Expedition's success. As much as the BMS was a coup for Sabine and elevated Ross to unrivalled celebrity for adding the south magnetic pole to his list of exploratory achievements, it was Fox's instruments that had taken centre stage on-board *Terror* and *Erebus*. Throughout the following decade, these Cornish-built dipping needles would continue to play an important role in the British Magnetic Scheme. In 1866 in his tenth series of 'Contributions', Sabine combined the results from *Terror*, *Erebus*, and *Pagoda* between 1839 and 1845 to give what he asserted was a complete magnetic survey of the south hemisphere, largely conducted with Fox's dipping needles.[129] Two years later, he published charts of these observations, the sum of 'the great national undertaking, the Magnetic Survey of the South Polar

---

[125] Ibid. 120–1.
[126] Lefroy, quoted in Goodman, 'From "Magnetic Fever" to "Magnetical Insanity"', 17.
[127] Goodman, 'From "Magnetic Fever" to "Magnetical Insanity"', 18.
[128] Herschel, quoted in Carter, *Magnetic Fever*, 49.
[129] Edward Sabine, 'Contributions to Terrestrial Magnetism: No. X', *Philosophical Transactions of the Royal Society of London*, 156 (1866), 453–543, at 453 and 457.

Regions of the Globe', adding a map for the northern polar regions in 1872.[130] In total, Sabine published fifteen 'Contributions to Terrestrial Magnetism', the final one appearing in 1877.[131] It was during the 1840s and 1850s, however, that the majority of the magnetic data for these publications would be obtained. In this venture, Fox's dipping needles would travel the world, far beyond the Atlantic and Antarctic. Ross's expedition was just the beginning of the transformation of Fox's natural philosophy from a regional body of knowledge and instruments into a global system of scientific investigation.

---

[130] Edward Sabine, 'Contributions to Terrestrial Magnetism: No. XI', *Philosophical Transactions of the Royal Society of London*, 158 (1868), 371–416, at 371; Edward Sabine, 'Contributions to Terrestrial Magnetism: No. XIII', *Philosophical Transactions of the Royal Society of London*, 162 (1872), 353–433, at 353.

[131] Edward Sabine, 'Contributions to Terrestrial Magnetism: No. XV', *Philosophical Transactions of the Royal Society of London*, 167 (1877), 461–508+v.

5
Expedition and Experiment
The British Magnetic Scheme, 1841–1843

At Carclew House in late January 1844, Charles Lemon hosted an auspicious gathering. His guests of honour—Commander Owen Stanley (1811–50), his wife, and Lieutenant Hutchinson of the Royal Engineers—were all in Falmouth. After serving as the scientific officer on George Back's 1836 Arctic Expedition aboard HMS *Terror*, Stanley had sailed to Australia and New Zealand between 1838 and 1843.[1] Throughout these voyages, Stanley had performed magnetic observations with a Fox-type dipping needle. Lemon's dinner party was more than a casual get-together: it was an instrumental reunion. Robert Were Fox, his wife Elizabeth, and his children were also invited, eager to inspect the Fox type that Stanley had returned from his recent expedition. To his enthralled Cornish hosts, Stanley told of how, amid all the 'hardship in all quarters of the world', the refinement and 'simplicity' of Fox's instrument had been a constant reassuring presence. Caroline Fox noted how he 'told us that his hair became grey during six months of anxiety when he was in the Terror in the Northern regions'.[2] Stanley had actually struggled with the Fox circle on Captain Back's expedition to the Arctic on-board *Terror*. The instrument had been perfectly workable in the summer of 1836, but during the winter he had complained that 'the dipping needle gave me the most trouble the weights used in ascertaining the intensity being so small & delicate as not to be easily handled with cold fingers'.[3] It is unclear whether it was the Arctic, or the Fox type, that had so greyed Stanley's hair. Stanley proceeded to explain how, on his later voyage, this precious instrument 'had sat on his lap across the great desert, sailed in his cabin over the Atlantic, the Pacific & Indian Oceans & been his companion in his solitary home on the borders of Siam & on the banks of the Tenassein River'.[4]

Told in the comfort of Carclew, Stanley's tales portrayed Fox's magnetic dipping circle as an almost human symbol of Cornish science and British civilization, triumphing and proving true throughout the trials and tribulations of global naval exploration. No matter what challenges Nature or non-Western societies

---

[1] RS, Robert Were Fox Papers, MS/710/109: [Owen] Stanley to [Robert Were] Fox, [n.d.].
[2] CUL, Add.9942/27: untitled note dated '27th Jan., 1844' and '30th Jan., 1844'.
[3] Stanley, quoted in Trevor H. Levere, *Science and the Canadian Arctic: A Century of Exploration, 1818–1918* (Cambridge: Cambridge University Press, 1993), 155.
[4] CUL, Add.9942/27: untitled note dated '27th Jan., 1844' and '30th Jan., 1844'.

had thrown at Stanley, his Fox type had been a trusted 'companion' through all. This was music to Fox's ears: it was exactly how the instrument, ideally, should perform. But this was very much an ideal. For all Stanley's boasting of instrumental integrity, the dinner was to serve up an amusing warning of the dangers of human fallibility and the capricious nature of dipping circles. The following afternoon, a wooden box was found in Carclew, containing Stanley's Fox dipping needle. The captain was distraught. Caroline reported that he 'greatly distressed himself at missing on the last day's journey the square box containing "his darling child", which he said had never before been separated from him during fifty thousand miles of travels'.[5] As much as this anecdote reveals how personalized Fox's needles could become and how attached naval officers grew to devices that were ever present on their long voyages, it also shows that the system of global magnetic observations was never more than a moment's carelessness away from disaster. Luckily for Stanley, he had only left his 'darling child' in the safe hands of Lemon, but not all magnetic instruments lived such charmed lives.

Stanley's experiences were typical of an expanding global network of circulating instruments; as the Admiralty launched a series of discovering expeditions around the world, tasked with a range of objectives, magnetic data collection took on increasing prominence in this naval enterprise. Despite this, Stanley's anecdotes revealed the difficulties of using dipping circles in extreme climatic conditions, the challenges of years at sea conducting survey work, and the delicate nature of Fox's instruments. Ross's Antarctic Expedition won fame for Cornwall's leading magnetic apparatus but, during the early 1840s, Fox and Sabine struggled to establish a system for equipping and maintaining naval officers with the means to conduct the BMS. Between 1840 and 1843, they readied expeditions for magnetic research in North America, West Africa, the Atlantic, and Australia. In these diverse regions, the trials of survey work would include tropical disease, sub-zero temperatures, instrumental damage from transportation, exposure to the elements, and contact with animals, as well as technical errors with the magnetic equipment itself. But from Falmouth and Woolwich, Sabine and Fox looked to establish a system which would overcome these problems and produce accurate, reliable magnetic data. From the preparations for the 1841 Niger Expedition to the despatch of Henry Lefroy's Canadian survey of 1843, Fox and Sabine developed a standardized network of practices and materials for magnetic survey work. This consisted of written instructions for observers, training sessions with dipping circles at Woolwich and Falmouth, the constant checking and repairing of instruments and, in particular, the rigorous examination of dipping needles, returned from expeditions and tested for their magnetic integrity. Collectively, this regime allowed Fox and Sabine to process incoming magnetic data from all

[5] Ibid.

around the world and adjust the numerical figures for errors arising from climatic conditions, human mistakes, and instrumental failings.

The importance of a systematic approach to magnetic survey work became particularly apparent in our own magnetic expedition in 2020. Importantly, Crosbie Smith and I came to appreciate the importance of user experience in magnetic experiments with Fox's dipping needles: to obtain data on a global scale required the combination of user experience and systematized instrumentation and practices. From Lisbon to St Helena, as we approached the equator, we observed the Earth's magnetic influence on our needle diminish, which was to be expected. However, in the lower latitudes around St Helena and the equator itself, the needle became completely inert: it was difficult to be sure if the instrument had lost its magnetism through some sort of damage, or if that was an accurate measurement of the magnetic properties of the region. For us, in 2020, we had a good idea of what to expect, but for nineteenth-century experimentalists, such moments must have been alarming. In these cases, global inquiry was contingent on how effectively its practitioners, usually naval officers, could assess the state of their instruments: they had to judge when their needles were behaving efficiently and distinguish between instrumental errors, local magnetic disturbances, and the Earth's magnetic influence. Accumulated experience was, therefore, invaluable. Only by performing a wealth of magnetic experiments, and through a close acquaintance with their instruments, could naval officers develop an effective understanding of how accurate a set of experimental measurements were. This was not just a matter of practical abilities gained through experience; it also involved a considerable intuitive element, relying on emotional, subjective feeling rather than specific learning. Magnetic experiments involved a combination of tacit and intuitive knowledge, uniting experientially attained skill with a sense of how a magnetized needle was behaving that was often shaped by anxiety over the instrument's condition and uncertainty over the local influence of terrestrial magnetism. In this chapter, for all that we see Fox and Sabine endeavour to standardize their system of magnetic measurements, and stress these qualities in their published outputs, it is the growing confidence and intuition of the BMS's experimentalists that takes a central, if often unstated, role.[6]

5.1. The Niger Expedition

On 1 June 1840, the Fox family was in London for an auspicious gathering. In the brand new evangelical bastion of Exeter Hall, over 4,000 people assembled to

---

[6] Edward J. Gillin, 'The Instruments of Expeditionary Science and the Reworking of Nineteenth-Century Magnetic Experiment', *Notes and Records of the Royal Society*, 76/3 (Sep., 2022), 565–592, at 590–591.

witness the inaugural meeting of the African Civilization Society (ACS), established to promote Christianity and end slavery throughout Africa. Its plan was to secure government support for a naval expedition to the River Niger, where it would forge trade links with local rulers, encourage the abolition of slavery, and spread the word of God. No one in the Exeter Hall crowd was more excited by this scheme than Barclay and Caroline Fox, who eagerly watched on with their father, Robert Were, and sister, Anna Maria.[7] Among the speakers were Prince Albert, Samuel Wilberforce, and Robert Peel, but the Foxes had turned out in force to hear the words of fellow Quaker Sir Thomas Buxton (1786–1845), the ACS's founder. Buxton had already convinced the Foxes of the merits of the Niger Expedition over a dinner in May.[8] Privately, doubts abounded among the Foxes' social networks, as John Sterling (1806–44) warned that Buxton's 'Civilization Scheme' was doomed to fail unless there was 'the establishment of British Empire in Africa'. While gardening with Caroline Fox, he admitted to having 'little faith that the savages of Africa will perceive the principles of political economy, when we remember the fact that the highly educated classes of England oppose the alteration of the Corn Laws'.[9] However, among the Exeter Hall throng, and in the pages of the ACS's publicizing journal, the *Friend of Africa*, there was no sign of anxiety.

Buxton's abolitionist zeal found support with the highest echelons of government. Thanks largely to the Secretary of State for War and the Colonies, John Russell (1792–1878), Lord Melbourne's Whig government had adopted the expedition as part of its colonial policy in January 1840. After Buxton delivered a petition of over 1.5 million signatures to Parliament, calling for an end to West African slavery, the government agreed to finance the construction of three steamships, the *Albert*, *Soudan*, and *Wilberforce*, which would sail deep into Africa to fulfil the ambitions of the ACS. These vessels were to be operated by the Royal Navy, with Captain Henry Dundas Trotter (1802–59) commanding the expedition from the *Albert*, and commanders William Allen and Bird Allen respectively commanding the *Wilberforce* and *Soudan*.[10] Along with a body of scientific men, the government paid for each ship to be custom-built with an elaborate ventilation system, sucking in and filtering air from high above the Niger's atmosphere, supposedly to neutralize the threat of tropical fever, which was commonly thought to be spread by bad air.[11] This state support changed the

---

[7] R. L. Brett (ed.), *Barclay Fox's Journal* (London: Bell & Hyman, 1979), 197–8.
[8] Horace N. Pym (ed.), *Memoirs of Old Friends: Being Extracts from the Journals and Letters of Caroline Fox of Penjerrick, Cornwall from 1835 to 1871* (London: Smith, Elder, & Co., 1882), 107 and 102.
[9] Ibid. 124.
[10] Parliamentary Papers, *Papers Relative to the Expedition to the River Niger*, No. 472 (1843), 83.
[11] John Daniell, 'On the Waters of the African Coast', *Friend of Africa*, 1/2 (15 Jan. 1841), 18–23; John Daniell, 'A Probable Cause of Miasma', *Friend of Africa*, 1/3 (1 Feb. 1841), 40.

nature of Buxton's project, representing a chance to extend Sabine's magnetic survey, already underway at the Antarctic.

By the time of the ACS's Exeter Hall gathering, this was no longer just a Quaker-driven mission, but a state-sponsored scheme for the promotion of free trade, broad Christianity, and opposition to slavery. But it was also to become part of Sabine's systematic surveying of the Earth's magnetism. The importance of an expedition to the African interior was, Sabine believed, obvious to all, opening up a part of the world otherwise 'regarded as inaccessible'. On learning of the government's support of the Niger scheme, Sabine was quick to conscript the venture within the BMS. He convinced the Royal Society's president, the Marquess of Northampton, to press the government over the importance of magnetic observations being made along the coasts and rivers of western Africa. Likewise, the expedition's officers would have to be persuaded of the 'peculiar scientific value at this time' of their measurements, while the Royal Society would prepare instructions and instruments for the expedition.[12] The government agreed, ensuring that the Niger Expedition would contribute to the 'investigation into the nature and causes of the *magnetic perturbations*, which have excited so great an interest in the last few years'.[13] Writing to William Allen in December 1840, Sabine explained that, although there was evidence of these disturbances being 'general and synchronous over the whole extent of Europe', there was little to suggest this extended to North America. Africa, on the other hand, was a completely unknown magnetic entity. 'We may infer', Sabine continued,

> that the causes of the perturbations are less distant from the earth than was at first apprehended; and they may possibly, therefore, be more easily sought out, especially by the extension of the stations, and by their being formed into groups. In this view the Expedition may afford a station of peculiar importance, as the central one of a group, of which the British Magnetic Observations at St Helena and the Cape of Good Hope, the Egyptian at Cairo, the French at Algiers, and the Spanish at Cadiz, may form the exterior stations. These are the *term-days* observations named in the subjoined instructions: they are made only on certain days, twelve in number, in each year, named by the Royal Society for general simultaneous observation at all parts of the globe; the instruments being observed exactly at every fifth minute, during twenty-four successive hours.[14]

Sabine thus made Allen very well aware of the expedition's place within the international study of terrestrial magnetism.

---

[12] Edward Sabine, 'Instructions for Magnetic Observations in Africa', *Friend of Africa*, 1/4 (25 Feb. 1841), 55–7, at 55.
[13] Ibid. 56.   [14] Ibid.

The *Friend of Africa* published Sabine's communication, bringing Britain's magnetic enterprise to a popular evangelical readership. The expedition would be part of the huge project to determine, as Sabine put it, 'the magnetic state of the whole globe at the present epoch, by systematic observations made nearly contemporaneously at almost every accessible part of its surface'.[15] Within this, Fox was to take a central role, providing magnetic instruments for Buxton's venture. William Allen wrote to Fox in October 1840 with news that 'we take a Dipping Needle on your excellent principle & Major Sabine has told me that I had better write to you on the subject'. Although short notice, the commander hoped a new device could be completed by the end of November, requesting Fox's 'directions for one to be made in the best manner for Her M. Service on the Niger Expedition'. Allen had past experience of a Fox type, but was excited at the prospect of a modified instrument. 'I think you said you have much improved on that which I have belonging to M^{ssr} Barclay', recalled Allen, but specified that 'as we only want it for Dip & Intensity we can dispense with the apparatus for the variation'.[16] Fox subsequently corresponded closely with Sabine to provide the Niger Expedition with reliable magnetized needles and dipping circles. Working to Fox's instruction, Jordan's foreman, William George, manufactured the needles, pivots, and balancing apparatus for the expedition's Fox types. Fox personally balanced the needles and advised William Allen to make out a table showing the relationship between weights and the angles of dip for each individual needle. Weights should be added, Fox recommended, beginning with 2.40 grains to give an index of 1.000 and then the mean angles from dip induced by each weight could be calculated in relation to this index. 'The formula for deducing the intensity from the angles I sent thee some time ago', wrote Fox, 'but if it be mislaid I can forward another copy.'[17] Along with a specially built Fox circle for William Allen on the *Wilberforce*, Fox and Sabine issued Bird Allen with an older instrument by Watkins & Hill for the *Soudan*.[18]

In addition to Fox dipping circles and needles from Falmouth, the Royal Society equipped the expedition with an impressive array of magnetic instruments, including a Robinson dipping needle, azimuth compasses, a transportable magnetometer built especially for the expedition under Weber's direction at Göttingen, Weber's absolute horizontal intensity apparatus, and Hansteen's absolute horizontal intensity apparatus. These were divided into two divisions 'to be embarked in different ships, and employed by different observers'. The first division, including Fox's apparatus, was to determine the

---

[15] Ibid. 55.
[16] RS, Robert Were Fox Papers, MS/710/11: William Allen to Robert Were Fox (1 Oct. 1840).
[17] TNA, Sabine Papers, BJ3/19/77–8: Robert Were Fox to Edward Sabine (13 Mar. 1841), 77–8.
[18] TNA, Sabine Papers, BJ3/19/106–7: Robert Were Fox to Edward Sabine (28 Dec. 1841), 106.

variation and *absolute horizontal intensity* by the transportable magnetometer, when the expedition may be stationary for three or four days; and when, for a single day only, the *variation* by the azimuth compass, the *dip* and *total intensity* by Fox's needles, and the *relative horizontal intensity* by Hansteen's apparatus.[19]

When the expedition was stationary on an appointed Term Day, the transportable magnetometer could be used to make term observations for variation and intensity. On these days, Sabine ordered that the crews unite to perform land-based measurements. As for the second division, Weber's apparatus would give absolute horizontal intensity when the expedition was stationary for three to four days. When only a single day on land was possible, the Robinson needle would take dip, an azimuth compass measure variation, and Weber's apparatus read intensity.[20]

Sabine urged the crew to avoid working on volcanic soil, especially on the expedition's initial stop on mainland Africa. This location, once selected, was to be revisited on leaving the Niger for three to four days of further measurements so as to provide a reference of how the instruments' magnetic character had changed throughout the voyage. On these observations, the expedition's subsequent readings for relative intensity could later be calculated. Likewise, on leaving and returning to England, 'the angles of deflection of Fox's intensity needles should be observed', as well as Hansteen and Weber's horizontal needles.[21] Along with this impressive instrumental package, Sabine issued details of how to use Fox's devices. Unlike the 1839 Antarctic Expedition, where Sabine and Fox could depend on Ross and Crozier's magnetic experience, the Niger Expedition entailed the training of naval officers and the production of printed instructions for conducting magnetic experiments. While Robinson's dipping needle required eight different positions of the circle for a complete observation to be made, with the poles of the needle reversed for each determination, Fox's instrument required no such adjustment. 'With Fox's dipping needle,' Sabine informed Allen, 'the poles are not reversed and two positions only are necessary.'[22]

With the expedition prepared for magnetic experiments, the *Albert* and *Soudan* sailed from Deptford, while the *Wilberforce* left Liverpool for Dublin, where she arrived on 17 February 1841, giving the officers a chance to consult with Lloyd.[23] The three ships then assembled at Devonport, where Fox himself inspected the expedition, having taken the mail packet to Plymouth on 25 April. He offered some last-minute magnetic instruction before the vessels departed for Africa.[24] Fully prepared to combine magnetic observations with the abolition of slavery, the expedition steamed from Plymouth to Madeira (Figure 5.1). At Porto Grande, Cape Verde, Allen took measurements with his dipping needle and magnetometer,

---

[19] Sabine, 'Instructions for Magnetic Observations in Africa', 56.   [20] Ibid.   [21] Ibid. 57.
[22] Ibid.   [23] Anon., 'Niger Expedition', *Friend of Africa*, 1/4 (25 Feb. 1841), 57–8, at 57.
[24] Brett (ed.), *Barclay Fox's Journal*, 227.

Figure 5.1 The Niger Expedition before its departure in 1841; engraved by E. Duncan (Author's image, 2021).

this being a location which could be correlated to the readings that Ross and Crozier had taken back in 1839.[25] After stopping at Sierra Leone in late June, the expedition reached the mouth of the River Niger in August. Here the real work began, as the three ships made their way cautiously into the interior of the West African coast. At first all went well, as the crews made contact with the river's rulers. Captain Trotter signed two treaties with African chiefs, the first with Obi Ossai of Aboh and the second with the Chief of Eggarah, the Attah of Iddah. Both kings agreed to trade in ivory and palm oil, end slavery, and allow Christian missionaries to enter their lands.[26] But then disaster struck, as fever broke out: by early September, over seventy out of 302 members of expedition were taken ill, and over a third of the European crew of 159 perished by the end of the mission, including Bird Allen.[27]

---

[25] William Allen and T. R. H. Thomson, *A Narrative of the Expedition Sent by Her Majesty's Government to the River Niger, in 1841; Under the Command of Captain H. D. Trotter, R.N.*, 2 vols. (London: Richard Bentley, 1848), i. 57.

[26] Parliamentary Papers, *Papers Relative to the Expedition to the River Niger*, No. 472 (1843), 141–4.

[27] See Edward J. Gillin, 'Science on the Niger: Ventilation and Tropical Disease during the 1841 Niger Expedition', *Social History of Medicine*, 31/3 (Aug. 2018), 605–26; Howard Temperley, *White Dreams, Black Africa: The Antislavery Expedition to the River Niger, 1841–1842* (New Haven: Yale University Press, 1991).

Back home in Britain, followers of the *Friend of Africa* continued to read of the expedition's inevitable success, but rumours were growing of an impending disaster. Things were equally alarming in terms of the expedition's magnetic results. In late 1841, Sabine received worrying reports over the integrity of both Fox types.[28] He wrote to Fox reporting on discrepancies in the African observations that had been sent back to England. Bird Allen had expressed doubts to Arthur Barclay, a relative of Fox's, over the method of mounting the needle within his circle, but Fox maintained that the well-travelled device gave uniformed readings.[29] As for William Allen's instrument, Fox reassured Sabine that the needles had all been thoroughly tested at all angles of deflection by magnets and weights, and were found to be extremely consistent in inclination. If the results from Africa were indeed producing discrepancies, he was adamant that this was not due to any defect or poor craftsmanship on George's part, but because the dipping circle had 'sustained some injury since'.[30] By early 1842, however, there was no doubt that the expedition was returning anomalous results. 'Has the cause of the discrepancies in the results out in Africa yet been discovered?', asked Fox in an anxious letter sent to Sabine in February.[31] Resolving this question would have to wait until the expedition's return, when the instruments could be properly tested and corrections made.

With uncertainty surrounding the incoming magnetic data, the true extent of the calamity became apparent when, in January 1842, *The Times* published a bleak account of the expedition, including reports of how tropical disease had forced its swift termination.[32] Seeking to limit the damage to public relations, Buxton and his collaborators claimed that the treaties represented real progress towards their aims. Barclay Fox, for one, took comfort from John Russell's upbeat address to the ACS's 1842 meeting, where he boasted that the 'natives having united with us to destroy the barracoons' marked a great step towards the abolition of slavery. It was, Barclay agreed, a qualified 'nonfailure'.[33] To more discerning audiences, however, there was no doubt that the expedition had been a complete catastrophe, as soon became clear to both Buxton and the Fox family. When the Foxes visited Buxton for dinner at his home in Cromer in September 1843, Caroline found him much changed, observing that he was no longer pleasant company, having 'never recovered his old tone of joyous mental energy since the failure of the Niger Expedition'.[34] The expedition's magnetic results appeared equally disappointing.

---

[28] For Allen's magnetic observations, see Allen and Thomson, *A Narrative of the Expedition*, 37–8.
[29] TNA, Sabine Papers, BJ3/19/103–4: Robert Were Fox to Edward Sabine (18 Dec. 1841), 103–4; BJ3/19/106–7: Robert Were Fox to Edward Sabine (28 Dec. 1841), 106.
[30] TNA, Sabine Papers, BJ3/19/106–7: Robert Were Fox to Edward Sabine (28 Dec. 1841), 106.
[31] TNA, Sabine Papers, BJ3/19/108–9: Robert Were Fox to Edward Sabine (14 Feb. 1842), 108.
[32] Anon., 'The Niger Expedition', *The Times*, 17887 (22 Jan. 1842), 5; 'A Surgeon', 'Niger Expedition.—Prevention of Fever', *The Times*, 17885 (20 Jan. 1842), 5.
[33] Brett (ed.), *Barclay Fox's Journal*, 274.   [34] Pym (ed.), *Memoirs of Old Friends*, 84.

Fox examined the returned African data in September 1842, but remained unable to account for the ongoing discrepancies. With Bird Allen's dipping needle reported to have been unsteady at Sierra Leone, where it gave a dip of 30° 16′, its readings off Liberia at Cape Palmas of 20° 40′ and 19° 16′ were, Fox wrote, 'also doubtless erroneous results being contradicted by six other nearly consistent observations'. While the cause could 'only be guessed at', he was adamant that the 'errors were not due to the instrument', asserting that the observations made off Cape Coast Castle on the Gold Coast showed the 'pivots were true'.[35] The Niger Expedition's anomalies remained a mystery.

Despite these instrumental concerns and although not ready in time for William Allen and Thomas Thomson's narrative of the Niger Expedition, published in 1848, Sabine publicly expressed satisfaction with the results brought back for the declination, intensity, and dip of magnetic force on the west coast of Africa. Thanks to Allen's 'scrupulous and persevering attention', these observations would form the 'principle part' of the maps for that quarter of the globe. For all that fever had ravaged the ship's crew, they continued to perform observations throughout. Given 'the peculiar objects and circumstances' of the expedition, as well as 'the sufferings which it underwent', Sabine thought Allen's 'careful attention to minutiae the more worthy of admiration'.[36] Not even the ravages of tropical disease had stopped Allen advancing the cause of magnetic science. These African results were of huge significance, Sabine argued, because of the expedition's location in relation to other magnetic stations across the world, including at Cape Town, Algiers, St Helena, and in Europe. 'The co-ordination of your observations on *Term Days*,' Sabine informed Allen,

> with the simultaneous observations made in different parts of the globe, is also in progress, and promises to afford some interesting and valuable results, particularly the Term-Day of April, 1842, which you observed at Fernando Po. It was a day of considerable magnetic disturbance apparently over the whole globe, and Western Africa seems to have had its fair share.[37]

This referred to the global observation of a magnetic storm in which dipping needles around the world had displayed extremely erratic magnetic behaviour. However, as much as Sabine was pleased to have magnetic data from the Niger Delta, the expedition had utterly failed to live up to expectations. Beyond the outbreak of tropical disease and lack of any genuinely tangible anti-slavery commitments from the Niger's rulers, the expedition had also raised concerns over the reliability of Fox's instruments. It was not that Sabine and Fox's system of global

---

[35] TNA, Sabine Papers, BJ3/19/132–3: Robert Were Fox to Edward Sabine (24 Sept. 1842), 132.
[36] Sabine, quoted in Allen and Thomson, *A Narrative of the Expedition*, p. ix.   [37] Ibid., p. ix.

magnetic surveying had broken down, but that the fragility of Fox types became increasingly apparent aboard *Soudan* and *Wilberforce*. Back in Falmouth and Woolwich, Fox and Sabine struggled to resolve these discrepancies.

5.2. Instrumental Upheavals

Despite the failings of the Niger Expedition, by 1842 there was a rich stream of magnetic data flowing back to Falmouth and Woolwich from around the world. Following his part on Back's Arctic Expedition of 1836 aboard *Terror*, Lieutenant Edward Barnett (1799–1879) had taken command of HMS *Thunder* in 1837, completing several years of survey work in the West Indies and Bahamas. Along with producing navigation charts for the navy, Barnett also conducted magnetic experiments. He had already learnt how to use a Fox type under Back's command, and put this experience to use, completing measurements with his own Fox circle between Nassau and England in 1841. That same year, Captain Francis Price Blackwood (1809–54) secured appointment to HMS *Fly* for its forthcoming expedition to the Torres Straits and northern coast of Australia. With the support of HMS *Bramble*, Blackwood left Falmouth in the spring of 1842, armed with a Fox type and orders to combine magnetic observations with his navigational survey work. Meanwhile, Henry Lefroy prepared to embark for an overland magnetic survey of northern Canada, keen to acquire the very best magnetic apparatus that Falmouth could supply. For Fox, the Niger Expedition was just one element of a rapidly expanding global network of magnetic surveys in which his instruments were taking a leading role. Awkwardly, however, these endeavours coincided with a reorganization of the personnel behind the production of Fox types.

In 1841, the Fox–Jordan collaboration that was so central to the production of the celebrated dipping needles of the 1830s and Ross's Antarctic Expedition came to an end. Fox wrote to Sabine in December 1840 with news that Jordan was about to leave Falmouth for London, where De la Beche had secured him a job as the keeper of mining records at the newly established Museum of Economic Geology on Jermyn Street. Although Jordan had agreed to complete his existing orders, Fox would have to find a new Falmouth-based instrument maker.[38] By March, he had found his man, following the provisioning of the Niger Expedition with dipping circles. 'His name is William George, lately TB Jordan's foreman; a superior mechanic, & the maker of Captain Allen's needles, pivots &c', he told Sabine, confessing that he did 'not know where another is to be found in whom

---

[38] TNA, Sabine Papers, BJ3/19/75–6: Robert Were Fox to Edward Sabine (2 Dec. 1840), 76; Anita McConnell, 'Jordan, Thomas Brown (1807–1890)', *Oxford Dictionary of National Biography*, https://www.oxforddnb.com/view/10.1093/ref:odnb/9780198614128.001.0001/odnb-9780198614128-e-15123, 23 Sept. 2004, accessed 24 Mar. 2020.

I can feel the same confidence as it respects the more delicate parts of the instrument'.[39] Although Fox did not want to press the mechanic's credentials so far 'as to interfere unsuitably with T. B. Jordan', he reassured Sabine that any order placed with George would be well justified. The supply chain of Cornish-built Fox-type dipping needles was, in this way, secured. Sabine, however, was loath to dispense with Jordan's services, seeing no reason why the instrument maker's relocation to London was not, in fact, advantageous. After all, this removed the need to communicate instructions to Falmouth and saved newly built needles the journey between Cornwall and the capital. Jordan wrote to Fox in March 1841, informing him that Sabine had confided to him that he did 'not wish to order those [dipping needles] of any one else', so long as Jordan continued to undertake orders. As Sabine thought 'it rather an advantage to the public service that they sh^d be made in London', Jordan was 'obtaining estimates from some of the London workmen'.[40] At stake here was not just Falmouth's position as the leading producer of dipping needles, but Fox's own autonomy over the manufacture of the instruments which bore his name. With Sabine and Jordan in London, the Cornish natural philosopher was at risk of becoming isolated from the production of magnetic apparatus.

Sabine was hesitant to entrust the manufacture of future dipping needles to a new instrument maker but, in Jordan's absence, Fox relied on George's mechanical skill and drew him increasingly into the process of instrument production and regulation. As Jordan's foreman, no one else had the same experience of crafting Fox types as George. Along with instruments for the Niger Expedition, George's workmanship had already impressed in assisting Jordan with needles for Blackwood's forthcoming voyage to Australia on the *Fly*. Fox expected to have the completed device ready for the captain's visit in March 1841 and boasted to Sabine that he would be shocked if Blackwood did 'not find the inst^{ts} answer completely both at sea & on shore'. George had two new dipping circles nearly ready, though they had been somewhat delayed by the artisan's recent 'sickness'. Despite poor health, Fox testified that George had worked hard to craft an accurate machine and had spared 'no pains to make good axles'.[41] This promotion of the new mechanic's credentials to Sabine continued when, in May, Sabine sent needles to Falmouth that had recently been returned from Ross's Antarctic Expedition. Without the actual Fox circle in which to mount these needles, which was still in use on *Erebus*, Fox doubted he could accurately test them. To trial Ross's needles, George would have to build a device with which to model the dipping circle that Ross carried on *Erebus*. Fox advised that George make a trial bracket

---

[39] TNA, Sabine Papers, BJ3/19/77-8: Robert Were Fox to Edward Sabine (13 Mar. 1841), 78.
[40] TNA, Sabine Papers, BJ3/19/79-80: Robert Were Fox to Edward Sabine (15 Mar. 1841), 79-80.
[41] TNA, Sabine Papers, BJ3/19/81-2: Robert Were Fox to Edward Sabine (17 Mar. 1842 [dated 1842, but certainly 1841]), 82.

for experiments on one of his new circles to see if Ross's needles could be mounted, balanced, and vibrated in the same manner as with Ross's 7-inch dipping circle. Through such a trial, Fox hoped that he 'might then ascertain any loss of force, & perhaps obtain data for getting proximate estimates of the intensity', providing Sabine forwarded some of Ross's maximum- and medium-intensity measurements.[42] Again attempting to ease Sabine's anxieties over workmanship, Fox reiterated that he had no 'doubt of the good quality of George's instruments; & he thinks that after he has completed his present orders, he will be able to turn out one a month' at £18 per device. With Jordan completing a third Fox type, intended to 'have silver links, & to be more elaborate than the instrument which is to cost about £18', Fox and Sabine now had two artisans delivering a regular line of dipping circles and needles.[43]

However, this new arrangement with George in Falmouth and Jordan in London somewhat weakened Fox's authority over the production of new devices. He was far from happy to have an instrument maker so far beyond his own direction of the construction process. Throughout 1841, Fox gradually undermined Sabine's confidence in Jordan, all the while promoting George. In June, Fox updated Sabine on George's work, with two new needles almost complete 'in good style', and a third in progress. Importantly, Fox emphasized his own role in collaborating with George to ensure quality. In answer to Sabine's concerns over the accuracy of recent intensity measurements on *Erebus*, Fox claimed that if 'both Capt Ross' needles had been thoroughly tested before they were sent away & tables of the intensity indicated by them made out, the intensity might be accurately deduced from his observations.'[44] On the contrary, Fox promised to perform this service in Falmouth. Fox's experimental testing of the magnetic force of new needles in his gardens at Rosehill and Penjerrick would be a real boon for George, collaborating to produce 'tables of the properties of every instrument, which will give the intensities by inspection, & give confidence to observers in the correctness of their deductions'.[45] This clearly gave George's products an edge over Jordan's. Later that summer, Fox completed his rout of Jordan's role in the dipping-needle production system. As Sabine wanted Jordan to finish and check George's Fox types, they were despatched on completion to Jordan's new workshop in London where they would be graduated and examined. This process should have taken five to six days, but on deciding that George's instruments were out of order, Jordan had kept possession of the new circles for almost six weeks. While this delay annoyed Fox, he used it to serve a useful lesson to Sabine of the dangers of removing construction work from Fox's direct instruction at Falmouth.[46]

---

[42] TNA, Sabine Papers, BJ3/19/83–6: Robert Were Fox to Edward Sabine (13 May 1841), 84–5.
[43] Ibid. 85.
[44] TNA, Sabine Papers, BJ3/19/87–8: Robert Were Fox to Edward Sabine (8 June 1841), 87–8.
[45] Ibid. 88.
[46] TNA, Sabine Papers, BJ3/19/89–90: Robert Were Fox to Edward Sabine (June or July 1841), 89.

Following this sullying of Jordan's reputation, Fox and George worked together through the summer of 1841 to examine Ross's returned needles in a new instrument, similar to the one in use on *Erebus*. On 28 August, Fox was able to recommend a correction for Ross's measurements of 6.5′ for every 10°F recorded at the time of observation. The problem was due, Fox surmised, to variations in temperature, which he had deduced through experiments with the needle 'by placing the instrument in a double case of sheet copper' and circulating hot water around it.[47] This was not just an effort to correct Ross's observations for Sabine's calculations, but a demonstration that George could replicate the precision of the Jordan-built 7-inch Fox type employed in the Antarctic. Evidently, the instrument he had made had given Fox confidence that his experiments on Ross's needles reliably replicated Jordan's method of suspending and balancing a needle on jewelled pivots. Sabine was also content, placing an order worth £25. 7s. 6d. with George for new instruments.[48] On 25 September, Fox reported that George had completed a new dipping circle that would be forwarded to London by steamer, along with a plan for specially designed gimbals for observations at sea.[49]

George was now firmly incorporated into the instrumental regime that was at the centre of the BMS: Fox and Sabine had secured their supply of dipping circles and magnetized needles. When reports reached Sabine in early 1841 that Captain Barnett's Fox type was producing anomalies, Fox speculated that the cause had 'arisen from injuries' sustained during travel, confidently vouching 'for the accuracy of the instrument…before it went from Falm[outh]'. Indeed, he had deemed it on trial to be decidedly 'superior to the instruments supplied to Captain Ross', while its pivots and jewels could not, he claimed, 'be injured by changing, as they must go in to their places'.[50] Nevertheless, Sabine and Fox put their needle conveyance network into action and, by September, two new needles had arrived in London, destined for the Cape of Good Hope. Of these, Fox had found one 'not quite satisfactory to me', having too small a pivot for the jewelled holes, but he had ordered it to be refashioned.[51] Consequently, he believed both needles to be of equal accuracy to the best that Gambey or Robinson could provide. Likewise, on 5 October, Fox again highlighted his own role in the production of magnetic instruments. On examination, he 'did not quite like one of George's needles, the axis of which did not appear to coincide with its magnetic axis, & he has therefore been engaged to day in making a new one'. George could not be accused of lacking commitment and resolve in the face of Fox's stringent demands. Labouring to

---

[47] TNA, Sabine Papers, BJ3/19/93–4: Robert Were Fox to Edward Sabine (28 Aug. 1841), 93; BJ3/19/91–2: Robert Were Fox to Edward Sabine (26 July 1841), 91–2.
[48] TNA Sabine Papers, BJ3/19/95–6: Robert Were Fox to Edward Sabine (16 Sept. 1841), 95.
[49] TNA, Sabine Papers, BJ3/19/97–8: Robert Were Fox to Edward Sabine (25 Sept. 1841), 97–8.
[50] TNA, Sabine Papers, BJ3/19/81–2: Robert Were Fox to Edward Sabine (17 Mar. 1842 [dated 1842 but certainly 1841]), 81–2.
[51] TNA, Sabine Papers, BJ3/19/97–8: Robert Were Fox to Edward Sabine (25 Sept. 1841), 98.

make a new axis, the mechanic broke two in the process, probably due to too hard a temper, but managed to complete the order on schedule for Barnett's use on *Thunder*.[52] With George manufacturing new needles to Sabine's orders before Fox subjected them to rigorous trials, a powerful system of quality control was in operation to sustain the BMS's investigations.

Unfortunately, the pivots of Barnett's new needle were damaged due to a 'mishap' in transit. Although Fox reassured Sabine that all had been in good order when the instruments left George's workshop, he promised that the mechanic would remedy any defects without charge.[53] George was sure that the damage had happened since leaving Falmouth, confiding to Fox that the needle 'must have been injured by some one before it came into his [Sabine's] hands'. He had been 'careful in packing it', confident that it had been 'perfect when it left Falmouth'. George guaranteed that he could repair the device within two weeks and was 'persuaded it is better adapted for sea service than any one has yet seen'.[54] Fox agreed, having found it a superb instrument on inspection and admitting distress at seeing such well-made pivots and jewels damaged through careless handling. He had, after all, ordered George to reject any axle that did not give 'coincident results', and took confidence from the mechanic's assurances 'that he takes even more pains than formerly to make them as perfect as he can'.[55] Fox believed that George had skill and experience equal to that of Jordan. Nevertheless, George found it hard to secure payments for his completed products, which Fox attributed to the instrument maker being 'rather short of friends'.[56]

On 25 February 1842, the broken dipping needle arrived at George's workshop. He took it to Rosehill 'to be examined before he took it out of the box', where Fox 'found that the needle for reversing, had pivots broken, & that those of the other two needles were sound'. With both jewels smashed and the pivot snapped, the instrument was 'useless', giving only 'sluggish movements'.[57] This was far from the perfect device that Fox had seen before its despatch to London the year before. As George had no jewels with larger holes, they together resolved on the solution of adding rings to help direct the pivots into place for changing needles. Fox believed the needle would be fully repaired and on a steamer for London within a week.[58]

However, on trialling the faulty instrument, Fox and George found that the pivots had been scratched by the sharp edges of the broken jewelled pallets and, as a result, needles A and B required repair. Once George had refashioned the

---

[52] TNA, Sabine Papers, BJ3/19/99–100: Robert Were Fox to Edward Sabine (5 Oct. 1841), 99–100; BJ3/19/101–2: Robert Were Fox to Edward Sabine (13 Oct. 1841), 101.
[53] TNA, Sabine Papers, BJ3/19/108–9: Robert Were Fox to Edward Sabine (14 Feb. 1842), 108; BJ3/19/106–7: Robert Were Fox to Edward Sabine (28 Dec. 1841), 107.
[54] TNA, Sabine Papers, BJ3/19/110: William George to Robert Were Fox (14 Feb. 1842), 110.
[55] TNA, Sabine Papers, BJ3/19/112–15: Robert Were Fox to Edward Sabine (19 Feb. 1842), 113–14.
[56] TNA, Sabine Papers, BJ3/19/103–4: Robert Were Fox to Edward Sabine (18 Dec. 1841), 103–4.
[57] TNA, Sabine Papers, BJ3/19/116–17: Robert Were Fox to Edward Sabine (26 Feb. 1842), 116.
[58] Ibid. 117.

pivots, Fox rebalanced needle B in the circle, but found that 'A did not require any alteration in this respect', which Fox believed to be 'striking proof of accurate workmanship'.[59] Fox retested the instrument, which now gave excellent results. This was, he contended, unsurprising, given the measures recorded were not any more accurate 'than the results obtained by George before the Inst^t was first sent away'.[60] The restoration of the dipping needle's physical integrity went alongside Fox's preservation of George's credentials as a trustworthy mechanic. Falmouth's location, while ideal for departing and returning naval expeditions, was proving troublesome in terms of conveying delicate instruments to London. In this process, Fox's assurances of quality control were essential to sustaining the reputation of Cornwall's manufacturers. It would not be until 29 December 1842 that Fox could announce that the circle was ready for Barnett's ongoing North American and West Indian survey.[61] The instrument was soon despatched to the Admiralty, from where it was forwarded to *Thunder*. The affair had been a challenging and lengthy experience, but it won George trust and a string of orders, and not just from the Admiralty.

George had also enhanced his credentials in his work for Blackwood's forthcoming Australian voyage. In early March 1842, George was finishing the new instrument for this expedition complete with a gimbal action to produce an equipoise at sea. Fox thought it superior to the earlier device sent to the Admiralty for Blackwood's use and was delighted that the captain intended to stay in Falmouth later in the year to 'go through a complete series of observations' under the Cornishman's guidance.[62] At the start of April, Fox reported delays to the new circle's completion, but promised to 'ascertain by experiment the coefficient for changes of temp^e for the reciprocal action of the needle under coercion by weights' as soon as it was delivered to Rosehill.[63] Soon after, George sent Sabine weights for Blackwood that Fox had tested and deduced accurate.[64] Sabine responded by sending Blackwood's previous Fox type to Falmouth. Fox and Sabine were therefore able to conduct comparative trials between the old instrument and the device recently completed. Fox's first object was to determine the amount of magnetic change the needles of Blackwood's original Fox type had sustained since their construction. He began on 6 April, getting a dip of 69.3°, and then took measurements with the circle's 'Deflector S', positioned at 40° from the dip, to calculate a mean dip of 69° 1' 6". The next day he took six readings with needle A, producing a mean of 69° 1' 5", followed by a mean of 69° 3' with needle B on the eighth. He then compared both to measurements with the deflector, giving

[59] TNA, Sabine Papers, BJ3/19/163-4: Robert Were Fox to Edward Sabine [c.1842], 163.
[60] Ibid. 164.
[61] TNA, Sabine Papers, BJ3/19/136-7: Robert Were Fox to Edward Sabine (29 Dec. 1842), 137.
[62] TNA Sabine Papers, BJ3/19/118-19: Robert Were Fox to Edward Sabine (7 Mar. 1842), 118-19.
[63] TNA, Sabine Papers, BJ3/19/120-1: Robert Were Fox to Edward Sabine (2 Apr. 1842), 120-1.
[64] TNA, Sabine Papers, BJ3/19/122-3: Robert Were Fox to Edward Sabine (4 Apr. 1842), 122.

an overall mean of both needles of 69° 2′ 5″. At Woolwich the same instrument had given a mean of 69° 7′ 9″, suggesting considerable consistency. Fox then repeated these trials with George's new Fox circle for Blackwood, intended for HMS *Bramble*. The same combination of needles and deflectors gave a mean of 68° 58′ 8″ in Fox's garden, while George had found the instrument produced 69° 1′ 5″ when checked on completion.[65]

Fox forwarded these results to Sabine, along with his own assessment of how the two circles and various needles had performed. He explained that during these trials each needle, when showing the dip, had been deflected by another needle mounted behind the dipping circle and

> the weights required to bring the mounted needle back to the dip indicated the amount of their mutual repulsion when at the angle of 70° from each other. The needle at the back being then removed, the mounted needle was coerced by weights only to the same angle (70°) from the dip, which gave the force of the earth's magnetism acting on the needle at that angle. The relative positions of the needles were then changed; the mounted one being made the deflecting one, & vice versa; & the same operations were repeated.[66]

From these arrangements it seemed that 'the loss sustained by both needles when weights only are used, is equal to one half the sum of the loss of the two needles successively reflecting each other'. This implied 'a decided advantage in fixing the deflecting needle perpendicular to the mounted one, rather than parallel to it'. Importantly, by this method, Fox asserted that the

> ratio of the loss sustained by each needle may therefore of course be ascertained at any place by means of the weights only; for as the sum of the weights required to deflect both needles to the given angle, is to their actual loss of force, so is the weight required by one of the needles, to its loss of force. It is evident that any other angles may be taken as well as 70°.[67]

Furthermore, Fox claimed that if the intensities recorded with weights alone varied from those with deflectors and weights, then it was clear 'that the needle or deflector/or both of them have sustained a change of force'.[68] Should a needle lose magnetic intensity, this would be shown when the weights were added, but not when the deflectors were employed. The combination of deflectors and weights, therefore, provided naval officers with the means both to examine the Earth's

---

[65] TNA, Sabine Papers, BJ3/19/21–2, 'Instrument for Capt Blackwood, from Woolwich C.2. Observations in RW Fox's garden, April 1842', 21–2.
[66] TNA, Sabine Papers, BJ3/19/23–6: 'Remarks of Some Experiments Made in Order to Ascertain the Amount of Changes Sustained by Dipping Needles in Their Magnetic Force' [1842], 23.
[67] Ibid. 24.   [68] Ibid. 25.

magnetic properties and to analyse the changing character of their own magnetic needles: an invaluable operation for users of Fox's instruments.

Later that April, Blackwood and his officers visited Fox, staying in Falmouth for instruction in magnetic experiments. Fox enjoyed this experience immensely, expressing how it

> has afforded me great pleasure to become acquainted wth Cap Blackwood & his officers. Some of them, especially Lt Shadwell, spent a great part of five days in my garden making observations, & I intend to send thee [Sabine] in two three days, some particular of the results obs^d by them.[69]

These were despatched a week later, having been 'made by the officers of the Expedition' with the George-built circle. Lieutenant Sulivan was also present, and helped Fox to make comparative experiments with a Robinson circle, the results of which were sent to Sabine for comparison with Blackwood's two Fox types, the latter constituting George's latest construction. Based on the mean results from two days of comparative observations between the three dipping circles, Fox boasted that Robinson's measurements 'agree exactly wth those obtained with my Inst^{ts}'.[70] Fox was determined that his dipping circles be seen as every bit as delicate as Robinson's, even on land. He was pleased to note that George's needles had performed excellently, though his earlier instrument, that of Blackwood's, was not quite so well made in terms of the jewelled holes for the pivots. George 'promised improvement in future'.[71] Nevertheless, as HMS *Fly* and HMS *Bramble* departed Falmouth at the end of the month, they not only carried the most up-to-date Fox type yet produced, but a crew well versed in magnetic experiments. Compared with the Niger Expedition a year earlier, this expedition was exceptionally well prepared to extend the BMS.

5.3. Lefroy and the North American Challenge

Along with the Antarctic and magnetic equator, the most important regions for magnetic investigation were the Arctic waters surrounding the north magnetic pole. By 1842, Canada was already a centre of magnetic research, with Toronto being home to one of the government's four permanent physical observatories. However, the data produced so far was from either this fixed position or the coastal regions which naval expeditions had surveyed. What was wanted was a capable magnetic observer to conduct a survey over the vast expanse of northern

---

[69] TNA, Sabine Papers, BJ3/19/128–9: Robert Were Fox to Edward Sabine (16 Apr. 1842), 129.
[70] TNA, Sabine Papers, BJ3/19/122–3: Robert Were Fox to Edward Sabine (4 Apr. 1842), 123.
[71] TNA, Sabine Papers, BJ3/19/130–1: Robert Were Fox to Edward Sabine (23 Apr. 1842), 130–1.

America. With Royal Society support and Sabine's direction, the first director of Toronto Observatory, Charles Riddell (1817–1912), was initially considered for the venture, but was deemed too valuable to lose to survey work.[72] Instead, Sabine selected Lefroy for the task. Although enthusiastic to establish St Helena Magnetic Observatory, Lefroy was ever eager to conduct magnetic survey work, initially proposing to Sabine that he conduct an 'African survey' in late 1840. Sabine rejected this offer but, in 1842, Lefroy secured a transfer to Canada. His task was to mirror Ross's southern exploits in the complex northern polar regions. By now, Fox was adept at equipping naval expeditions with magnetic instruments, but to survey the Arctic would involve overland work, presenting a very different challenge. Having seen Crozier's Fox type in action on *Terror*, Lefroy wrote to Sabine in 1841 to request a similar instrument for himself with which to conduct a survey of St Helena.[73] Although impressed with Fox's instruments on *Terror* and *Erebus* in 1839 and keen to employ a Fox type in Canada, Lefroy had some experience of the instrument's mixed performance. When the *Terror* returned from the Arctic in November 1837, he had inspected Stanley's Fox type at Chatham. Lefroy recalled how, on-board *Terror*, he enjoyed 'some pleasant intercourse with her officers, particularly with the second lieutenant, Owen Stanley, who introduced me to an instrument I was afterwards destined to make extensive use of-Fox's dip circle. He had been the first observer to employ it in Arctic regions, or indeed on any distant voyage'.[74]

For Canada, Lefroy acquired the Gambey type that FitzRoy had employed on HMS *Beagle*, for which Robinson made two new magnetized needles, and a George-built 7-inch Fox type with three needles.[75] In June 1842, he went to stay with Fox at Penjerrick 'for the purposes of instruction in the manipulation of his dip circle', before departing for Quebec on the *Prince Regent*, using this Atlantic crossing to practise with his needles.[76] He arrived in Montreal on 15 September and reached Toronto a month later. In April 1843, Lefroy commenced his surveying of the Hudson Bay Territory, using canoes to move his instruments up the Ottawa River. Reaching Fort Chipewyan in September, he settled in for a winter of observations. From the start, it soon became clear that this work would be a very different challenge to Ross's in the southern hemisphere. Moving magnetic instruments in canoes and wagons, packed with straw, proved disastrous.

---

[72] Matthew Goodman, 'From "Magnetic Fever" to "Magnetical Insanity": Historical Geographies of British Terrestrial Magnetic Research, 1833–1857', PhD thesis, University of Glasgow, 2018, 116–20.
[73] Ibid. 120–2.
[74] Lefroy, quoted in Trevor H. Levere, *Science and the Canadian Arctic: A Century of Exploration, 1818-1918* (Cambridge: Cambridge University Press, 1993), 155; Levere, 'Magnetic Instruments in the Canadian Arctic Expeditions of Franklin, Lefroy, and Nares', *Annals of Science*, 43/1 (1986), 57–76, at 62.
[75] Levere, 'Magnetic Instruments in the Canadian Arctic Expeditions of Franklin, Lefroy, and Nares', 64–7; see also Christopher Carter, *Magnetic Fever: Global Imperialism and Empiricism in the Nineteenth Century* (Philadelphia: American Philosophical Society, 2009), 127–49.
[76] Lefroy, quoted in Goodman, 'From "Magnetic Fever" to "Magnetical Insanity"', 122.

Lefroy's Gambey and Fox types were 'shaken to pieces': the Gambey arrived completely broken, while needle C for the Fox circle lost its magnetic force due to the jolting of the carts.[77] Here, Fox's inappropriateness for land surveying become apparent, weighing 37 lb, including its box, compared to the Gambey at 27 lb.[78] Though broken, Gambey's circle with Robinson's magnetic needles seemed to outperform the Fox type. Throughout his journey to Fort Chipewyan, Lefroy observed a considerable alteration to the magnetic state of his instruments. On one occasion he almost lost his needles when they fell into a river, only to be recovered 3 miles downstream. Two days later the same needles fell out of his pocket, losing further magnetic strength. On 20 June 1843, a stray calf walked into his equipment, knocking over his Gambey, smashing its cover, and rendering it useless. Rust, arising from exposure to the elements, was a constant problem, while the axle of needle C for the Fox circles was out of shape, having to be replaced in August. Fox needle A worked well enough, but needle B soon ceased to oscillate. As historian Matthew Goodman has argued, the Fox circle failed to live up to expectations. For all that these Cornish instruments were celebrated for their robustness at sea, on land they proved fallible, with the delicate needle axles and jewelled pallets particularly vulnerable.[79]

After a winter of measurements, Lefroy departed Fort Chipewyan in March 1844 to conduct further survey observations, leaving behind his Fox type and taking only the lighter Gambey. Although the Fox circle had disappointed, by the time Lefroy returned to Montreal in November 1844, he had covered some 6,000 miles and observed terrestrial magnetic phenomena at over three hundred stations. Sabine subsequently published the results of the North American Survey in 1846, but with little reference to the varied performance of Lefroy's instruments.[80] As Jenny Bulstrode has observed, away from the discipline of naval ships and Fox's gardens at Rosehill and Penjerrick, his dipping circles showed an alarming tendency to be disrupted.[81] Fox's needles moved easily with the roll of the ocean swell, even in storms, but the vibrations of overland transportation ruined the workings of the instruments. Likewise, for land measurements, Sabine's observers generally relied on Gauss's more delicate magnetometer, in which a magnetic needle was suspended on knife edges which rested on agate planes. By the action of weights, adjusted with screws, Gauss's apparatus brought the needle from an inclined to a horizontal position. Lloyd developed this device, boasting that his magnetometer could detect a change of 1 part in 40,000 of total magnetic

---

[77] Goodman, 'From "Magnetic Fever" to "Magnetical Insanity"', 123–4.   [78] Ibid. 126.
[79] Ibid. 130–3.
[80] Edward Sabine, 'Contributions to Terrestrial Magnetism: No. VII', *Philosophical Transactions of the Royal Society of London*, 136 (1846), 237–336, at 238 and 240.
[81] Jenny Bulstrode, 'Men, Mines, and Machines: Robert Were Fox, the Dip-Circle and the Cornish System', Part III diss., History and Philosophy of Science Department, University of Cambridge, 2013, 33–5.

intensity.[82] Despite Lefroy's underwhelming experience of Fox's instruments, this did little to undermine their growing reputation. Given George's successful exhibition of workmanship in the construction and examination of dipping needles for Blackwood and Barnett, it is not surprising that he soon attracted orders for instruments from beyond the BMS and Royal Navy. In November 1844 he received a request for details of one of his new dipping circles, 'in a portable form', from the Austrian explorer Virgil von Helmreichen (1805–52), who was interested in acquiring a device with which to survey Brazil at the expense of the Austrian government.[83] Fox responded with a plan of George's latest Fox type and instructions, intended for both land and sea measurements.[84] Soon after, Sabine requested three new instruments on Helmreichen's behalf. Fox regretted that George could not sell these for any less than £54 to £53 in total, which would 'rather exceed V. Helmreichen's calculations, but this cannot be helped'.[85] Undeterred by the costs, however, Helmreichen proceeded with the order and Fox kept the Austrian explorer supplied with instruments and instructions throughout the 1840s via the Brazilian packet service, which provided a convenient connection between Rio and Falmouth.[86]

5.4. Conclusion

Between 1841 and 1843, Fox dipping needles were in use in the Antarctic on *Terror* and *Erebus*, in the West Indies and North America with Lefroy and Barnett, amid the tropical growth of the River Niger, as well as with Blackwood on his voyage to Australia. Across the vast expanses of the globe, naval officers were simultaneously measuring the Earth's magnetic field on the Indian, Atlantic, Arctic, Pacific, and Southern oceans, all united by Fox's standardizing instrumentation and Sabine's authoritative direction. From Falmouth and Woolwich, a worldwide experiment was underway. This was celebrated as a truly global undertaking and, as such, relied on a constantly developing system of supplying, regulating, and repairing dipping needles. The training of crews in magnetic experiments, the constant checking of instruments returned from expeditions, the exchange of written instructions from Sabine and magnetic data from naval officers, the network of agreed observation spots at Falmouth, Madeira, St Helena, and Cape Town were all intended to eradicate error from the investigation.

---

[82] Levere, *Science and the Canadian Arctic*, 156.
[83] TNA, Sabine Papers, BJ3/19/136-7: Robert Were Fox to Edward Sabine (29 Dec. 1842), 136; BJ3/19/138-40: Robert Were Fox to Edward Sabine (11 Mar. 1843), 138.
[84] TNA, Sabine Papers, BJ3/19/149: Robert Were Fox to Edward Sabine (18 Nov. 1843), 149.
[85] TNA, Sabine Papers, BJ3/19/151-2: Robert Were Fox to Edward Sabine (25 Nov. 1844), 151-2.
[86] RS, Sabine Correspondence, Vol. 2, D–J, MS/258/575: Robert Were Fox to Lt Col. Edward Sabine & Elizabeth Sabine, 12 Sept. 1846; TNA, Sabine Papers, BJ3/19/155: Robert Were Fox to Edward Sabine (13 Nov. 1845), 155.

This was an elaborate system that promised Fox and Sabine some sort of control over their instruments and observers as they travelled the globe. And while the instruments themselves often failed to live up to expectations, this system represented an effort to overcome the considerable challenges of expeditionary life, from extreme climatic conditions around the poles and equator to tropical diseases, the trials of transportation, and even mischievous cows.

Along with this elaborate infrastructure for magnetic data collection, the problem of securing good observers remained. When Fox's instruments did work, it was crucial that naval officers knew how to use them. Sabine's drafting of instructions on magnetic measurements for the Niger Expedition had demonstrated that to transform the BMS into an effective global network of observers required clear instrumental and experimental guidelines. Ideally, naval officers could visit Fox in Falmouth for a period of training, but this was not always possible. Therefore, an important element in establishing a worldwide system of magnetic data collection involved the production of printed instructions. To this end, Fox and Sabine produced increasingly refined manuals for naval and army officers equipped with Fox types. In early 1842, Sabine took the earlier works that Jordan and Fox had published during the 1830s, revised them, and sent them to Falmouth for proofing. Fox replied in April, recommending the addition of information on temperature corrections and pre-prepared tables for intensities, as well as formulae for calculating the diminished magnetic force of needles.[87] By March 1843, they had put together an impressive new set of instructions with elaborate details of how to take observations at sea. Subsequently, officers were issued with printed manuals in the use of Fox types. It was recommended that twelve observations be made, including an equal number with the circle facing east and then west.[88] As well as instruments and training, Fox and Sabine provided future naval expeditions with instruction through these detailed publications. Their evolving magnetic regime was becoming increasingly sophisticated at overcoming the capricious nature of surveying the Earth's magnetic phenomena and exerting experimental control over vast geographical space. Throughout the remainder of the 1840s, this enterprise would expand into the most ambitious scientific venture of the age.

---

[87] TNA, Sabine Papers, BJ3/19/128–9: Robert Were Fox to Edward Sabine (16 Apr. 1842), 128–9.
[88] CUL 9942/34: 'Instructions for Using Mr. Fox's Instrument for Determining the Magnetic Inclination and Intensity' (c.1842–3), proof with Fox's annotations, 2.

6
Discovery, Disaster, and the Dipping Needle
Britain's Global Magnetic System, 1843–1850

At the BAAS meeting of 1845, a special Magnetic Congress was held to celebrate and promote the advancement of the science. At that very moment, Royal Navy ships were carrying Fox dipping circles around the world's oceans, from London and Cape Town to the Arctic, Far East, and Pacific. The 1845 Magnetic Congress represented a chance to reflect on the BMS's progress and examine its place in the broader cultivation of science. With Alexander von Humboldt, Gauss, Hansteen, and Weber all invited, John Herschel heralded the magnetic survey as a great civilizing mission. Herschel declared that a permanent physical observatory, such as the ones at Cape Town or Rossbank, was 'part of the integrant institutions of each nation calling itself civilized'.[1] With news that the government would continue the BMS until at least 1848, he lauded the central role of Parliament in advancing magnetic science, which he construed to be evidence of the British state's wisdom. Although personally exhausted with the magnetic campaign, Herschel hoped all of Britain's colonies would establish observatories, providing a global network of scientific centres. The BAAS and Royal Society would encourage the government to found new magnetic observatories in Ceylon, New Brunswick, Bermuda, and Newfoundland.[2] Within this increasingly extensive philosophical programme, Fox's instruments were playing a crucial part. He was particularly delighted to learn of the BMS's 1845 extension, believing that the data produced so far showed 'unequivocally the importance of continuing the system of simultaneous observations in different parts of the world'.[3] As to the form this systematic surveying took, he advised Sabine that he would rather see the number of observatories diminished than have 'any limit now put to the duration of the combined system which has been so auspiciously commenced, & the results of which <u>must increase</u> in value every year'. This was hardly surprising, given that Fox's dipping needles had become the standard instruments for naval expeditions but were not the

---

[1] John Herschel, 'The President's Address', *The Athenaeum*, 921 (21 June 1845), 612–18, at 614.
[2] Christopher Carter, *Magnetic Fever: Global Imperialism and Empiricism in the Nineteenth Century* (Philadelphia: American Philosophical Society, 2009), 122–5.
[3] TNA, Sabine Papers, BJ3/19/153–4: Robert Were Fox to Edward Sabine (3 Mar. 1845), 153.

preferred apparatus for permanent observatories, where accuracy was prioritized over robustness, more delicate needles could be used to measure dip, and magnetometers were favoured for measures of absolute intensity. Keen for an enlargement of the BMS's expeditionary component, Fox suggested the project might be expanded to have simultaneous observations made on daily changes in declination, dip, and intensity at the greatest accessible 'depths in the earth, & also at considerable elevations above its surface, in order to ascertain if the apparently more local magnetic phenomena are modified by these circumstances'.[4] The Cornish natural philosopher again looked to the mine as a source of new scientific knowledge.

Fox and Sabine's systematic production of magnetic data was well underway. Between 1843 and 1850, the provisioning and training of British naval expeditions for magnetic survey work became highly standardized after the trial and error of the late 1830s and early 1840s. Fox and Sabine had learnt much from their experiences of the Antarctic, Canada, and the Niger. By 1843, their regime of instruments, training, and instruction manuals was firmly established. And yet, as this systematization extended the reach of the BMS ever further around the world, challenges remained. As the Franklin Expedition to the Arctic in 1845 demonstrated, disaster was never far away. But more problematic were the human and natural trials expeditions faced as they travelled to unmeasured regions.

Given that to completely understand the Earth's magnetic field the entire world had to become a site of experimental investigation, the use of Fox's instruments drew British naval crews and expeditions into previously uncharted territories. This has important implications for our understanding of the relationship between nineteenth-century science and colonialism. It is often assumed that colonialism followed commerce, with Britain gradually expanding its political and military grip over regions, having initially been predominantly interested in a territory's trade. Yet during the voyages of naval ships like HMS *Samarang* and HMS *Rattlesnake*, as well as overland expeditions, a range of extra-imperial spaces became sites of British scientific activity. Places like Borneo, Japan, Polynesia, the Arctic, and Korea were not yet, and would never become, British overseas territories. Magnetic data collection often preceded colonial expansion, which suggests that such colonizing processes were immensely complex and cannot be reduced to simple questions of commercial interest. More than this, the interactions between the expeditionary crews and varying local inhabitants that the BMS initiated reveal the differing character of mid-nineteenth-century colonial, and non-colonial, exchanges. From the townsfolk of Manila and ruling elites in Japan to pirates off Borneo's coast, the BMS's expeditions witnessed a diverse response to its magnetic experiments and, frequently, Fox's dipping needles in particular.

[4] Ibid.

6.1. Science on HMS *Samarang*

By 1843, Sabine had accrued a wealth of magnetic data from the Arctic and Antarctic, the Atlantic, Pacific, and Indian oceans, as well as from Britain's global network of physical observatories. There was, however, still one geographic region that required urgent investigation if the BMS was to complete its aims, and that was the magnetic equator, the movement of which varied unpredictably over time and space. The only colonial territories which Britain possessed for its constant examination were St Helena and Singapore, and these were, by the mid-1840s, well-established centres of magnetic science. Yet Sabine recognized the importance of a magnetic investigation of the seas around Singapore and in the Far East, which were poorly charted, let alone magnetically measured. Between 1843 and 1847, under the command of the unscrupulous but industrious Captain Edward Belcher (1799–1877), HMS *Samarang* would voyage to Hong Kong and, from there, magnetically survey Borneo, the Philippines, Korea, and Japan. It helped that Belcher was already vastly experienced in magnetic experimentation, as well as in cultivating goodwill with local dignitaries to facilitate such measurements; he also knew Fox's instruments well, having used them throughout his previous expedition along the coast of America from Vancouver to Panama. Eager to leave England following a scandalous court case against his wife, who had accused Belcher of 'knowingly and wilfully infecting her with a disease he had contracted as a result of ante-nuptial irregularity', his opportunity came in November 1835, when he was commissioned to replace the invalided Frederick William Beechey (1796–1856) as captain of HMS *Sulphur*, presently surveying North and South America's Pacific coast.[5]

On arriving at Panama in the spring of 1837, Belcher took command, along with possession of Beechey's 6-inch Robinson circle and several magnetized needles to complement a set of improved quality that he had taken out with him.[6] On 20 March, Belcher identified a spot for his first observations, ordering the *Sulphur*'s crew to set up a temporary magnetic observatory on an island just off Panama's Pacific coast. Here, Belcher performed measurements which 'were principally magnetic' and, in recognition, 'this island received the name of Magnetic Island'.[7] It was to be a vital location for his forthcoming survey work. Belcher compared his extensive collection of magnetic needles, employing the island as

---

[5] J. K. Laughton and Andrew Lambert, 'Belcher, Sir Edward (1799–1877)', *Oxford Dictionary of National Biography*, https://www.oxforddnb.com/view/10.1093/ref:odnb/9780198614128.001.0001/odnb-9780198614128-e-1979, 23 Sept. 2004, accessed 24 Mar. 2020.

[6] Edward Sabine, 'Contributions to Terrestrial Magnetism: No. II', *Philosophical Transactions of the Royal Society of London*, 131 (1841), 11–35, at 11.

[7] Edward Belcher, *Narrative of a Voyage Round the World, Performed in Her Majesty's Ship Sulphur, During the Years 1836–1842, Including Details of the Naval Operations in China, from Dec. 1840, to Nov. 1841*, 2 vols. (London: Henry Colburn, 1843), i. 25.

an index for subsequent intensity observations: Magnetic Island was given the relative intensity of 1.000, with measurements subsequently made between Panama and Cocos Island, from there to the Sandwich Islands, and then back to America's north-west coast, Vancouver, San Francisco, and Mazatlán.[8] Following Belcher's preliminary readings, he took increasing delight in his magnetic inquiries, learning much about the effects of local interference on needles. The western end of Cardon Island, for instance, provided challenging conditions, being 'of volcanic origin, and the beach contains so much iron, that the sand, which probably is washed up, caused the magnetic needle to vibrate 21° from zero'.[9] Belcher resolved this by carrying his needle to the summit of the island, where successful observations could be performed with reduced local interference. The *Sulphur* returned to repeat magnetic experiments on Magnetic Island in 1839, where the expedition's instruments could be checked to see how their magnetic character had altered over the previous two years. Later that year, Belcher received orders to return to England by a western route across the south Pacific, surveying and observing island groups along the way. The *Sulphur* reached Singapore in October 1840, but the recent outbreak of hostilities with China saw him diverted back east to the Canton River. Nevertheless, Belcher despatched his magnetic data from North America's western coast to Beaufort at Greenwich in time for inclusion within Sabine's Royal Society paper of February 1841. Sabine praised Belcher's 'zeal' in the pursuit of magnetic science, publishing his results in the *Philosophical Transactions*.[10]

After participating in the Anglo-Chinese Opium War, the *Sulphur* finally returned home in January 1842, having performed magnetic observations along the way in Ceylon, the Seychelles, Madagascar, Cape Town, St Helena, Ascension Island, and Cape Verde. At Cape Town, Belcher took the *Sulphur* to Simon's Bay, where the ship's 'needles were again tested at the position occupied by Captains James Clark Ross, and Crozier'.[11] On reaching the isolated Atlantic island of St Helena, Belcher again sought out the observation locations from Ross's Antarctic Expedition, pitching his tent at the precise position of measurement. He visited Plantation House, calling on the governor, who accompanied the *Sulphur*'s officers to Longwood House where they met the observatory's new director, Lieutenant William Smyth (1816–87). However, the *Sulphur*'s next port of call, Ascension Island, provided a less amicable scene: arriving in May, they found a grim sight—the *Albert*, 'one of the unfortunate Niger expedition, commanded by Commander Fishbourne, awaiting further orders'.[12] With its crew depleted and weakened by yellow fever and malaria, this relic of the Niger Expedition provided

---

[8] Sabine, 'Contributions to Terrestrial Magnetism: No. II', 12–13 and 16.
[9] Belcher, *Narrative of a Voyage Round the World*, i. 27.
[10] Sabine, 'Contributions to Terrestrial Magnetism: No. II', 11.
[11] Belcher, *Narrative of a Voyage Round the World*, ii. 294.   [12] Ibid. 296.

an ominous warning of the perils of mid-nineteenth-century exploration. On Belcher's return to Britain, Sabine had the expedition's needles tested at Woolwich, finding that in the three years since leaving Panama in March 1839, these delicate instruments had lost much magnetic force, especially between the Seychelles and Madagascar.[13] Duly corrected, Belcher's magnetic observations provided rich data for the fourth edition of Sabine's 'Contributions to Terrestrial Magnetism', published in the *Philosophical Transactions* in 1843. Along the north-west coast of America, the Pacific, and the China Sea, no fewer than thirty-two stations had been determined to complement the readings Belcher had made at sea.

Belcher did not linger long in England. After just four months back home, in November 1842, he was appointed to the *Samarang* for a new expedition with orders to survey Borneo, the Philippines, Taiwan, and the coast of China. In this task, Belcher's magnetic research was to continue, with Fox playing a pivotal role. Sabine ordered a new Fox circle and needles for the expedition. Despite 'the past discouraging experience' with George, Fox and Sabine persisted with the Falmouth mechanic. 'I really expect it will be ready in about the time mentioned, for I sh^d be very sorry for such a first rate observer as CaB[elcher] to sail without it', reflected Fox.[14] Importantly, he was no longer dependent on Jordan for jewelled pallets on which to balance the needles' pivots (Figure 6.1). 'George has a good stack of jewel'd holes from which to make his selection,' wrote Fox to Sabine, 'instead of having to wait for supplies from London.'[15] George fulfilled this promise, having the device ready for despatch by 30 December, but Belcher requested the instrument remain with Fox for collection when the *Samarang* called at Falmouth. Fox was delighted, boasting 'that for once, George has been punctual', and predicted that Belcher would be impressed with the circle.[16] While still in possession of this new instrument, Fox took the chance to experiment with the device in early 1843. He told Sabine that this investigation had helped him determine 'the influence of the earth's magnetism in modifying the force of the deflectors'. Each one of these cylindrically encased magnets was 3¼ inch long by ¼ inch diameter (Figure 6.2). Fox calculated that Belcher's new 'deflector S' had 1/345 the influence of the Earth's magnetism acting on a needle, compared to 'deflector N's' 1/342. The effect in modifying the force of the needle was thus 'only about half as great as it was with the deflector', which suggested great accuracy.[17]

After setting sail from Spithead, the *Samarang* arrived at Falmouth on 5 February 1843, where Belcher visited Fox for tea at Rosehill. Barclay excitedly noted that the captain arrived 'in the *Samarang*, Sloop of War, on his way to

---

[13] Edward Sabine, 'Contributions to Terrestrial Magnetism: No. IV', *Philosophical Transactions of the Royal Society of London*, 133 (1843), 113–43, at 113, 117–18, and 121.
[14] TNA, Sabine Papers, BJ3/19/134–5: Robert Were Fox to Edward Sabine (23 Nov. 1842), 134.
[15] Ibid.
[16] TNA, Sabine Papers, BJ3/19/136–7: Robert Were Fox to Edward Sabine (29 Dec. 1842), 136.
[17] TNA, Sabine Papers, BJ3/19/141–2: Robert Were Fox to Edward Sabine (16 Mar. 1843), 142.

Figure 6.1 The jewelled pivots holding in place the needle of a George-built Fox type, dating from around 1845. Note the groove around the edge of the needle's central wheel (Author's picture, 2020).

China on a surveying voyage. His object in touching at Falmouth is one of my father's instruments, which he is busily learning.'[18] Along with obtaining final onshore readings with his Hansteen-designed needles for land-based measurements, Belcher took possession of the new Fox dipping circle and an assortment of magnetic needles. In Fox's garden, he performed experiments with both of his magnetic instruments, expressing his personal gratitude to the Cornish natural philosopher and gaining some final advice regarding the use of the device.[19] Fox and Belcher exchanged experiences over the use of silk to suspend weights. Fox told Belcher that the 'advantage of using a single fibre of silk instead of many combined in all experiments is so great' that it was best to adopt the practice, to give more uniformed results owing to the regularity of friction a single thread produced.[20] Under Fox's gaze, Belcher performed this technique, producing a

---

[18] Barclay, quoted in, R. L. Brett (ed.), *Barclay Fox's Journal* (London: Bell & Hyman, 1979), 302.
[19] Edward Belcher, *Narrative of the Voyage of H.M.S. Samarang, During the Years 1843–1846; Employed Surveying the Islands of the Eastern Archipelago*, 2 vols. (London: Reeve, Benham, and Reeve, 1848), i. 3.
[20] TNA, Sabine Papers, BJ3/19/138–40: Robert Were Fox to Edward Sabine (11 Mar. 1843), 138–40.

DISCOVERY, DISASTER, DIPPING NEEDLE 195

Figure 6.2 The back of a Fox type showing the insertion of one of the cylindrically encased deflector needles. At the top of the circle is a second hole for the second deflector, which can be screwed in. Note that both sit on a disc that can be rotated to an angle in reference to dip (Author's image, 2020).

graph which charted results of tremendous consistency. They then took the new instrument on-board *Samarang* and readied the ship and its dipping needles for the expedition. Together, they deduced that the 'local attraction on board the Samarang was very great' and tested the various needles in comparison to their earlier trials at Rosehill. As Belcher explained,

> in instances where the Dipping Needle and other delicate instruments are to be observed on the quarter-deck, and the results of vessels equipped as the 'Samarang' to be regarded of authority, it is imperative that the ship be reduced to the same condition as she would be when such observations are to be made at sea; that all chance of error may be removed, or means pursued to arrive at data by which any may be corrected. These observations referring to the magnetic dip, intensity, and variation of the compass, must be taken throughout the voyage from a fixed point; prior to sailing, therefore, the ship is to be reduced to her sea footing; that is to say—to that state in which she would be found under ordinary breezes, away from land and out of soundings. The cables, as well as

her chain-messenger, should be unbent and below; all her boats up and stowed, and if provided with chain-topsails-sheets, particularly on the mizen-mast, they should be *home*. At Falmouth all this was easily provided for...and the ship being secured by hawsers to the mooring buoy was swung to thirty-two points of the compass for Magnetic Variation, and to sixteen for Dip and Intensity. At other parts I should deem it sufficient to unshackle and pay down the after part of the chain-cable, and to hang the ship 'before all' at the bitts—all other precautions observed.[21]

Yet again, Falmouth's position as a centre of magnetic expertise proved invaluable, providing Fox not only with a place of instruction for his instruments, but also with a site of preliminary experiments where the *Samarang* could be prepared for future magnetic survey work.

The *Samarang* sailed from Falmouth on 9 February, having enjoyed four days of Fox's hospitality and scientific assistance. Attempting to repeat observations at similar locations to previous survey expeditions, including Ross's Antarctic voyage, Belcher anchored off the island of St Jago at Cape Verde. The magnetic character of the eastern and western sides of this volcanic island were well known, providing an excellent site of comparison. Leaving Porto Praya on 7 March, Belcher followed an easterly, 'almost abandoned track' to the Cape of Good Hope, so as to obtain magnetic data from a previously unexamined region.[22] At Cape Town, Belcher visited the Magnetic Observatory, after which its director, Lieutenant Clerk of the Royal Artillery, came aboard the *Samarang* to see it being swung. More magnetic observations were made in Simon's Bay, where Belcher selected the exact location that Ross had employed in 1840 and Belcher had used in 1842. Though overgrown by grass and shrubs, this spot was well known and, by now, marked by a wooden post.[23] Belcher despatched the expedition's preliminary measurements back to Sabine who, in November 1843, forwarded them to Fox. On comparing Ross's intensity results by deflector from 1840 with Belcher's new experiments in the same location with weights, Fox saw that the intensity measurements differed by just 0.016 relative to London. Despite a few anomalies, Fox was staggered at the consistency between Ross's and Belcher's results and was pleased to learn that 'the influence of induced magnetism on the ships iron does not keep pace with the changes in force producing it'.[24] Belcher's observations with needle C at Cape Town, however, alarmed Fox, causing him to speculate that the instrument had sustained damage during the voyage.[25] In November 1843, he wrote to Belcher, requesting an update on how this needle was behaving but,

[21] Belcher, *Narrative of the Voyage of H.M.S. Samarang*, i. 3–4.
[22] Ibid. 5–6 and 8.   [23] Ibid. 11.
[24] TNA, Sabine Papers, BJ3/19/145–7: Robert Were Fox to Edward Sabine (18 Nov. 1843), 145.
[25] RS, Robert Were Fox Papers, MS/710/18: Edward Belcher to Robert Were Fox (4 Sept. 1845).

although the letter would eventually reach the *Samarang*, there was to be no response from Belcher for over two years, as the expedition sailed for the Far East.

Having crossed the Indian Ocean, with stops at Singapore, Macao, and Hong Kong, the *Samarang* anchored off Manila, in the Spanish Philippines, on 16 March 1844 (Figures 6.3 and 6.4). Here, it is apparent that while colonial entrepôts provided valuable staging posts, the increasing emphasis of the BMS was on obtaining data from beyond Britain's imperial spaces. Anxious to make observations in a magnetically unsurveyed territory, Belcher secured permission for experiments from Captain-General Alcala, the Spanish governor of the Philippines, who provided an exercising ground south of the city where the expedition could erect the tents of its portable magnetic observatory. Alcala also appointed an armed guard to keep away the fascinated local crowds, as Belcher suspected 'from the fearfully superstitious character of the natives, that some mischief might result from our being taken for necromancers'.[26] Although the crew accidentally erected the magnetic tents in the wrong location, selecting a 'green spot within the Southern Bastion', the experiments provided a unique opportunity for Belcher to share his enthusiasm for magnetic phenomena with the local population. A crowd of townsfolk made their way down to the tents to witness the experiments, including Alcala, who 'appeared to be much gratified by viewing the instruments, as well as by an explanation of the portable Magnetometers, Fox's needle, &c'. In discussions with Alcala, Belcher explained 'the nature of our intended operations within his government, and the advantages which would result to the civilized world', to which the Filipino governor consented, so 'that in every place under his jurisdiction orders would be given permitting us to make any that might be thought desirable'.[27] In Belcher's narrative, Fox's dipping needle appeared to have impressed Manila's society and helped the captain secure access to this magnetically uncharted region. The *Samarang* pressed on to Spanish-held Samboanga, where the local governor, Colonel Figueroa, granted permission for the temporary magnetic observatory to be erected and the experiments duly proceeded with the Fox type.[28]

On 16 April 1844, the expedition reached Sulu, under the rule of Sultan Jamalul-Kiram I, where Belcher received a far less amiable welcome. On landing to conduct magnetic observations, the *Samarang*'s crew encountered 'several small parties, armed with spears and krises'. Belcher reported that, 'as neither party understood Malay, nor we their language (Bisayan), explanation, or remonstrance, was impossible'. This was not just a threatening position for Belcher's men, but for the Fox dipping needle itself, with Belcher detailing how the 'proximity of their weapons much disturbed our magnetic observations'.[29] An armed guard was

---

[26] Belcher, *Narrative of the Voyage of H.M.S. Samarang*, i. 100.
[27] Ibid. 101.   [28] Ibid. 106–7.   [29] Ibid. 114.

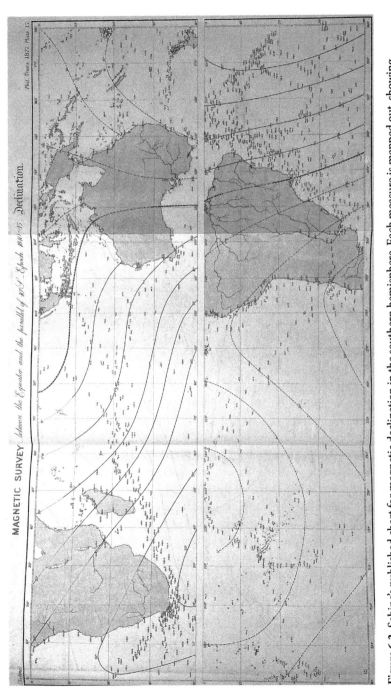

Figure 6.3 Sabine's published chart for magnetic declination in the southern hemisphere. Each measure is mapped out, showing Belcher's Indian Ocean crossings of 1842 and 1843, his *Sulphur* voyage across the Pacific, as well as those of Ross between 1839 and 1843 (Author's image, 2021).

Figure 6.4 HMS *Samarang*, capsized in the Sarawack, Borneo, on 17 July 1843, before the crew were able to refloat her (Image in author's possession, 2021).

called out to protect the ensuing magnetic observations, but the iron weapons continued disrupting the Fox's oscillations. Similar difficulties were subsequently encountered off the east coast of Borneo, on Tawi-Tawi Island, a pirate stronghold. Belcher accordingly selected Unsang for magnetic readings, this being the nearest relatively safe land within sight of Borneo, allowing him to complete his magnetic surveying of the eastern extremity of the coastline.[30] Often, Belcher and his crew were obliged to manage with make-do observatory locations, transforming the local environment into a suitable place for magnetic experiments. When the *Samarang* anchored in the Strait of Patientia, within sight of Gilolo, a detachment occupied a small island, little more than a reef, and reorganized the coral flat into a site of experimental observation. Belcher ordered the construction of 'an islet for my instruments, by piling up coral slaps upon the reef. It proved to be very fragile, being as it were an underwood of coral trees into which we sank, by our own weight, about eighteen inches'.[31] Nevertheless, with the aid of pounding crowbars, the crew secured the instruments firmly in place: this was a far cry from the stable conditions of Rosehill garden. Forenoon observations were taken, before Belcher indulged in a little shell-collecting about the reef. However, on returning to the magnetic instruments for noon readings, the captain and his detachment were surprised to see 'several natives' fleeing the experimental reef platform. Evidently, the mysterious magnetic observatory, complete with the Fox type, had aroused the curiosity of the local inhabitants. Soon after, a band of some forty men appeared about a quarter of a mile off, armed with bundles of spears

---

[30] Ibid. 118–19.   [31] Ibid. 135.

and dressed in scarlet red clothing, revealing them to be local 'pirates'. Encircling the observatory, they began throwing spears, 'some of which fell very near the instruments'.[32] Belcher's men retreated to their gig, clutching their scientific apparatus. Once on-board, they fixed bayonets and delivered a few volleys of musketry, forcing the pirates to flee. It made for a sensational story in Belcher's subsequent narrative of the voyage, published in 1848 after his return to Britain, but this had been a close-run thing for the prized Fox.

On Christmas Day 1844, the *Samarang* sighted the mouth of the Pantai River, Pulo Panjang, on the coast of Borneo. In his search for an observation spot, Belcher could identify nothing but loose mud, making a magnetic experiment at this 'almost unknown river' a difficult prospect. Undeterred, the expedition's crew took advantage of the mangrove trees that grew along the coast, selecting the largest, of which 'the trunk was sawed off... [at a] height above high water as to serve for the Azimuth circle'. A second nearby tree was also cut down and wooden planks from the ship were used to construct a platform where the artificial horizon could be positioned. As this lacked stability, three spike nails were driven in around the instrument's circumference and a 12-inch observing slate was balanced on top of them to form a level plinth for the horizon. From the muddied riverbanks, a magnetic observatory had been built, on which the Fox type was mounted. Belcher later boasted how his crew had, effectively, fashioned from 'an almost inaccessible marine position, a dry, comfortable, and efficient observatory'.[33] When an account of this featured in Belcher's published narrative, the implications were subtle but clear for British audiences: the scientific investigation of terrestrial magnetism appeared to bring order to the most untamed natural environments.

The *Samarang*'s difficulties continued into 1845, as the magnetic observers contended with natural and social challenges alike. The expedition found itself at Balambangan Island, off Borneo's north coast, for its subsequent monthly magnetic Term Day, but Belcher felt the risk of attack from the local populace too great for his crew to disembark and erect the observatory's tents and equipment. His men might have gone armed, but this, Belcher believed, would have been counterproductive due to the magnetic interference from the 'iron implements of war' which the locals brandished, not to mention the expedition's muskets.[34] The problem was solved through careful diplomacy, as Belcher secured permission from the Sultan of Gunang Taboor, the ruler of the territory, to perform experiments on the grounds of the local aristocrat Datoo Danielle, an old ally of the English. Well enclosed and about half a mile from the town, Danielle's home provided fine hospitality and greater comfort than the tents of the portable observatory. Eager to facilitate Belcher's work, Danielle invited his relatives to witness the experiments who, Belcher reported,

[32] Ibid. 136.   [33] Ibid. 210–12.   [34] Ibid. 257.

exhibited great anxiety to view our instruments,...they were much delighted by the beauty of the instruments; more particularly by Fox's dipping needle, placed beyond the limits of influencing the more delicate magnetometers. Of the uses of this instrument they appeared to comprehend more than I had given them credit for, although I have remarked, as a general feature amongst the better educated Malays, as well as Chinese, that they understood more of the properties of the magnet than many educated Europeans. I expected to excite their surprise by the reversal of the poles of the Dipping Needle, and I was assisted by my very intelligent friend Mr. Wyndham, as interpreter; but the better informed of Datoo Danielle's family, gave me to believe that they understood it perfectly. Indeed, I was told that the younger brother, Udin, was an ingenious mechanic, and could take a watch to pieces, and clean or repair it.[35]

Belcher and his officers were impressed by the philosophical knowledge of their hosts, reflecting that the Sultan of Sulu would do well to raise a man such as Danielle to the position of chief or prime minister. But what is most interesting about this exchange is the way in which such cultural encounters blurred and problematized traditional attitudes towards non-European inhabitants. That Belcher thought the people of Malaya and China exhibited a superior level of philosophical understanding of magnetic science to 'many educated Europeans' suggests that the face-to-face experiences of naval crews often failed to live up to traditional colonial ideologies.

After Borneo and the Philippines, the *Samarang* sailed through the Korean Archipelago. In August 1845, the *Samarang* approached its final destination, Japan, anchoring in the port of Nagasaki in search of supplies. The local authorities sent out boats and provided a friendly welcome. Belcher met with Nagasaki's dignitaries and explained the purpose of his expedition. While they agreed that the ship could continue to Loo-Choo, exploring the Ryukyu Islands, the Japanese were concerned over the magnetic survey work, informing the captain 'that it was forbidden to measure the land in Japan' and that to enter the country's towns would, as Belcher put it, 'offend their prejudices'.[36] Discussions remained polite and civil, and eventually the Japanese officials secured permission from the Emperor for provisions to be taken on board, providing that the crews respect Japanese customs and not remain on land at night to observe the stars. Magnetic measurements were to be confined to the ship to respect Japanese customs.[37] Nevertheless, Belcher later boasted that his 'ideas as to the practicability of overcoming the prejudices of Loo Choo, Japan, & Korea are partly successful'.[38] The investigation of terrestrial magnetism had fostered unprecedented cultural

---

[35] Ibid. 258–9.   [36] Ibid. ii. 6.   [37] Ibid. 6–10.
[38] RS, Robert Were Fox Papers, MS/710/18: Edward Belcher to Robert Were Fox (4 Sept. 1845).

exchanges between British naval officers and the Japanese authorities. Belcher later told Fox that the Japanese had

> treated me well, allowed me to remain on shore until midnight, longer had I wished, to determine the Latitude—but only on an uninhabited island at Nangasaki—This is more than they ever granted to any preceding one for more than 40 years—They looked to my <u>return</u> with <u>satisfaction</u>, and told me 'that it was probable I would be permitted to land at Nagasaki' & amongst the Korean Group; I did as I pleased; keeping out of their villages; which I knew they disliked,—ascended their mountains, and had my <u>observatory on shore</u>—at Loo Choo, I was treated very handsomely—They gave permission to travel over their islands and make their survey and furnished my party with <u>horses</u> to <u>view</u> the <u>country</u> ascending the highest mountains, which commands a view of both seas, and their city beneath—So long as you respect their towns, and give women time to hide themselves, they have no objections to entering their country.[39]

It is very evident that Belcher's efforts to respect Japanese customs actively shaped where and when he was able to take magnetic measurements. But more than this, it is clear that Anglo-Japanese cooperation in the pursuit of magnetic science had actually opened up a land little known by British explorers: these early British–Japanese interactions were, therefore, the direct consequence of the surveying of terrestrial magnetism, preceding the signing of the first Anglo-Japanese Friendship Treaty in 1854.

Having reached Japan, and with the expedition's main objectives largely fulfilled, Belcher returned to British-held Hong Kong to refit and take on provisions. Belcher wanted to continue his magnetic research in Japan, but all his efforts 'were frustrated by being compelled to return to Hong Kong, the admiral having no vessel to bring me provisions'.[40] Despite this diversion, he left Japan armed with a wealth of magnetic data from some two years of exploration. With Hong Kong in sight, on 4 September 1845 Belcher finally sent a detailed account of the dipping needle's performance, responding to Fox's letter from November 1843. The captain apologized for lack of communication, but enclosed valuable information on Fox's instrument. Calming Fox's worries over needle C, expressed two years earlier, Belcher reported that it had since proved reliable. Fox supposed that the 'error at the Cape probably arose from some mistake of mine'. Since then, the needles had rarely differed by more than seven minutes, although Belcher noted that C did not oscillate 'well and requires a heavy dose of the magnet to cause it to act properly'.[41] Although back in April 1844, at Manila, Belcher's three Fox needles had coincided extremely closely, at Borneo the dips for needles A, B, and C

---

[39] Ibid.   [40] Ibid.   [41] Ibid.

were, respectively, 2° 32' 55", 2° 33' 53", and 2° 36' 53". Providing needle C was kept magnetized, Fox's instruments produced consistent measurements. He experienced further inconsistencies on an island in the Korean Archipelago, 34° 16'N and 127° 13'E, where C gave 12° 30' to A's 48° 42' and B's 48° 45' 37". This was wildly inconsistent. Then at Nagasaki, A and B gave 44° 44' 25" and 45° 6' 2" respectively, before differing by over four minutes in the Japanese Rykukyu Islands. All of this illustrated 'that the differences are not steady', causing Belcher to doubt that deflector S had worked as it was intended to. He suspected that the extreme heat of the tropics might have affected its magnetism. At the same time, the deflector's screw cap kept falling off, which Belcher had secured 'by jamming it in with paper'.[42] Throughout the trial's expeditionary survey work, Belcher's well-crafted Fox circle had assumed an increasingly ragged appearance.

Along with needle C, Belcher also had growing doubts over needle A and relied almost completely on needle B throughout the voyage. 'The results—with B, which is my pet needle, I am certain of', he reported. Nevertheless, Belcher assured Fox that he would be 'pleased to see the state of the needles—The electro-typed A very little changed and A & B as bright as when I left Falmouth'. Physically, Fox's needles had endured well, despite having 'had their share of exposure on half tide rocks and positions where the spray of the sea dimmed the glass of the circle'.[43] As for the weights, Belcher had managed not to mislay a single one since leaving Falmouth, but the silk thread for intensity experiments had required replacing several times. The captain claimed to have become highly skilled at using this apparatus for intensity observations, detailing how he 'generally test[ed] the pulley, & silk, by noticing whether on lifting the hook, without weight, whether takes the 1°. 2°. or 3° to draw the silk over it—at times the hook alone will effect it'. As for using the weights in combination with the deflectors, Belcher assured Fox that he had 'paid attention to your wish relative to the Intensity by the Deflectors but until they give me a little more time for play I despair of performing any thing properly'.[44] What he required was more experience with the instrument, but finding time for this was tricky, given Belcher's duties as a captain.

For all the challenges, Belcher thought Fox should be delighted with the performance of his circle and needles throughout *Samarang*'s voyage. 'You will be pleased', wrote Belcher, 'to find that your circle has determined the first well authenticated Dip action in Japan or the Korea.'[45] Once at Hong Kong, Belcher again wrote to Fox, sending copies of his most recent dip observations. This correspondence reached Fox on 29 November, with the Cornishman ecstatic to hear news of his instrument's performance on *Samarang*. He wrote immediately to John Phillips, enthusiastically announcing that he had 'some very good results of magnetic dip &c. obtained by Sir E Belcher at Japan & in the China seas.

[42] Ibid.   [43] Ibid.   [44] Ibid.   [45] Ibid.

He writes me that the needles retain their polish well'.[46] Belcher's expedition continued to perform magnetic observations after leaving Hong Kong, keeping Term Days as usual, with the expedition recrossing the Sulu and Mindoro seas and following Borneo's northern coast in late 1845. On the apparently unclaimed island of Cagayan Sulu, Belcher pitched another temporary magnetic observatory on a small rocky islet, perched on a reef at the 'entrance of a most romantic circular basin', which appeared to be an old crater. Belcher's crew were, by now, well versed in constructing makeshift magnetic observatories. Here among the island's dense vegetation, Belcher and the officers took dip and intensity measures, before visiting some nearby houses and introducing themselves to the island's inhabitants.[47] When the *Samarang* travelled back to England, it stopped along the way at Simon's Bay, in October 1846, to remeasure Ross's celebrated magnetic post.[48] Once home, Belcher received yet another hero's welcome, before publishing an account of the expedition, complete with extensive magnetic data.

6.2. Magnetic Miseries

The *Samarang*'s voyage needs to be seen in the context of the BMS's increasing occupation with extending its investigation beyond the familiar routes of imperial communication and trade, particularly those to India, Hong Kong, Singapore, and Australia, via the Atlantic and Cape of Good Hope. Belcher's expedition had helped do this, obtaining magnetic measurements from the unmeasured coasts of Borneo, Korea, Japan, and the Philippines. However, to understand the Earth's magnetic behaviour in equatorial regions, it was as important to acquire data overland as from a fixed location over a considerable duration of time. Both of these tasks fell, not to a naval expedition, but to Lieutenant Charles Elliot (1815–52), of the Madras Engineers. Tasked with establishing a physical observatory at Singapore, Elliot had travelled out to the Straits Settlement in 1840 on the East India Company's *Minerva*, armed with a selection of scientific instruments, including a Fox type. With these, he would produce a huge volume of measurements, which were eventually published in 1851. After setting up the East India Company's Magnetic Observatory at Singapore and completing a period of observations, Elliot applied to the Royal Society to conduct survey work in Borneo in 1842. This suggestion was approved and he subsequently organized a two-month-long expedition to the region, before returning to Singapore to resume his work at the observatory.[49] After his travels in Borneo, Elliot soon grew

---

[46] OUM, Papers of John Phillips, Phillips 1845/36: Robert Were Fox to John Phillips (29 Nov. 1845).
[47] Belcher, *Narrative of the Voyage of H.M.S. Samarang*, ii. 112–14.   [48] Ibid. 209.
[49] Fiona Williamson, 'Weathering the Empire: Meteorological Research in the Early British Straits Settlements', *British Journal for the History of Science*, 48/3 (Sept. 2015), 475–92, at 483–5.

restless at his observatory, writing to Sabine in March 1845 with dire warnings over his health. 'I have no organic disease that I can make out past a gradual loss of health and stamina', which, he claimed, resulted from five and a half years of magnetic investigations in the tropics. Requiring a 'total absence of all excitement', Elliot speculated 'that 5 years of observations would be sufficient for this observatory', as few changes to or abnormalities in the Earth's magnetic field were discernible. As for his magnetic apparatus, all of 'Lloyd's instruments in India for determining the changes of vertical force have proved failures'. In contrast, he reported that 'Fox's dip circle is a very nice instrument—the needles are very well finished, but appears to me to be too heavy for a portable instrument. The dip found from deflectors never comes out equal to the dip found directly being always more And the difference being constant'.[50] Nevertheless, for all their difficulties of transportation, Fox's instruments had provided a rare source of magnetic consistency. Elliot was sorry to report that a recent misfortune had befallen the treasured device, explaining how he 'had a set of most valuable needles stolen from me a short time ago but I never could find out the thieves. It must doubtless have been done by some of the people about me in the house who were struck at the care I took, always cleaning them myself'. Elliot felt this misfortune all the more, having 'now nothing but Fox'.[51] As ever, the BMS's systematized data collection was vulnerable, be it to human frailties or acts of sabotage.

By June 1845, the BAAS was satisfied with Elliot's measurements and terminated the Singapore operation. He subsequently secured Sabine's permission to resume his overland survey work in Borneo, 'Java, Sumatra, and the whole of our Eastern possessions in the bay of Bengal'.[52] Following the recent theft of his dipping needles and poor state of his equipment, Elliot requested Sabine send him 'a portable magnetometer, and a Fox dip circle—the latter I consider as especially useful in these seas, as for every minute of latitude from this to Java there are two minutes of dip'.[53] Such rapid fluctuations in magnetic phenomena made these equatorial regions especially important to the BMS, complementing those from St Helena. However, for such work, he found the recommended needles of Lloyd's design and construction unreliable, reporting that

> Hansteen's needles are proposed as a substitute to Lloyd's Intensity needles in low magnetic latitudes, yet I have got out some very accordant results being exactly the same as the total intensity derived from the 15 inch magnets and the

---

[50] RS, Sabine Correspondence, Vol. 2, D–J, MS/258/485: Elliot to Edward Sabine (18 Mar. 1845).
[51] Ibid.
[52] RS, Sabine Correspondence, Vol. 2, D–J, MS/258/496: Elliot to Edward Sabine (24 Mar. [c.1845]); Williamson, 'Weathering the Empire', 488.
[53] RS, Sabine Correspondence, Vol. 2, D–J, MS/258/496: Elliot to Edward Sabine (24 Mar. [c.1845]).

dip at Sarawak in Borneo as compared with Singapore. Hansteen's needles have not yet arrived at a permanent state of magnetism.[54]

Equatorial magnetic measurements required exceptionally delicate needles to detect minor changes in magnetic force but, for this, Elliot thought Fox's instrument superior to its rivals.

Elliot departed Singapore for Java in 1845, undertaking a gruelling regime of magnetic observation. On expedition, he would erect a temporary observatory and take several months of readings, which could then be compared to results from Singapore's permanent observatory, before moving on to a new location. Waking each day at 5.30 a.m., he arrived at his station for 6 a.m. to take dip, intensity, and variation measurements, and continued experiments throughout the afternoon. Elliot despatched regular reports and requests for new equipment back to Sabine, noting enviously how Belcher had 'assistance which of course I could never hope for', as well as a 'suite of tents' with which he could 'take every precaution in placing the instruments at proper intervals'.[55] Continuing his magnetic expedition throughout 1846, Elliot reached the Dutch East Indian capital, Batavia, in July 1847 where he received his new Fox dipping circle the following month. By November, after several months of travel, this device was in a poor state due to rough handling during transportation. Along with losing a set of Royal Society thermometers, Elliot struggled to get consent from the local authorities for magnetic observations, but was still determined to produce a magnetic report for Sabine, following a tough regime of magnetic data collection.[56] Accounts of fatigue and tropical disease filled Elliot's miserable letters back to Woolwich. Between 8 and 9 December 1847, he endured an awful night in which his assistant suffered a four-hour 'attack of fever'.[57]

In late 1847, Elliot's expedition commenced its survey of the Coco Islands and, in August 1848, he returned eight months of results from his observations at Java back to London (Figure 6.5). He then ventured eastward towards the western regions of New Guinea, so as to cross the magnetic equator, before returning to Singapore at the end of the year. From there, he hoped to push on to Bengal and Calcutta to secure results to compare with those of the observatories at Madras, Trivandrum, and Simla, all of which were important sites for observing equatorial terrestrial magnetism.[58] In particular, Trivandrum Observatory was an especially industrious site of magnetic data production, positioned on the tip of India, in what is today Kerala. Since 1840, the Raja of Travancore, Swathi Thirunal Rama Varma (r. 1829–46) had offered his astronomical observatory to the BAAS for

---

[54] Ibid.
[55] RS, Sabine Correspondence, Vol. 2, D–J, MS/258/485: Elliot to Edward Sabine (18 Mar. 1845).
[56] RS, Sabine Correspondence, Vol. 2, D–J, MS/258/486: Elliot to Edward Sabine (2 Nov. 1847).
[57] RS, Sabine Correspondence, Vol. 2, D–J, MS/258/487: Elliot to Edward Sabine (9 Dec. 1847).
[58] RS, Sabine Correspondence, Vol. 2, D–J, MS/258/488: Elliot to Edward Sabine (4 Aug. 1848).

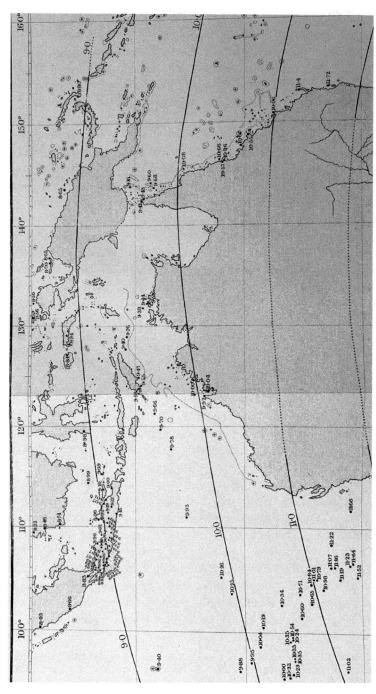

Figure 6.5 Sabine's chart for magnetic intensity between 1840 and 1845, published in 1876, including Elliot's measures from Java and the Far East (Author's image, 2021).

inclusion within the BMS. Given Trivandrum's proximity to the magnetic equator, the BAAS obliged, sending out magnetic instruments and a director, John Caldecott (1800–49), to oversee magnetic observations. Exceptionally well equipped thanks to their princely financing, Caldecott and his successor, Allen Brown, performed some 350,000 observations over fourteen years, constituting one of the BMS's most substantial data sets.[59] Elliot's surveying of the land around the Bay of Bengal provided invaluable measurements to complement this network of Indian magnetic observatories. Throughout these measurements, Elliot continued to be impressed with the reliability of his Fox dipping needle. 'I have taken an immense number of observations of dip with your Fox,' he told Sabine, 'which is a beautiful instrument and works remarkably well—but I cannot say the same of the gymbal stand. It is by no means of the substantial description requisite for board ship work—it is more a toy than a solid piece.'[60] The Fox type and carrying box were heavy enough on their own. Nevertheless, Elliot had found the circle itself invaluable, even managing intensity measurements with both weights and deflectors on the Coco Islands. In August 1849, an exhausted Elliot reached Madras, having 'undertaken the magnetic survey of the Indian Antipoleys at the recommendation of the Royal Society'.[61] Overall, he had taken magnetic observations at some fifty stations in Borneo, Singapore, the Cocos Islands, and India, employing four different dipping needles.

6.3. The Voyage of HMS *Rattlesnake*

Just as Belcher's *Samarang* voyage was coming to an end, so Sabine and Fox set about equipping a new venture, tasked with circumnavigating the world and surveying its magnetic properties along the way. Between December 1846 and 1850, Owen Stanley's *Rattlesnake* expedition sailed east, via Cape Town, to the northern coast of Australia and New Guinea, before returning across the Pacific Ocean, stopping at the Falkland Islands along the way. Having seen action in the Opium War of 1839 to 1842, the *Rattlesnake* had been converted into a survey ship in 1845 and was equipped with an array of magnetic apparatus, including several of Fox's construction. Responsibility for magnetic experiments fell to Lieutenant Joseph Dayman, who had experience of a Fox type from his time on *Erebus* with Ross, while the experienced Captain Stanley would provide constant scrutiny and assistance. Armed with a 6-inch inclinometer of Robinson's construction and a Fox-type dipping circle, Dayman took regular magnetic observations on land with both instruments, as well as at sea with the Fox. He carried two needles for

---

[59] Jessica Ratcliff, 'Travancore's Magnetic Crusade: Geomagnetism and the Geography of Scientific Production in a Princely State', *British Journal for the History of Science*, 49/3 (Sept. 2016), 325–52.
[60] RS, Sabine Correspondence, Vol. 2, D–J, MS/258/488: Elliot to Edward Sabine (4 Aug. 1848).
[61] RS, Sabine Correspondence, Vol. 2, D–J, MS/258/491: Elliot to Edward Sabine (6 Aug. 1849).

Robinson's instrument, labelled 'A1' and 'A2', and three for the Fox, listed as 'Fox A', 'Fox B', and 'Fox C'. Comparative readings ensured that if any one of these suddenly lost magnetic intensity, the error would be immediately apparent.

Dayman's results were to provide a valuable comparison to those of Ross and Crozier taken between 1839 and 1843. After arriving at Madeira, he recorded dips of 59° 41' 7" and 60° 40' 2" respectively with his Robinson and Fox instruments in Funchal, before hauling the Fox to the summit of the Pico dos Bodes where it gave 64° 10' 5" (Figure 6.6). After observations with the Robinson type on Rat Island at Rio, the *Rattlesnake* sailed to Cape Town. There, Dayman recording a dip of 53° 40' with Fox A, working at the same observation spot in the dockyard of Simon's Bay that Ross and Crozier had employed in 1840 and that Belcher had since remeasured twice. On Tonnelier's Island at Port Louis on Mauritius, Dayman compared his two Robinson needles: A1 gave 53° 48' 9" to A2's 53° 48' 8", suggesting superb integrity between the instruments. By the time the *Rattlesnake* reached Rossbank at Hobart, the same two needles had diverged somewhat, with A1 giving 70° 36' compared to A2 at 70° 41' 5". Dayman continued to use the Robinson for land measurements along the northern coast of Australia, where the difference between the two needles varied between one and two minutes.[62]

Figure 6.6 Owen Stanley's depiction of Dayman performing magnetic measurements with his Fox type at the summit of the Pico dos Bodes on Madeira (Public Domain, 2022).

---

[62] Joseph Dayman, 'Observations of the Mean Magnetic Inclination Made on Shore in the Voyage of H.M.S. Rattlesnake, by Lieut. Joseph Dayman, R.N.; Instruments Employed: Robinson's 6-Inch Inclinometer; Fox's Dipping Apparatus', in John MacGillivray, *Narrative of the Voyage of H.M.S. Rattlesnake, Commanded by the Late Captain Owen Stanley, R.N., F.R.S. &c. During the Years*

Writing from the *Rattlesnake* on 6 April 1847, Stanley despatched an early report back to Fox of the expedition's initial magnetic observations, thanking the Cornishman for the 'complete list' of instruments for the dipping needle he had despatched from England. Due to heavy weather, he was sorry that the Fox type had at times been unmanageable, making measurements for 'both Dip and Intensity with weights & deflectors' impossible.[63] But Stanley was confident that the observations that had been taken were done in the manner 'particular to Colonel Sabine', who would 'make good use of them'. As for Dayman, he had nothing but praise for the officer's experimental skill. Along with emphasizing the competence of the *Rattlesnake*'s naturalist, John MacGillvray (1821–67), Stanley provided Fox with a gripping account of the expedition's three-week stay at the Cape of Good Hope and acquaintance with the local Kafu people. The colony suffered a 'want of money', making the purchase of victuals expensive. Within three weeks, the £50,000 worth of gold that the ship had conveyed to the colony had been completely spent, this being a 'drop in the ocean'. Along with extravagant prices, Stanley reported of murders and attacks from the Kafu and speculated that the improvement of the region would be a slow process. Evidently, magnetic science in embryonic colonies was an expensive and dangerous business.[64]

After Madeira and Cape Town, it was Dayman's preference to use the Robinson type onshore and the Fox at sea. But by April 1848, as the expedition traversed Australia's north coast past Barnard Island, off Cape York, he performed further comparisons between the Robinson and Fox apparatus, finding Robinson A1 gave 33° 10′ 2″ to Fox C's 33° 8′ 4″. At Port Essington, on Australia's northern coast, Dayman compared Fox C on the *Rattlesnake* with the same needle on land, giving respective readings of 35° 14′ 6″ and 33° 48′. The error was not, he reported, due to the ship's iron, but to local interference, asserting that the 'observations on board the ship at this station are the nearest to the truth, there being much iron-stone strewed over the country about the observation spot on shore'.[65] At Coral Haven, 'On a patch of Coral near Pig Island', he trialled the Robinson type against Fox, replicating this on the uninhabited Duchâteau Islands, in the Coral Sea. Dayman used Fox B along the south coast of New Guinea and for the remainder of the voyage, but took thorough comparative land-based observations between his three Fox needles on Garden Island at Port Jackson, the Bay of Islands at New Zealand (near Kororareka Bay), and on the eastern Falkland Island close to the chaplain's house in Port Stanley (Figure 6.7). This allowed the lieutenant to

---

1846–1850; *Including Discoveries and Surveys in New Guinea, the Louisiade Archipelago, etc. To Which Is Added the Account of Mr E. B. Kennedy's Expedition of the Cape York Peninsula*, 2 vols. (London: T. & W. Boone, 1852), i. 337–42, at 337–8.
[63] RS, Robert Were Fox Papers, MS/710/110: Owen Stanley to Robert Were Fox (6 Apr. 1847).
[64] Ibid.
[65] Dayman, 'Observations of the Mean Magnetic Inclination Made on Shore in the Voyage of H.M.S. Rattlesnake', 339–40.

Figure 6.7 Owen Stanley's depiction of inhabitants from Brumi Island, New Guinea, performing a dance on HMS *Rattlesnake* in 1849 (Public Domain, 2022).

monitor the integrity of his on-board measurements as *Rattlesnake* crossed the Pacific. Throughout his voyage the three needles increasingly diverged, with each giving an identical dip of 62° 44′ 9″ at Port Jackson, but with Fox A's 59° 37′ 6″ differing from B and C's respective measures of 59° 44′ 2″ and 59° 28′ 1″ at New Zealand. On reaching Port Stanley in 1850, Fox A gave 52° 19′ 6″ to B's 51° 43′ 3″ and C's 50° 58′ 8″. After repeating observations in Berkeley Sound, which the crews of HMS *Terror* and HMS *Erebus* had measured in April 1842, *Rattlesnake* returned to Britain via the Azores, where Dayman made brief comparative observations between Fox A and B, confirming the divergence recorded on the Falklands. With A giving 67° 26′ 9″ to B's 66° 58′ 4″, the former needle was tending to a slightly higher degree of dip.[66]

Taking into account these anomalies, but with sufficient data to make the necessary corrections, Sabine published the *Rattlesnake*'s results as he received magnetic reports throughout the expedition. Appearing in his ninth edition of 'Contributions to Terrestrial Magnetism' in 1849, Sabine was able to compare Dayman's results from Britain to the Cape of Good Hope between 1846 and 1847 with those of Ross, Sulivan, Lefroy, Clerk, and Moore. Providing a magnetic declination map for the Atlantic from the latitude of 60° north to 60° south, Sabine included few references to Fox's needle, with the instruments now well established

---

[66] Ibid. 341.

as the Royal Navy's standard for measurements taken at sea.[67] In 1852, MacGillvray published the complete data set from Dayman's onshore observations, but was unable to share the on-board measurements, these requiring 'several corrections before they are fit for publication.'[68] Nevertheless, the *Rattlesnake* expedition had proved another successful operation for Fox's experimental, instrumental regime.

6.4. Relics and Rejects: The Franklin Expedition

Not all expeditions worked out so well as those of the *Samarang* and *Rattlesnake*. The most famous of all magnetic undertakings was, predictably, a complete failure. After a miserable governorship in which his and his wife's reforming zeal had alienated local settlers, Sir John Franklin left Van Diemen's Land in November 1843. Back in England, he secured appointment to command a new expedition to the Arctic in 1845, in what was hoped would be the last of Barrow's expeditions, locating the North-West Passage and completing the magnetic survey of the northern polar regions. Barrow had actually wanted the young and energetic commander James Fitzjames (1813–c.1848) to lead this venture, having served on the Euphrates Expedition of 1834 to 1837 and in the Anglo-Chinese Opium War. Though experienced in the science of terrestrial magnetism, some suspected that Franklin was, by now, in a condition too physically poor to direct the Admiralty's endeavour.[69] Echoing the 1839–43 Antarctic voyage, Franklin's expedition would consist of HMS *Terror* and HMS *Erebus*, with Crozier serving as second-in-command. Franklin was 'really flattered' at having such an experienced captain for *Terror*, despite having heard rumours that 'the Board contemplate having a commander only on the 2ᵈ ship...there are two persons of that rank whom they have in their eye, and are said to possess considerable scientific qualifications, besides having the advantage of youth.'[70] However, no amount of 'youth' or 'scientific qualifications' could rival Crozier's magnetic experience, especially with the Fox dipping circles.

Sabine and Lloyd took responsibility for the expedition's magnetic preparations, opting for the ever-reliable Fox type for on-board measurements and a Robinson for land observations.[71] Having learnt of the forthcoming expedition, Fox wrote

---

[67] Edward Sabine, 'Contributions to Terrestrial Magnetism: No. IX', *Philosophical Transactions of the Royal Society of London*, 139 (1849), 173–234, at 173 and 175–6.
[68] Dayman, 'Observations of the Mean Magnetic Inclination Made on Shore in the Voyage of H.M.S. Rattlesnake', 337.
[69] Andrew Lambert, *Franklin: Tragic Hero of Polar Navigation* (London: Faber and Faber, 2009), 153; B. A. Riffenburgh, 'Franklin, Sir John (1786–1847)', *Oxford Dictionary of National Biography*, https://www.oxforddnb.com/view/10.1093/ref:odnb/9780198614128.001.0001/odnb-9780198614128-e-10090, 23 Sept. 2004, accessed 24 Mar. 2020.
[70] Scott, Ms248/316/17: John Franklin to James Clark Ross (9 Jan. 1845).
[71] Lambert, *Franklin*, 153.

to Sabine on 3 March 1845 with news that while 'George has nearly completed a dip circle which promises to turn out well', he was 'about to give up his present employment so that this may be the last of his make'. Changing the subject altogether, Fox enquired whether a George-built Fox type would be wanted for the Arctic Expedition.[72] Sabine recognized the auspicious timing of the completion of this new instrument, precisely when one was required for *Terror* and *Erebus*, and duly requested Fox prepare it for London. Within five days of his initial query, Fox wrote again, delighted that George had agreed to give him the new circle to examine within a week. 'If the jewel'd holes & pivots, or most of them should be found satisfactory the inst^t may be forwarded on the following week', Fox confirmed, while warning that 'it is not to be expected that all the pivots will be passed'. Furthermore, if a second Fox type was required, Fox promised to 'get another made by an ingenious mathematical instrument maker in this neighbourhood', offering to personally 'finish the balancing of the needles'.[73] Fox was already working to secure a replacement supplier of magnetic apparatus with George's impending departure.

George left for London by steamer on 21 March, taking with him the new Fox type for Captain Crozier. Eager not to lose future commissions, George had promised Fox 'to do some of the nicer parts of dipping needles' in his spare time, providing he could 'make arrangements with an instrument maker to undertake the heavier part of the work'. Fox had examined George's new circle and its needles, finding Fox B 'very good' and C 'satisfactory', but conceded that 'A gives too little dip by 6' or 8' both direct & reversed, & had time allowed I sh^d have recommended a new axle in this case, as I am anxious to see the Arctic Expedition provided with the very best needles'. Nevertheless, he suspected this might be an error on his own part, given that George thought that Fox had overestimated the error, confessing that 'it may be so, as I was far from well when I examined it, & c^d not spend much time about it'.[74] He concluded that the new circle for the Franklin Expedition had all the characteristics of a well-built instrument.

Terror and *Erebus* also carried the most up-to-date magnetic instructions for both. Charles Riddell, assistant superintendent of ordnance magnetic observatories at the Royal Military Repository in Woolwich, fellow of the Royal Society, and a Royal Artillery lieutenant, wrote this extensive text under Sabine's guidance in 1844. Riddell had considerable experience of magnetic science, having founded Toronto Magnetic Observatory. His *Magnetical Instructions* became the standard Royal Navy textbook for exploratory expeditions tasked with magnetical survey work. Here, Franklin's officers could read an account of the magnetic phenomena

---

[72] TNA, Sabine Papers, BJ3/19/153–4: Robert Were Fox to Edward Sabine (3 Mar. 1845), 154.
[73] RS, Sabine Correspondence, Vol. 2, D–J, MS/258/572: Robert Were Fox to [Edward Sabine] (8 Mar. 1845).
[74] RS, Sabine Correspondence, Vol. 2, D–J, MS/258/573: Robert Were Fox to [Edward Sabine], 22 Mar. 1845).

of declination, dip, and intensity, with the last defined in terms of the force exerted on the needle 'to return it in its position of equilibrium, or the power with which, when withdrawn from this position, it tends to revert to it'.[75] Along with mathematical formulae for calculating this quality, instructions to keep to Göttingen time, and details of Term Days—which were appointed to Fridays preceding the last Saturday of the months of February, May, August, and November, and the Wednesday nearest the twenty-first of the other eight months—*Magnetical Instructions* gave information on how to construct a temporary magnetic observatory. Riddell included prepared tables to be filled in during the voyage. Most importantly, however, Franklin's crew could read of the very latest prescribed regime of magnetic observations. With an azimuth compass to give absolute declination, the Fox dipping circle would give sea measures for dip and intensity. In good weather, the ship's magnetometer could confirm the Fox's readings.[76] As with earlier instructions, Riddell and Sabine demanded a routine of aiming the Fox needle north, taking three measurements, then turning it 180° to repeat these readings, giving observations with the circle facing both east and west. The needle could then be taken out, reversed, and the experiment repeated, giving a batch of results from which means for dip and intensity could be calculated. These changes in the direction of the instrument's face would

> eliminate the error arising from the deviation of the line joining the zero points of the circle from the horizontal position and the error arising from the deviation of the magnetic axis from the axis of form; the object of repeating these observations with the poles of the needles reversed is to eliminate the error arising from the non-coincidence of the centre of gravity with the centre of motion.[77]

To an aspiring magnetic observer on *Terror* or *Erebus*, there was no shortage of guidance in the unlikely event that Franklin and Crozier were not on hand to give instruction. While the expedition made its final preparations, Franklin's officers practised with their magnetic instruments at Woolwich. Fitzjames, in particular, spent a considerable period of time with Sabine, learning how to manage Robinson and Fox types alike, although he would certainly have encountered a Fox circle while on the earlier Euphrates Expedition. During the voyage, Fitzjames would take responsibility for the daily magnetic observations at sea on *Erebus*, while Franklin and Crozier would orchestrate the onshore elements of the investigation.[78]

---

[75] C. J. B. Riddell, *Magnetic Instructions for the Use of Portable Instruments Adapted for Magnetical Surveys and Portable Observatories, and for the Use of a Set of Small Instruments Adapted for a Fixed Magnetic Observatory; With Forms for the Registry of Magnetical and Meteorological Observations* (London: Her Majesty's Stationery Office, 1844), 3.
[76] Ibid. 7 and 84.   [77] Ibid. 84–5.   [78] Lambert, *Franklin*, 154–8.

On 19 May 1845, Franklin's expedition set sail for the Arctic from Greenhithe. During the early stages of the expedition, Fitzjames kept up regular reports for Sabine. On leaving Stromness on 3 June, he drafted an account of his first experiences with his Fox dipping needle. Writing off the northern coast of Scotland, the officer detailed of how he had taken 'a complete set of observations with Fox using both deflectors separately and together and the 3 weights'. While on land at Stromness, he had also trialled *Terror*'s unifilar magnetometer, but informed Sabine that although the deflections were satisfactory, the vibrations were not, due to a small bank of sandstone under the instrument affecting the needle. Fitzjames's problems did not end there: he was also struggling with *Erebus*'s Fox type. 'I am terribly disappointed with my Fox which is *rotten*,' he lamented; 'it is sunk so low in the gunwale stand that the line exactly comes on with the point of the needle at 70° and at 80°. I shall not be able to set it. So we *raise* the cable and we waited to try the Unifilar on it *at sea*. Crozier has been try[ing] vibrations with Hansteen at sea very well.'[79] These difficulties were soon to get worse. On 11 July, Fitzjames wrote the expedition's final magnetic report to reach England, detailing readings made during the voyage to the Whalefish Islands, off the west coast of Greenland. He included a table of measurements from 1 and 4 July, using two different needles, but the report was ominous. Fitzjames described the magnetic observations, 'or rather those we have not made', for the 'weather during our passage to this has been anything by pleasurable'. The Fox needle in particular was causing Fitzjames great trouble, as he explained:

> No observation either with Fox or anything else—Not that *my* observations with Fox can even be of much use for the instrument being cut only to degrees could give a reading of any intrinsic value—I imagine I appreciated to every 5′—and even if I could estimate to two minutes I would put no value on the results.[80]

His Fox dipping needle had not been finished and lacked the precision of earlier models. Back in London, a disappointed Sabine read of how 'the vertical is about ¼ of a mile from the limb—and in fact the instrument is rotten'. The poor workmanship depressed Fitzjames, preventing him from 'taking the great interest I otherwise should in the observations and Experiments I perceive might be made with a good instrument'.[81] He could not understand how the oversight had occurred. 'Why the Admiralty should have palmed off a rotten affair like this on us I am at a loss to determine,' wrote Fitzjames; 'I can only suppose that it was supposed my observations would not be of use as compared to Capt Crozier's.'[82] Be it the fault of its mechanic or of the Admiralty, Fitzjames was resolved that the

---

[79] TNA, Sabine Papers, BJ3/17: James Fitzjames to Edward Sabine (3 June 1845).
[80] TNA, Sabine Papers, BJ3/17: James Fitzjames to Edward Sabine (11 July 1845).
[81] Ibid.   [82] Ibid.

blame was not due to his own lack of skill with the Fox type. Despite this, he claimed that the 'Fox observations' he had taken 'present a tolerably good series of means'. Although lacking accuracy and precision, the needle was at least consistent, if unreadable.

In contrast, Fitzjames's Robinson circle excelled. On arriving at the Whalefish Islands, he had been anxious to calculate an index of error for his Fox dipping needle,

> and for this purpose got up on 'Boat H' and the 'Robinson' which is the most beautiful instrument – and has cost me some humble for I lost the final series of observations—having dropped my book on the hills and—in taking second series I observed that the needle A1 gave a much smaller result than the other. I consequently repeated the observations three times with all the needles and came to the conclusion that the Magnets supplied with the instrument were not strong enough to magnetize the needles properly.[83]

Fitzjames repeated his observations on 11 July, having remagnetized his needles with strong magnets that Lloyd had supplied. This achieved more equal results, but A2 still lacked magnetic intensity. For this, Crozier had provided Fitzjames with instruction, allowing him to make 'observations with Lloyd's intensity instrument which however good the theory of it, so much for Magnetics'.[84] Despite these problems, he thought the Robinson type 'beautifully delicate', giving a good measure of absolute dip to which the Fox could be compared and its error calculated. Bound for Lancaster Sound, west of Baffin Bay, Fitzjames promised Sabine he would persevere with the magnetic investigations, but the final news from *Erebus* was bordering on disastrous for the expedition's magnetic ambitions.

The final correspondence received from the expedition, dispatched from the Whalefish Islands, revealed similar magnetic anxieties on-board *Terror*. Crozier updated Ross on the magnetic observations, having found—like Fitzjames—his own Fox type difficult to handle due to rough weather. Crozier's time had 'been a good deal occupied with dips', of which he had sent Sabine an abstract, but explained how 'our passage across was very unfavorable for observing such constant heavy Sea and a great deal of wet—very uneasy days and C^d not manage azimuths'. This was turning out to be a very different expedition to Ross's voyage. Crozier complained of the unsteady motion of his dipping needle and regretted that he did not have his trusted instrument from the Antarctic Expedition. He was, nevertheless, under no obligation to make magnetic observations at sea, but rather to provide guidance, having found 'by the instructions that Fitzjames is appointed to superintend the Mag. Observations I will therefore take just so

---

[83] Ibid.   [84] Ibid.

much bother as may amuse, without considering myself as one of the staff'.[85] Crozier was feeling far less involved with the magnetic investigations than he had been with Ross and was beginning to doubt Franklin's leadership. To Ross, he confided that the expedition had arrived in the polar regions too late and would blunder into the ice, risking 'a second 1824', recalling Franklin's grim Mackenzie River experiences. 'How do I miss you,' he wrote to his ex-commander; 'I cannot bear going on board Erebus.' Franklin was kind, but would have *Terror*'s captain to dinner every night if Crozier consented. 'I wish you were here, I would then have no doubt as to our pursuing the proper course', wrote a distraught Crozier, disenchanted with both Franklin and the magnetic enterprise.[86]

On *Erebus*, Franklin had no such doubts, either in his own judgement or in the progress of his expedition's magnetic investigations. To William Edward Parry he boasted that his ships had three years of provisions and reported that the 'magnetic men were landed with their instruments' on 'the Boat Island', this being the same position that Parry had selected for his measurements back in the 1820s. Unaware of Crozier's growing isolation, Franklin was impressed with the zeal of his officers and reported that, 'Knowing what an excellent instructor and fellow worker Crozier was & will prove to Fitz-James, I have left the magnetic observations of the Erebus to the latter—who is most assiduous respecting them'.[87] Fitzjames's enthusiasm was heartening and he eagerly encouraged his fellow officers to take an interest in terrestrial magnetism. Yet this delegation to Fitzjames of the responsibilities for *Erebus*'s observations, under Crozier's superintendence, was in stark contrast to Ross's four years of constant management of his Fox type during his voyages to the Antarctic. The instruments, leadership, and observers all appeared to be struggling to fulfil the high expectations of the venture.

After the Whalefish Islands, Franklin's expedition sailed into the Arctic and was never seen or heard from again. Throughout 1846, fears grew back home that Franklin had met with calamity, with John Ross voicing concerns that the expedition had become trapped in the ice (Figure 6.8). In 1847, he advised the Admiralty to dispatch a rescue mission with all haste, but Parry and James Clark Ross felt such alarm to be premature. With the Admiralty slow to take action, Lady Franklin campaigned to initiate a search for her husband. Eventually, the Admiralty conceded that *Terror* and *Erebus* almost certainly required help and sent an expedition to find Franklin in 1848. Having promised his fiancé, Ann Coulman (1817–57), that he would retire from polar exploration, rejecting invitations to lead the 1845 expedition, James Clark Ross came out of retirement to command the Admiralty's rescue mission. A year later, in 1849, he arrived back in England

---

[85] Scott, Ms248/364/26: Francis Crozier to James Ross (9 July 1845).   [86] Ibid.
[87] Scott, Ms438/18/7: John Franklin to Edward Parry (10 July 1845).

Figure 6.8 Edwin Landseer's romanticized portrayal of the fate of Franklin's expedition, *Man Proposes, God Disposes*, completed in 1864 (Public Domain, 2023).

with no news of the missing expedition.[88] Between 1846 and the late 1850s, the search for Franklin gripped the nation. With support from Charles Dickens, Lady Franklin raised funds for further expeditions, with John Ross undertaking a voyage of his own in 1850. Like his nephew, he returned disappointed. In 1851, Stephen Pearce (1819–1904) painted a fictionalized study of what became popularly known as the 'Arctic Council', a group of leading naval authorities tasked with finding Franklin (Figure 6.9). This painting toured Britain to raise money for more ships to be sent in search of *Terror* and *Erebus*. Among the naval officers that Pearce depicted were Francis Beaufort, John Barrow junior, William Parry, James Clark Ross, and Sabine. On the wall behind them were portraits of John Barrow senior, Fitzjames, and Franklin. This was an august, if romanticized, gathering of patrons and practitioners of magnetic science brought together in the hunt for the lost expedition.[89]

The search for Franklin stimulated new scientific inquiry. On hearing of his relief effort in December 1847, Charles Darwin wrote to James Ross asking if the captain would procure him botanical samples during his search. Darwin requested that Ross 'preserve in spirits the northern species of cirripedia or Barnacles, noting the latitude under which found', from both coastal rocks and floating timber. This would be invaluable for his forthcoming monograph on cirripedia and, as 'Barnacles are so easily scraped off the rocks & put into spirits', it would cause Ross 'but little trouble'.[90] It was not just botany that Franklin's disappearance helped to promote, but magnetic science too. In his search for

---

[88] Elizabeth Baigent, 'Ross, Sir James Clark (1800–1862)', *Oxford Dictionary of National Biography*, https://www.oxforddnb.com/view/10.1093/ref:odnb/9780198614128.001.0001/odnb-9780198614128-e-24123, 23 Sept. 2004, accessed 24 Mar. 2020.

[89] Elizabeth Baigent, 'Arctic Council (act. 1851)', *Oxford Dictionary of National Biography*, https://www.oxforddnb.com/view/10.1093/ref:odnb/9780198614128.001.0001/odnb-9780198614128-e-95281, 28 Sept. 2006, accessed 24 Mar. 2020.

[90] RS, Sabine Correspondence, Vol. 2, D–J, MS/258/384: Charles Darwin to James Ross (31 Dec. 1847).

Figure 6.9 Stephen Pearce's *The Arctic Council Planning a Search for Sir John Franklin*, 1851. From left to right, Franklin, Fitzjames, and John Barrow are included as portraits on the wall, with George Back, Edward Parry, Edward Bird, James Clark Ross, Francis Beaufort (sitting), John Barrow junior, Sabine, William Hamilton, John Richardson, and Frederick Beechey (sitting) all present (Reproduced by permission of the National Portrait Gallery, 2023).

Franklin in 1852, Belcher conducted extensive magnetic observations, as did Leopold McClintock (1819–1907) on the *Fox*.[91] Along with new magnetic observations, Sabine hoped that, if McClintock found *Terror* and *Erebus* trapped in the ice, he would also retrieve the expedition's magnetic data. While the crew had almost certainly perished, Sabine was confident that there would

> be found in one or the other, or in both ships, the records of observations in at least two intermediate localities, in which the Expedition may have been stationary in different years; because both ships were furnished with the proper instruments, and some of the Officers had attended at Woolwich to practice with them before the Expedition sailed: to this it may be added, that both Sir John Franklin and Captain Crozier were strongly impressed with the desirability of making the observations, and letters are extant from both, written from Davis Strait, after they had sailed from England, expressing their full intention to set up the instruments wherever the ships should be detained for a sufficient period to give the observations value.[92]

While the British public worried that Franklin's expedition had been frozen in, Sabine saw such a misfortune as a magnetic opportunity, effectively transforming the two ships into a stationary magnetic observatory. Sabine emphasized the possibility of these records existing to McClintock so that he would be aware of 'the scientific importance' of locating them. Though it was unlikely the crews would be found alive, a retrieval of magnetic data would provide some philosophical consolation.

Magnetic relics were to be found, but not the data that Sabine hoped for. In April 1859, McClintock's expedition made a grim discovery, purchasing debris from Franklin's expedition from local Inuit. A month later they found the remains of an encampment, which appeared to have been a temporary magnetic observatory, at Cape Felix on King William Island. After following the coast south to Point Victory, McClintock's men then found the beach covered in the eerie remains of the last desperate stages of Franklin's expedition, including stores and clothes. It seemed the crews had abandoned their ships on 22 April 1848 and tried to escape on foot. Then, in June 1859, McClintock's crew discovered a cryptic note, hidden in the Ross Cairn that had been erected at Victory Point on King William Island in 1830, revealing that Franklin had perished on 11 June 1847, at which point Crozier had taken command. Near the cairn, McClintock's men also

---

[91] Edward Sabine, 'Results of Hourly Observations of the Magnetic Declination Made by Sir Francis Leopold McClintock, and the Officers of the Yacht "Fox," at Port Kennedy, in the Arctic Sea, in the Winter of 1858–9; and a Comparison of These Results with Those Obtained by Captain Rochfort Maguire, and the Officers of Her Majesty's Ship "Plover," in 1852, 1853, and 1854, at Point Barrow', *Philosophical Transactions of the Royal Society*, 153 (1863), 649–63, at 649.

[92] Ibid. 650.

found a 6-inch Robinson dipping needle, taken when the crews abandoned *Terror* and *Erebus* and not discarded until late into the disaster (Figure 6.10). Built in 1840 at his London workshop at 38 Devonshire Street, Portland Place, Robinson's brass dipping needle and silver vertical scale were to become iconic of the heroism of Franklin's sacrifice. Even as calamity beckoned, it appeared that magnetic observations remained a priority for Franklin's starving men. In taking the Robinson type, Crozier or Fitzjames had selected the superior circle for land readings, presumably leaving the disappointing Fox type on the ships.[93]

It took Sabine three months to 'prepare & unify' the magnetic apparatus McClintock retrieved, the delay causing some delay to Lady Franklin's plans to have them exhibited to raise fresh public enthusiasm to renew the search.[94] They were, however, ready for the private inspection of Queen Victoria at Windsor Castle by October 1859; the queen thought the surviving material and confirmation of

Figure 6.10 The Franklin Expedition's 6-inch Robinson dipping needle that McClintock retrieved, now in the National Maritime Museum at Greenwich (Author's Image, 2019).

---

[93] Lambert, *Franklin*, 61 and 279–80; this dip circle is currently held at the National Maritime Museum, Greenwich, collection reference AAA2223.

[94] Scott, Ms248/469/2: Edward Sabine to Lady Franklin (Sept. 1859).

Franklin's death would provide Lady Franklin with some comfort.[95] After their royal inspection, the remains of the expedition would take on a huge cultural significance and almost sacred value in Victorian society. Having appeared in the *Illustrated London News*, the objects went on exhibition, becoming emblematic of the heroic scientific civilizing search in the face of barbaric nothingness (Figure 6.11). Franklin's 'relics'—including instruments, a medicine chest, a prayer book, spectacles, and the dipping needle—at once linked Christianity, science, and imperialism for enthralled audiences.[96] For all that Franklin's expedition had been one of scientific exploration and magnetic inquiry, the result was that the Arctic increasingly became understood as a place of isolation and desolation in the Britain public imagination. The irony of the retrieved dipping needle was that the instrument, intended to reveal the complexities of Nature and magnetic phenomena, was conscripted within a heroic narrative in which Franklin and his crew had perished in an empty, desolate land.[97] McClintock's subsequent *The Voyage of the 'Fox' in the Arctic Seas* was a huge success, selling over seven thousand copies.

6.5. Conclusion

The expansion of the BMS accompanied a broader flourishing of the survey sciences within the Royal Navy. Parliament and the Admiralty tasked naval expeditions with observing an ever-wider range of natural phenomena, as the navy's officers made a growing contribution to the natural sciences. The study of terrestrial magnetism had, throughout the 1840s, been the dominant enterprise of British science, but it had fostered a naval culture of scientific investigation, from meteorology and botany to astronomy and geology. Given that Britain's expeditions had shown how they could effectively obtain magnetic measurements from around the world, it was clear that they could also provide data on a broad range of natural questions. Yet not all naval officers were natural philosophers. Acknowledging this limitation, Herschel produced the *Manual of Scientific Enquiry* for the Royal Navy in 1849. Intended as a thorough guide to the philosophical investigation of natural phenomena for travelling naval officers, Herschel organized an impressive array of authors, including Whewell on tides, Charles Darwin on geology, and Airy on astronomy. Effectively, this work was a handbook on how to perform global studies of natural phenomena. Unsurprisingly, magnetic science featured prominently. Herschel commissioned Sabine to redraft

---

[95] Scott, Ms248/469/4: E. Becker to Edward Sabine (26 Oct. 1859).
[96] Adriana Craciun, *Writing Arctic Disaster: Authorship and Exploration* (Cambridge: Cambridge University Press, 2016), 35–49.
[97] Ibid. 7–10.

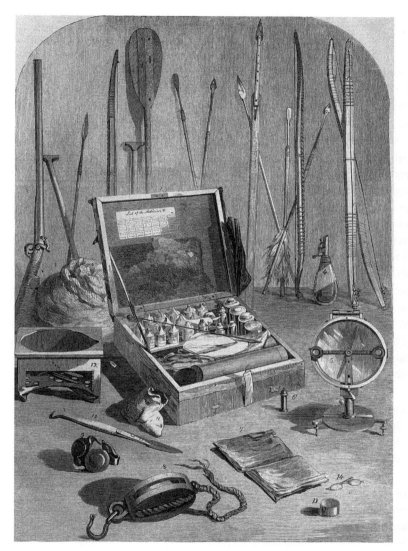

Figure 6.11 The *Illustrated London News*'s depiction of the relics recovered from the Franklin Expedition, including the Robinson dipping needle to the right (Author's image, 2023).

the Royal Navy's standard instructions for magnetic observations for the *Manual*'s chapter on 'Terrestrial Magnetism', which included a summary of the BMS to date. Acknowledging the Royal Navy's crucial role in the surveying of the Earth's magnetic field on the world's oceans, constituting three-quarters of the globe, Sabine praised 'the zealous and unwearying assiduity of British officers' in taking a lead in the scientific enterprise. He cited expeditions completed, including

Ross's to the Antarctic, Sulivan's to the Falklands, Belcher's on the *Samarang*, and the Niger venture of 1841, as well as voyages still underway or whose whereabouts were unknown, notably Franklin's.[98] As much as the 1849 *Manual* set out a programme of future expeditionary science to be undertaken, it was also something of a celebration of what had already been achieved through collaboration between the government, the Royal Navy, and Britain's leading philosophical authorities in the investigation of terrestrial magnetism.

Sabine's revised magnetic instructions took an eminent position in this catalogue of expeditionary sciences. He explained how declination—the angle of difference between geographic and magnetic north—could be mapped, producing isogonic lines that were valuable for navigation. Isodynamic and isoclinal charts for dip and intensity were useful in 'theoretical respects', with these three maps providing the 'basis of a systematic view of terrestrial magnetism as it manifests itself to us on the surface of the globe'.[99] The charts produced, in this way, appeared to represent Nature's magnetic system itself. However, Sabine warned his readers of the BMS's limitations, reflecting that all measures of intensity at sea were relative, giving only a ratio to absolute observations on land. The importance of known values for onshore stations was therefore paramount to examining how intensity varied relatively over space, but this still provided an important picture of how the Earth's magnetic force operated.[100] In this task, Sabine ascribed a central role to Fox's instrument. To give a ratio of force for a geographical position of a ship at sea, a base station had first to be measured,

> where the instrument has been landed and used in precisely similar observations to those made on board ship. The instrument is the well-known apparatus devised by Mr. Fox, which has contributed more to a knowledge of the geographical distribution of terrestrial magnetism than any other recent invention.[101]

For all that on-board observations of absolute magnetic intensity remained unobtainable, Fox's dipping circle was, Sabine alleged, of the foremost calibre for maritime readings. Fox's system of a grooved wheel and suspended weights allowed the needle to be deflected from its position of rest in the magnetic direction to a new position in equilibrium between the opposing forces of the Earth's magnetism and the deflecting weight.[102] A series of added weights and observations could produce a mean. Sabine surmised that 'the needles made by Falmouth artists, under Mr. Fox's own superintendence, have generally proved the most remarkable in preserving their magnetism unchanged for years and in all

---

[98] Edward Sabine, 'Terrestrial Magnetism', in John F. W. Herschel (ed.), *Manual of Scientific Enquiry; Prepared for the Use of Her Majesty's Navy: And Adapted for Travellers in General* (London: John Murray, 1849), 14–53, at 29–32.
[99] Ibid. 27.   [100] Ibid. 14.   [101] Ibid. 19–20.   [102] Ibid. 20.

DISCOVERY, DISASTER, DIPPING NEEDLE 225

Figure 6.12 An intensity experiment with a George-built Fox type, showing the suspension of weights by silk thread from the grooved wheel of the dipping needle. Weights could be added to the hanging hooks to coerce the needle to its original dip (Author's image, 2020).

climates'.[103] Sabine included an appendix on how to manage a Fox type, recommending the deflectors be applied at a 30° angle from the dip, and ½ grain weights be added gradually to coerce the needle back to its dip (Figure 6.12). As a general

[103] Ibid. 21.

routine, he advised three observations of dip, three observations with deflectors N and S to measure the angle of deflection from dip, and then both steps to be repeated with a second needle.[104] Fox's dipping needles therefore appeared in Herschel's *Manual* as the most eminent magnetic devices available for expeditionary science.

Between Ross's departure for the Antarctic and the publication of Herschel's *Manual*, a decade had passed. Expeditions carrying Fox's dipping circles had criss-crossed the world from the North to the South Pole, throughout the Far East and Indian Ocean, on the Atlantic and Pacific, around Africa and Europe, and along the Earth's equatorial regions. The use of these instruments, so vaunted in Sabine's account of the BMS's progress, had initiated diverse cultural exchanges. Due to the necessity of land-based measurements, expeditions explored new lands, which they would have had no other interest in exploring beyond this scientific imperative. The building of temporary observatories in tropical and polar wildernesses clearly had a symbolic value, appearing to be 'civilizing' acts. The study of terrestrial magnetism and the use of Fox's dipping needles fostered new exchanges between naval expeditions and diverse societies around the world, from the townsfolk of Manila to Japanese dignitaries. On occasions, such as when encountering pirates off Borneo, this interaction was hostile, but some audiences showed interest in Fox's magnetic devices, exhibiting knowledge of magnetic phenomena. Fox and Sabine's system of magnetic data production had, inadvertently, initiated new multicultural interactions. It had taken a central part in the rapidly escalating processes of European colonialism.

---

[104] Ibid. 38–42.

7
The Twilight of Cornish Science and the Systematization of Oceanic Navigation, 1850–1907

By the 1850s, there were few who could rival Robert Were Fox's authority in the science of terrestrial magnetism. His Falmouth-built instruments had achieved international acclaim, while his experimental skill and experience in the study of magnetic phenomena had been at the centre of the world's foremost scientific enterprise for over a decade. However, after the flurry of naval surveys that followed Ross's 1839–43 Antarctic Expedition, the heat of the BMS was, by the 1850s, dissipating. With the investigation largely complete, Sabine found himself swamped with data and completely occupied with publishing the enterprise's results. Throughout this, Fox provided advice and interpretations over measurements and instruments, but the industrious work behind building and regulating a global network of dipping needles and observers was practically complete. As the BMS's attention turned to calculations, corrections, and charting, Falmouth's position as a centre of British magnetic science diminished: now came the turn of Sabine and the number crunching of his Woolwich assistants. This work would not be concluded until 1870 but by then it was very clear that the Earth's magnetic properties changed in such a manner as to defy mapping. Yet the inquiry also had two more tangible outcomes. The first was to establish a link between solar flares and terrestrial magnetism. Sunspots were a well-known phenomenon, but Sabine's investigation had revealed coincidences between solar activity and the fluctuations of the Earth's magnetism: solar spots, releasing immense heat and charged particles across the universe, increased the intensity of magnetic phenomena. There was another product of all this data collection: around Britain's empire, a network of observatories had been established. Here was a system, under government control, which could be adjusted for the scientific analysis of a diverse range of natural phenomena, from astronomical events to oceanic currents and weather patterns. This was an unprecedented resource for the state-supported scrutiny of Nature. In Victorian Britain, these observatories took on more than just a purely scientific character: they embodied imperialistic notions of progress and modernity. At the Magnetic Congress of 1845, Herschel heralded the magnetic survey as a great civilizing mission. A scientific observatory, he declared, such as the ones at Cape Town, Singapore, or Hobart, was essential for

An Empire of Magnetism: Global Science and the British Magnetic Enterprise in the Age of Imperialism. Edward J. Gillin,
Oxford University Press. © Edward J. Gillin 2024. DOI: 10.1093/oso/9780198890959.003.0008

any nation with aspirations to appear 'civilized'.[1] For newly established colonies, he continued, such institutions would mark them out as places of enlightenment.

These fixed places of magnetic inquiry were of little value to Fox: the cessation of expeditions reduced demand for his robust instruments, which had been so carefully crafted for survey work. Nevertheless, Fox's study of terrestrial heat, electrical currents, and terrestrial magnetism had come a long way since his mine experiments in the 1820s. Through his partnership with Sabine, he had transformed this Cornish science into a global investigation which mobilized the naval and colonial resources of Europe's foremost imperial power. His knowledge and instruments had underpinned a worldwide experiment in which the state had taken an unprecedented role. Much of this had built on the lessons learnt during Sabine's BAAS magnetic survey of the British Isles, back in the 1830s. In many ways, this had provided a model that could be scaled up and organized into a global study. Yet before the end of the 1850s, there was to be one final Indian summer for Cornish natural philosophy, as Britain's leading magnetic specialists returned to the subject of the original BAAS survey. This would provide Fox and his dipping needles with a short-lived renaissance, but it would also reveal how out of touch Cornwall's leading magnetic authority had become within the space of just a few years. In 1845, this was all to come: Fox first faced the challenge of restoring confidence in his dipping circles, following the failure of the Franklin Expedition.

7.1. Trusting Wilton

The confirmation of Franklin's death might have provided closure for Lady Franklin, but at Penjerrick the sadness of this news was amplified by the growing realization of how disastrous the expedition had been for Fox's dipping circle.[2] The final voyage of HMS *Terror* and HMS *Erebus* was more than just a personal tragedy for the Fox family: it was a catastrophe for Falmouth's scientific instrument makers. With George having moved out of Falmouth and retired from constructing entire dipping circles, Fox and Sabine desperately wanted a new mechanic. Given Fitzjames and Crozier's lamentable experience on the Franklin Expedition, it was unlikely that the Admiralty would return to George, even if he had not retired. With the final magnetic observations from *Terror* and *Erebus* lost, it remained unknown just how poorly built George's final Fox type had been, but the reports from the Whalefish Islands were alarming enough. In August 1846, Fox proposed a new artisan altogether. Fox assured Sabine that, should he require

---

[1] John Herschel, 'The President's Address', *The Athenaeum*, 921 (21 June 1845), 612–18, at 614.
[2] Horace N. Pym (ed.), *Memoirs of Old Friends: Being Extracts from the Journals and Letters of Caroline Fox of Penjerrick, Cornwall from 1835 to 1871* (London: Smith, Elder, & Co., 1882), 321.

new dipping needles, they could still be 'well made by an ingenious instrument maker in this neighbourhood'. He recalled that in the early 1830s, before his partnership with Jordan, he had been able to have dipping needles built for his work on the British Isles survey by bringing together artisans with different skills. A local mechanic had produced the magnetic needles while a Falmouth watchmaker crafted the pivots in which to balance them. Fox could return to this system of divided labour, should nobody of George or Jordan's skill be found. However, he confidently informed Sabine that he had a craftsman in mind who 'makes superior instruments for the miners, & has a good dividing angle made by himself'.[3] Sabine was cautious, but Fox reiterated the credentials of the new artisan in December 1847, declaring that there 'is an instrument maker in this neighbourhood who is quite approved, & I think that we could together concoct a good dip & intensity apparatus. He succeeds well with theodolites, & I can balance the needles'.[4] His name was William Wilton (1801–60).

Sabine agreed to give Wilton a chance, requesting a new Fox type early in 1848. Though anxious that he could not complete one within the specified six weeks, Wilton was delighted. Fox shared in Sabine's fears, especially if he had 'as heretofore, to reject some of the pivots'. Despite this, by February he was pleased to see the pivots of the new circle's third magnetic needle 'promise well even beyond my expectations'.[5] Fox suggested to Sabine that the instrument be sent out for Dayman's use on HMS *Rattlesnake*, presently off the coast of Australia. Wilton had already overhauled Stanley's previous dipping needle the preceding April, which he had returned to the Admiralty in October 1848. This 'was one of the very first; & made by Watkins & Hill', having two pillars to support the axle, instead of one, as newer models possessed. Fox thought it unlikely 'that the needle will now give a correct indication of dip, but it may be adjusted by means of the instrument...should it be required'.[6] Wilton was not only building new circles, but also maintaining those of earlier construction.

Fox's dipping needles had secured such international attention that he was also receiving regular foreign orders. In October 1849, the French explorer Antoine Thomson d'Abbadie (1810–97) was in London in search of magnetic instruments. Eager for an alternative to the 'cumbersome' Gambey types in use in France, d'Abbadie had enquired at Watkins & Hill about a more portable Fox dipping circle, but they demanded '31£ and a month's delay' for the construction of a new instrument. After meeting Lloyd, d'Abbadie wrote to Fox, having learnt 'that your

---

[3] RS, Sabine Correspondence, Vol. 2, D–J, MS/258/574: Robert Were Fox to [Edward Sabine] (20 Aug. 1846).
[4] RS, Sabine Correspondence, Vol. 2, D–J, MS/258/576: Robert Were Fox to [Edward Sabine] (28 Dec. 1847).
[5] RS, Sabine Correspondence, Vol. 2, D–J, MS/258/577: Robert Were Fox to [Edward Sabine] (26 Feb. 1848).
[6] RS, Sabine Correspondence, Vol. 2, D–J, MS/258/578: Robert Were Fox to [Edward Sabine] (28 Oct. 1848).

instrument is only made by a Falmouth artist under your own immediate direction'. Setting out his credentials, he explained how,

> In 1836 I made a voyage to Olinda in the Brazils for the express purpose of observing daily variations of the horizontal needle which I registered every 20 minutes day & night for two months. I have since observed in Italy, Egypt & Abyssinia. I mention this not to boast of my magnetic exploits but merely to show you that in forwarded my present request you will not throw away your kindness on an observer totally ignorant or inactive. I wish to multiply observations of the dip & intensity in France and diffuse there a knowledge of your instrument which is all but unknown on the other side of the channel and would I am sure, if very portable, render good service.[7]

D'Abbadie regretted not having a Fox type on his 1837 Ethiopian Expedition, 'as I could have determined several positions of the magnetic Equator in the heart of Africa while the unwieldly dimensions of Gambey's *bounole l'inclinaison* forced me to leave it at Gondar in Abyssinia'.[8] Since then, he had visited Barrow, on Oxendon Street, who had shown him two Fox types, but these were very large and not for sale.

D'Abbadie now planned 'a magnetic tour' through southern France and northern Spain, and wanted a small Fox type and explanatory pamphlet with which to complete the task. He was impressed that Fox's instrument measured both intensity and variation, but confessed he had not seen 'this admirable improvement in the old instrument of yours lying at Barrows'.[9] Fox replied in early November, despatching one of his dipping circles to London by mail coach soon after, but d'Abbadie had already left London. He was extremely grateful to Fox, and had his landlady on Manchester Street forward the device to him in Paris. Given that 'the instrument which you destine for me cannot have a telescope at the back for observing the Variation', he requested Fox's

> instructions in order to get it added here, and, according to your experience, the possible error incurred by determining that difficulty of ascertaining the azimuths where the dipping needle is vertical, at least within the Tropics. You mention 'many checks and modifications in order to ensure the truth of the results' & I would request you to mention them to me, that I may not remain an unworthy & partially ignorant possessor of your instrument.[10]

---

[7] RS, Robert Were Fox Papers, MS/710/1: Antoine Thomson d'Abbadie to Robert Were Fox (29 Oct. 1849).
[8] Ibid.   [9] Ibid.
[10] RS, Robert Were Fox Papers, MS/710/2: Antoine Thomson d'Abbadie to Robert Were Fox (11 Nov. 1849).

Before he left London, he had also met Sabine and

> learnt from him that you had already made the magnetic tour which I projected through the South of France. However as I may be able to observe at new points within the marker of your magnetic net, I would request you to inform me in which paper those observations of yours have been published.[11]

Delighted with Fox's swift delivery of a new circle, d'Abbadie paid the £16 invoiced. The potential crisis in confidence that could have undermined Fox's instruments following the disastrous Franklin Expedition appeared to have averted.

Fox's reputation remained high into the 1850s. When the Swedish government looked to equip the frigate *Eugenie* for magnetic survey work in 1852, they requested a Fox type from Sabine, who duly conveyed the order to Falmouth.[12] Promoting Sweden's trade, *Eugenie* would become the first Swedish warship to circumnavigate the globe. Three years later, William Scoresby travelled to Australia on an iron ship to examine how the vessel's hull affected its on-board compasses. Ever eager to advance his practical understanding of how the Earth's magnetic field interacted with a ship's iron, the heroic whaling captain acquired a Fox circle and needles from the Admiralty. All appeared well, except that the intricate weights were missing. He immediately wrote to Fox, ordering a new set be made. Sailing on the *Royal Charter*'s maiden voyage to record the change of its magnetic character over time and space, Scoresby went on to observe a drastic alteration in the manner in which the vessel's iron acted on its compass, but died before he could publish his findings. Scoresby's use of Fox's instrument, as well as requests from d'Abbadie and the Swedish government, testified not only to the Cornishman's credentials but also to the ways in which his wares transcended the BMS's remit, assisting both private and international magnetic enterprise.

At the same time as producing new magnetic devices, Fox and Wilton continued to check and repair instruments, providing a regular service for Sabine, who continued directing global magnetic observations until 1854.[13] Fox was even consulted over the nation's foremost engineering ventures. In September 1855, an anxious Isambard Kingdom Brunel sought Fox's advice on a magnetic problem concerning his enormous iron steamship, the 'Leviathan', which was under construction on the Isle of Dogs. Having already seen his previous ship, the *Great Britain*, run aground amid rumours of compass error resulting from magnetic disruption, Brunel was desperate to avoid a repeat with what would eventually become the SS *Great Eastern*. Was there 'any application', Brunel enquired, that

---

[11] Ibid.
[12] RS, Robert Were Fox Papers, MS/710/90: Edward Sabine to Robert Were Fox (26 June 1852).
[13] RS, Robert Were Fox Papers, MS/710/91: Edward Sabine to Robert Were Fox (15 Dec. 1855).

Fox knew of, which could be implemented as 'the work proceeds' and which would neutralize 'the changes on the current of local influence and the promotion of the plan of minimum disturbance'. Although he reported that the 'body of my big ship is completed or nearing so', Brunel vowed to adopt any measures Fox thought prudent.[14] This engineering enquiry was somewhat vague but, nevertheless, demonstrated that Fox had ascended to a trusted position in matters magnetic, including those related to the perilous task of building iron ships. As engineers and naval architects like Brunel and John Scott Russell turned out increasing numbers of iron leviathans, Fox provided instruments and scientific experience that promised to make oceanic navigation safe in this new age of naval architecture.

Nevertheless, Wilton struggled to build the same degree of trust that his predecessors, George and Jordan, had attained. As late as March 1850, Fox sent a letter to Sabine warning that although Wilton had sent his latest instrument to Woolwich, he had done so before it could be checked at Rosehill or Penjerrick. The weights in the box had 'not therefore been corrected by me which is necessary, as Wilton's balance is not sufficiently delicate for this purpose', explained Fox, promising to send a proved set of weights within the week.[15] Further concerns surfaced in 1853 when Fox suggested Wilton's practice of polishing the terminal axles of his needles had been a mistake. For all this, Wilton and Fox continued meeting Sabine's orders well into the 1850s.[16] At the Great Exhibition of 1851, it was a Wilton-built Fox dipping circle, alongside one of Wilton's theodolites for underground mine surveying, that represented Falmouth's contribution to British science and instrumentation (Figure 7.1). However, this auspicious occasion did mark something of a climax to Cornwall's part in Victorian magnetic science.

7.2. Reporting the BMS

Since Ross had sailed for the South Pole in 1839, the BMS had gone from strength to strength. After the government's initial investment, Sabine secured renewed funding from Parliament in 1842 and again in 1845. Although far from enthusiastic, Herschel requested that Prime Minister Robert Peel extend the investigation until 1848, asserting that the project was a valuable use of taxpayers' money.[17]

---

[14] RS, Robert Were Fox Papers, MS/710/21: Isambard Kingdom Brunel to Robert Were Fox (27 Sept. 1855).

[15] RS, Sabine Correspondence, Vol. 2, D–J, MS/258/579: Robert Were Fox to [Edward Sabine] (9 Mar. 1850).

[16] TNA, Sabine Papers, BJ3/19/156–7: Robert Were Fox to Edward Sabine (23 Nov. 1853), 156–7.

[17] John Cawood, 'The Magnetic Crusade: Science and Politics in Early Victorian Britain', *Isis*, 70/4 (Dec. 1979), 493–518, at 516.

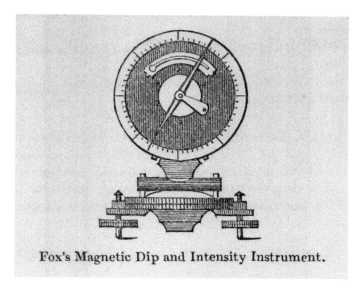

Fox's Magnetic Dip and Intensity Instrument.

Figure 7.1 Fox's Wilton-built dipping needle that went on display at the 1851 Great Exhibition (Author's image, 2021).

It did, however, take extensive time and labour to convert taxes into magnetic data and then reduce these into publishable results. Working under an avalanche of incoming observations, Sabine and his Royal Artillery assistants of the Magnetic Department, established at Woolwich Arsenal in 1841, converted the reports of naval and army officers and magnetic practitioners from all over the world into the printed maps and tables that constituted the Royal Society's extensive series of 'Contributions to Terrestrial Magnetism', all published in the *Philosophical Transactions*. Although magnetic observations officially ended in 1854, the Magnetic Department operated until 1857. Along with reducing observations, producing means for dip, intensity, and declination, making corrections for temperature variations, local disturbances, user errors, and instrumental failings, Sabine orchestrated an intensive bureaucratic machine in which his hard-worked artillery officers processed the incoming magnetic data (Figure 7.2).[18]

Yet for all this apparent progress, the BMS had been dogged throughout by controversy and division. With Lloyd somewhat isolated at Dublin, Sabine had taken almost complete control over the processing of survey results from the start, thanks to the support of Richard Hussey Vivian (1775–1842), Master

---

[18] Matthew Goodman, 'Follow the Data: Administrating Science at Edward Sabine's Magnetic Department, Woolwich, 1841–1857', *Notes and Records*, 73/2 (2019), 187–202; Matthew Goodman, 'From "Magnetic Fever" to "Magnetical Insanity": Historical Geographies of British Terrestrial Magnetic Research, 1833–1857', PhD thesis, University of Glasgow, 2018, 200–37.

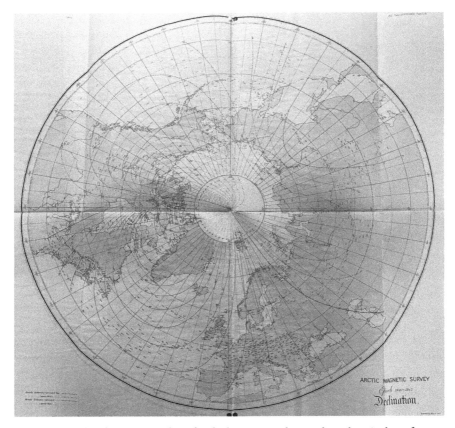

Figure 7.2 Sabine's magnetic chart for declination in the northern hemisphere for 1840–5, published in 1872 (Author's image, 2021).

General of the Ordnance, who provided Woolwich staff for calculations.[19] When the question of a second government grant arose in 1842, Sabine was adamant the BMS be continued, but Herschel was critical of such endless numerical gathering, laconically describing this pursuit of data for charts as 'chartism', invoking the politically radical Chartist movement. At the same time, when Herschel asked Gauss for advice over how to make the collection of data more efficient, the German magnetic specialist expressed doubts over the accuracy of Sabine's results. This strain in German–British communication reflected what was a growing bone of contention between Sabine and Herschel over the extent to which the BMS should cooperate with its Continental counterparts. When Wilhelm Weber delivered a report to the BAAS on the renewal of the BMS in 1845, his stressing of the importance of international standardized practices for precise measurements

---

[19] Cawood, 'The Magnetic Crusade', 513–14.

raised further questions over European collaboration.[20] These anxieties were reflected in the low attendance of Continental delegates at the BAAS's Magnetic Conference later that year.

Herschel was equally dismayed at Sabine's dictatorial manner over the question of where to locate the scheme's new London-based magnetic observatory. In 1841, Sabine persuaded the Royal Society to finance the establishment of a magnetic observatory at Kew, which was to become the central geophysical observatory for the British Empire.[21] Originally built for King George III in 1769, the building had fallen into disuse, but Sabine oversaw its purchase and revival as a site of specialized magnetic research. After a year of Royal Society patronage, the BAAS took over the new institution in 1842, as it expanded into the broader study of meteorology and, eventually, solar astronomy.[22] But this brought Sabine into direct rivalry with the Astronomer Royal, Airy, who had only just established a Magnetic and Meteorological Department at the Royal Observatory. Since his appointment to Greenwich in 1834, Airy had promoted magnetic research, obtaining advice from Lloyd, James Forbes, and Gauss concerning instruments and practices. Gauss responded enthusiastically, recommending a Göttingen-built magnetometer and that Greenwich join the Magnetischer Verein's regime of coordinated observations. Airy then requested £600 from the Admiralty towards building a magnetic observatory.[23] However, the grant was not approved until April 1837, and further delays in construction meant that by 1841 Greenwich had not secured the magnetic prominence that the Astronomer Royal had hoped for. Similarly, Sabine was reluctant to delegate authority over the scheme's magnetic observatory to Airy, who would, at the 1845 Magnetic Conference, oppose the BMS's renewal. Despite all this, though historians have emphasized Sabine's autocratic control over the BMS, his partnership with Fox demonstrated the considerable extent to which he relied on collaboration throughout the enterprise.

By the 1850s, with the survey work largely complete, Sabine turned to the question of analysing all this extensive magnetic data. In particular, it was the influence of the Sun on the Earth's magnetic field that drew his attention. After publishing images of the movement of needles throughout the day at observatories in Toronto, Hobart, Cape Town, and St Helena in 1851, Sabine demonstrated how

---

[20] Diane Greco Josefowicz, 'Experience, Pedagogy, and the Study of Terrestrial Magnetism', *Perspectives on Science*, 13/4 (Winter, 2005), 452–94, at 453–4 and 487.

[21] Cawood, 'The Magnetic Crusade', 514.

[22] Lee T. Macdonald, *Kew Observatory and the Evolution of Victorian Science* (Pittsburgh: University of Pittsburgh Press, 2018), 24–49; Lee T. Macdonald, 'Making Kew Observatory: The Royal Society, the British Association and the politics of early Victorian Science', *British Journal for the History of Science*, 48/3 (Sept. 2015), 409–33.

[23] Lee T. Macdonald, 'The Origins and Early Years of the Magnetic and Meteorological Department at Greenwich Observatory, 1834–1848', *Annals of Science*, 75/3 (2018), 201–33, at 219–22.

crucial solar activity was to magnetic phenomena.[24] Since 1843, he had been convinced that magnetic storms would eventually be found to 'have a character of *periodicity*'. The BMS had, by 1851, demonstrated that such magnetic disturbances were far from irregular.[25] Between 1843 and 1848, the BMS had measured a progressive increase in the mean diurnal range for declination, dip, and intensity. The cause was uncertain, but in 1852 Sabine posited that he could not overlook 'the possibility of the discovery of some cosmical connection which may throw light on a subject as yet so obscure'.[26] Asserting that the Sun was clearly the '*primary* source of all magnetic variations which conform to a law of local hours', it occurred to Sabine that the Sun might also be the cause of variations over a longer period of time. The phenomenon of solar spots was already well known. The earliest recorded observation of this activity was in 28 BCE in China, while in Europe Galileo had first noticed dark spots on the Sun's surface in 1610. William Herschel (1738–1822) had continued this study in the 1770s, believing that sunspots increased the output of solar heat on the Earth. He later claimed that there was a correlation between the number of sunspots and historic wheat prices, suspecting that increased solar activity enriched agricultural produce with warm weather and good harvests. Though poorly received when Herschel first posited this idea in the 1800s, his son, John, took up the theory and proceeded to investigate solar phenomena during the 1830s at the Cape of Good Hope. In his 1847 *Results of Astronomical Observations*, he speculated on the nature of sunspots and hoped that new photographic apparatus would advance knowledge of this subject.[27] Sabine was equally aware of this broader cosmological question.

Looking for coincidences between the results of the magnetic survey and solar phenomena, he found that the epochs of solar-spot variation that German astronomer Heinrich Schwabe (1789–1875) had previously observed, were 'absolutely identical with those which have been here assigned to the magnetic variations'.[28] Sabine's wife, Elizabeth Juliana Leeves (1807–79), had recently translated Alexander von Humboldt's *Cosmos* (1845), which included Schwabe's claims, made in 1843, of a ten-year solar-spot cycle, with 1848 identified as a solar maximum. This corresponded precisely with the extremity of magnetic disturbance observed throughout the BMS. Furthermore, 1843 appeared to mark the points both of

---

[24] Edward Sabine, 'On the Annual Variation of the Magnetic Declination, at Different Periods of the Day', *Philosophical Transactions of the Royal Society of London*, 141 (1851), 635–41.

[25] Edward Sabine, 'On Periodical Laws Discoverable in the Mean Effects of the Larger Magnetic Disturbances', *Philosophical Transactions of the Royal Society of London*, 141 (1851), 123–39, at 123–4.

[26] Edward Sabine, 'On Periodical Laws Discoverable in the Mean Effects of the Larger Magnetic Disturbances: No. II', *Philosophical Transactions of the Royal Society of London*, 142 (1852), 103–24, at 120–1.

[27] Lee T. Macdonald, '"Solar Spot Mania": The Origins and Early Years of Solar Research at Kew Observatory, 1852–1860', *Journal for the History of Astronomy*, 46/4 (2015), 469–90, at 471–2.

[28] Sabine, 'On Periodical Laws Discoverable in the Mean Effects of the Larger Magnetic Disturbances: No. II', 121.

least magnetic disturbance and of least solar-spot activity.[29] Given this coincidence, Sabine speculated that the gaseous envelope of the Sun appeared to give 'rise to sensible *magnetical* effects at the surface of our planet, without producing sensible *thermic* effects'.[30] He therefore believed that he had pinpointed the physical cause of magnetic storms and, with it, the BMS could claim a prized philosophical discovery. Sabine's enterprise appeared to have vindicated John Herschel's belief that sunspots were related to the aurora borealis and suggested that electric currents in space were 'auroralized' by the Sun's outer atmosphere. Here was a sparkling example of how state-funded scientific inquiry could advance philosophical knowledge of Nature. At Kew Observatory, a rigorous programme of solar investigation followed, with the building of a new solar telescope in response to calls to investigate solar flares. Begun in 1854, it came into use in 1858, as the science of terrestrial magnetism morphed into the wider cultivation of solar astronomy.[31] In many respects, there was a neat circularity to this for Fox. Although interested in magnetic phenomena throughout the 1820s, his curiosity had grown through his observation of the aurora borealis at Falmouth between 1830 and 1831. His instrument had helped demonstrate how such aesthetic marvels of Nature, witnessed on Earth, were the products of the Sun's activity far away across the galaxy. By the 1850s, the magnetic data obtained with his dipping needles was taking Britain's leading natural philosophers back to the examination of solar phenomena.

7.3. The British Isles Survey of 1857

There was to be one final act in the drama of Sabine's magnetic programme. In 1857, the BAAS appointed a commission to evaluate a proposed continuation of the global system of magnetic observations. Reporting in December, Sabine summed up the committee's conclusions. The BMS had, he asserted, been a great success, especially in having demonstrated that the 'sun's regular action on the magnetism of the globe is determined by a law of no small complexity and intricacy' and traced this relationship 'with precision and certainty'.[32] With the discovery of a correlation between solar spots and terrestrial magnetic phenomena, Sabine was sure that a new network of observatories should be established to renew the BMS. The BAAS committee recommended that Vancouver Island, Newfoundland, the Falkland Islands, and Pekin be adopted as new sites of fixed

---

[29] Ibid.; Macdonald, '"Solar Spot Mania"', 474.
[30] Sabine, 'On Periodical Laws Discoverable in the Mean Effects of the Larger Magnetic Disturbances: No. II', 122.
[31] Macdonald, '"Solar Spot Mania"', 475–8.
[32] Edward Sabine, 'Report of the Joint Committee of the Royal Society and the British Association, for Procuring a Continuance of Magnetic and Meteorological Observatories', *Proceedings of the Royal Society of London*, 9 (1857–9), 457–74, at 459.

magnetic observation. The Falklands would offer a rich comparison to the old observatory at the Cape of Good Hope, while Pekin would effectively 'complete and carry round the globe the chain of northern middle latitude stations', including those in Toronto and Russia.[33] Despite Sabine's continuing global aspirations, British enthusiasm for such magnetic enterprise was exhausted. The zeal of the 1830s and 1840s was long gone. Far from orchestrate a renewed global investigation, Sabine was already struggling to organize a repeat of the survey of the British Isles.

At the BAAS's 1856 meeting in Cheltenham, there was much self-congratulating among Britain's magnetic promoters. It was noted that the next meeting, in 1857, would mark the twentieth anniversary of the BAAS's original domestic magnetic survey. Reflecting on this, Sabine vainly boasted that this project represented 'the first complete work of its kind planned and executed in any country as a national work, coextensive with the limits of the state or country, and embracing the three magnetic elements'. Evidently, he considered the start of the BMS to be not 1839 but the British Isles survey two years earlier. Not short of nationalistic hubris, Sabine lauded the subsequent mobilization of Britain's naval, military, and colonial resources to extending the British survey, the example of which 'was speedily followed by the execution of similar undertakings' around the world. The magnetic maps, charting the isoclinal and isodynamic lines of terrestrial magnetism, marked the fulfilment of the venture's objectives. Yet even more remarkable than these apparently invaluable documents, was the worth of the data collected in itself. Beyond practical matters of navigation, such figures would help natural philosophers to work out the laws and cause of the '*secular change* in the distribution of the earth's magnetism, perhaps the most remarkable of the yet unexplained natural phenomena of the globe'.[34]

Sabine had, in fact, missed the Cheltenham meeting, instead visiting the Continent for his health. But if the BAAS attendants were eager to celebrate the success of the British Isles survey, it was not lost on anyone that its five investigators were, perhaps surprisingly, all still alive. With Sabine at work at Woolwich and Kew, Lloyd ensconced in Trinity College Dublin, Phillips now at Oxford, Ross retired to the country, and Fox as omnipotent as ever at Falmouth, this endurance seemed fortuitous. Seizing on this shared longevity, the BAAS appointed the five ageing philosophers to repeat the magnetic survey of the British Isles. Lloyd was keen, promising to perform the Irish element of the task, as was Sabine, who rushed back to England on learning of the commission.[35] Getting Ross, Phillips,

---

[33] Ibid. 464–6.
[34] Edward Sabine, 'Report on the Repetition of the Magnetic Survey of England, Made at the Request of the General Committee of the British Association', in *Report of the Thirty-First Meeting of the British Association for the Advancement of Science* (London: John Murray, printed by Taylor & Francis, 1862), 250–79, at 250–1.
[35] Ibid. 251.

and Fox on board was, however, another matter. Although he initially agreed to contribute, Fox played only a small part in the proceedings. In 1855 his son, Barclay, died, followed shortly by Barclay's wife, and then Fox's own wife, Maria, in 1858. So within the space of three years, Fox found himself alone with his daughters Anna-Maria and Caroline at Penjerrick. Ross was equally inactive on the project. The Arctic veteran and recently promoted rear-admiral never recovered from the death of his wife in 1856 and lacked the energy of his early career. As for Phillips, he had wanted to help with the survey but, as in 1837, found his role constrained by instrumental problems. Having become keeper of Oxford's Ashmolean Museum in 1854 and director of the new University Museum in 1857, he had remained interested in magnetic science.[36] Although he was happy to partake in the BAAS's new magnetic survey of the British Isles, he ultimately did little. Ross instructed Phillips to take magnetic observations in northern England, but when the Oxford geologist insisted he use his old instruments from the 1830s, Sabine vetoed Ross's request. As a result, Phillips's contribution was confined to dip and intensity observations in York throughout the autumn of 1857, made with obsolete magnetic needles. When these were compared to Ross's dip results for the same location, Phillips's differed by ten minutes and were clearly unacceptable. Sabine omitted Phillips's measurements from the final report for the BAAS.[37]

Beset by personal tragedies, a lack of commitment, and dated instruments, the new BAAS survey made little progress. To complete the Scottish part of the investigation, Sabine appointed John Welsh (1824–59) of Kew Observatory, but he did not last long under the rigours of survey work, abandoning the project in 1858 due to an incurable lung disease. His death a year later had, Sabine speculated, been 'accelerated it is feared by his too persistent exposure in the second year of the Scottish survey'.[38] Kew's meteorologist, Balfour Stewart (1828–87), took over Welsh's work, as yet another name was added to the tally of casualties from Britain's global magnetic enterprise. Survey work was, it seems, a perilous business not only in the Earth's tropical and polar regions, but also much closer to home in the Scottish Highlands. It was soon clear that things were going badly. Lloyd wrote to Fox on 29 April 1857 with news that Sabine had called a meeting for 14 May at Kew Observatory for those involved 'in the new magnetic survey of the United Kingdom'. Both Lloyd and Sabine hoped Fox could attend, as the project was lacking direction and might have to be deferred until 1858.[39] The limited participation of Ross, Phillips, and Fox was here finally acknowledged, and Sabine resolved that he himself would have to perform the majority of the work.

[36] Jack Morrell, 'Phillips, John (1800–1874)', *Oxford Dictionary of National Biography*, https://www.oxforddnb.com/view/10.1093/ref:odnb/9780198614128.001.0001/odnb-9780198614128-e-22163, 23 Sept. 2004, accessed 24 Mar. 2020.
[37] Jack Morrell, *John Phillips and the Business of Victorian Science* (Ashgate: Aldershot, 2005), 298–9.
[38] CUL, Add.9942/33: Sabine, 'Report on the Repetition of the Magnetic Survey', 251.
[39] RS, Robert Were Fox Papers, MS/710/66: Humphrey Lloyd to Robert Were Fox (29 Apr. 1857).

Following this renewal of the survey's objectives, he began magnetic observations in the summer of 1858.

With Sabine having completed measurements for three locations in South Wales, and with five to six more planned to complete the region within a month, Fox sent dip observations from eighteen stations in England. For all his zeal, age was also catching up with Sabine, who confessed to Ross that for observatory and survey readings, he was now required 'occasionally to use spectacles'.[40] Nevertheless, with Ross and Fox failing to provide further data, Sabine took almost sole responsibility for the observations to produce isoclinal and isodynamic lines for most of England. Working alone, this was a slow task, consuming the summers of 1858, 1859, 1860, and 1861. Although far from active in the new survey, Fox offered some assistance. When Sabine toured the south-west of England in the summer of 1859, he went to stay with the Cornishman at Penjerrick. There they took readings of dip on 23 June and 4 July to go with intensity observations made on 30 June and 1 July.[41] Cornwall was an important location for comparing how the dip of England's west coast had changed between 1837 and 1860. Land's End and Lowestoft formed the western and eastern limits for references of variation over time, with Lowestoft's annual change of 2′36″ comparing to Land's End's 3′09″. In total, West Cornwall's dip had changed by over 1° 11′ in twenty years.[42] Along with his joint efforts at Falmouth in 1859, Fox provided data for only three other stations, including Eastbourne.[43]

This was a far cry from the industrious Fox of the 1830s and 1840s. Still, he was happy to help Sabine in procuring Falmouth-built dipping circles. In February 1860, Sabine asked Fox to

> tell the artist that we shall glad to receive at Kew the Circle for which he was paid last year: It is destined for the voyage to the Dutch East India Islands, and will be called for by Dr Bergama, who writes that he shall be here early in March to receive all the instruments prepared for his Magnetic Observatory and Survey.[44]

Although lamenting how distant and out of touch he had become from Fox, Sabine was pleased to report that Kew's 'self-recording magnetometers' were proving a 'most successful experiment', and that the results produced for 1858 and 1859 appeared '<u>much more</u> consistent than any of the years of eye-observation from any quarter'.[45] Fox must have read this with at least some regret. For observatory work, the business of magnetic observation was becoming increasingly

---

[40] Scott, Ms1226/27/1: Edward Sabine to James Clark Ross (24 July 1858).
[41] CUL, Add.9942/33: Sabine, 'Report on the Repetition of the Magnetic Survey of England', 254 and 265.
[42] Ibid. 259.   [43] Ibid. 272.
[44] RS, Robert Were Fox Papers, MS/710/92: Edward Sabine to Robert Were Fox (10 Feb. 1860).
[45] Ibid.

automated. Here there is a clear sense that Falmouth had fallen behind as a centre of magnetic instrumentation. The dream of automatic magnetic observation had been a long time coming and, when it did, Fox's devices were made redundant for all but survey work. Fox had known of the prospect of this change since 1839 when Charles Wheatstone had written to him, excitedly speculating on the possibilities of photography for registering the indications of metrological and magnetic instruments. The 'same idea' had, Wheatstone believed, 'also occurred to Major Sabine', but he thought Fox might already be 'actually constructing an instrument on this principle'.[46] However far Fox got with this invention is unclear, but by the late 1850s he was quite disconnected from the development of the latest magnetic devices.

Nevertheless, Fox could still advise on the more traditional manual dipping needles on which survey work depended. In March 1860, with the new investigation into the British Isles' magnetism nearing completion, Sabine assessed the results of his experiments at Penjerrick. He had since been making observations with two 9-inch circles at Kew, both built to Gambey's pattern. The mean dip of eight observations with the first, which Robinson had built, was 68° 20′, compared to the second, an instrument of Barrow's construction, which gave 68° 19′ from six observations. In comparison, back in March 1859, Sabine had performed dip experiments with six different circles and twelve needles that produced results ranging from 68° 21′ to 68° 22′. This suggested an annual decline in dip of two to three minutes. Fox had recently sent a new circle to Kew which had given 68° 13′ dip in the garden back in June 1859. Sabine was uncertain how to explain this discrepancy. Even allowing for two minutes' annual change, the variance was hard to account for, with Sabine wondering if the different measurements were due to instrumental errors. He feared that his long-trusted supply of instruments from Falmouth was no longer producing the reliability of old.[47] Indeed, Fox's mechanic of choice, Wilton, died in 1860, leaving his son William Wilton junior to take over the business. Before the year was out, Fox and Wilton junior ended their partnership amid claims of poor workmanship.[48]

In the end, Fox's presence in the ensuing BAAS report on the state of the British Isles' magnetic character was minimal. The increasingly bespectacled Sabine published the findings in 1861, consisting mostly of his own labours and Lloyd's Irish contribution. Rather than naming magnetic instruments by maker, they appeared numbered, subject to Kew Observatory's organization of its dipping circles. For dip observations, Sabine selected a circle 'of the well-known English pattern which has been adopted some years past at the Kew Observatory; the circle was

---

[46] RS, Robert Were Fox Papers, MS/710/119: Charles Wheatstone to Robert Were Fox (13 Mar. 1839).
[47] RS, Robert Were Fox Papers, MS/710/93: Edward Sabine to Robert Were Fox (23 Mar. 1860).
[48] Jenny Bulstrode, 'Men, Mines, and Machines: Robert Were Fox, the Dip-Circle and the Cornish System', Part III diss., History and Philosophy of Science Department, University of Cambridge, 2013, 5 and 38–40; see also Robert Wilton, *The Wiltons of Cornwall* (Chichester: Phillimore, 1989).

6 inches in diameter, fitted with both verniers and microscopes, and with two needles, each 3½ inches in length. An account of the results 'obtained with twelve circles of this pattern with their 24 needles in 282 determinations made by different observers at the Kew Observatory' which appeared in the eleventh volume of the *Proceedings of the Royal Society*, concluded that Circle no. 30 was the most distinguished.[49] Sabine therefore selected this device for all of his dip observations between 1858 and 1860. For his readings on the east coast of England, made in the summer of 1861, he used Kew Observatory's Circle no. 33, which was of equal accuracy. Sabine boasted that Circles 30 and 33 gave very superior results and were in a class 'unsurpassed by any other form of instruments in use either in our own or in any other country for the determination of the magnetic dip'.[50]

With the Magnetic House of Kew Observatory as the reference point for checking the integrity of these instruments, not only had the Fox type been replaced, but so too had Fox's gardens as leading magnetic test venues. As Sabine surveyed Wales in 1858, the south-west and north of England in 1859, eastern England between Cambridge and Margate during 1860, and then Kent in 1861, he returned regularly to Kew for dipping-needle trials.[51] Comparing the isoclinal lines produced for 1860 with those from 1837, it seemed that there had been a change of 6° 17' across the British Isles. The isoclinal lines had shifted to between N71° 22'E and S71° 22'W in 1860 from N65° 05'E and S65° 05'W in 1837. Sabine explained that 'a real change has taken place', with the lines of dip having 'increased the angle which they make with the geographical meridians; a change implying that in the interval the secular diminution of the dip has been greater on the *West* than on the *East* side of the island'.[52] From north to south the change was also apparent. Over the previous twenty-three years, the south coast's mean annual secular change was 2' 68" compared to the northern border of England's 2' 05".[53]

For intensity measurements, Sabine had employed a collimator magnet, with the rate of a chronometer compared to the arc of a magnetic needle's vibration.[54] Unlike with a Fox type, he had preferred to use separate instruments for dip and intensity, hoping to achieve superior precision. His subsequent isodynamic map demonstrated a very distinct line of intensity of 10.322 running through the British Isles. As for how this compared to the 1837 survey, Sabine regretted that the 'method of determining the value of the force in absolute measure had not then been introduced, and the values of the force at the different stations are expressed in that Table (as was then the custom) relatively to the force in London expressed by 1.000'. This measure of relative intensity, made with an index like London or Falmouth, had been the limit of Fox's instrument. Sabine's 1861 report concluded that the angle at which the isodynamic lines running through England intersected with the geographical meridian had increased since 1837. This alteration

---

[49] CUL, Add.9942/33: Sabine, 'Report on the Repetition of the Magnetic Survey of England', 252.
[50] Ibid. 252.   [51] Ibid. 255–6.   [52] Ibid. 258.   [53] Ibid. 260.   [54] Ibid. 261.

was 'similar in character to the change in the direction of the isoclinal lines in the same interval, but somewhat less in amount'.[55] The data for the final magnetic quality, that of declination, came from the Royal Navy's superintendent of the Compass Department, Frederick John Evans (1815–85). With these figures, Sabine produced the maps of isogonic lines which completed his BAAS report.[56] Sabine presented his final comparative results from the surveys of the 1830s and 1850s in the *Philosophical Transactions* in 1870.[57]

Sabine never returned to the rigours of survey life. And there would not be another naval voyage of magnetic investigation until the *Challenger* expedition of 1872. As for Sabine's collaborators, they did not return to the study of terrestrial magnetism either. Having never recovered from his wife's death, Ross died in 1862, just a year after publication of the second BAAS report on the magnetism of the British Isles. Phillips met his end after dinner at All Souls College in 1874, having fallen down the stairs after a presumably indulgent meal. Likewise, after the second BAAS survey of the British Isles, Fox withdrew from philosophical investigations, retiring to his country residence at Penjerrick. The deaths of his wife, Maria, and son, Barclay, robbed him of much of his scientific enthusiasm, as well as removing his familial support; Catherine Fox's death in 1871 exacerbated this, leaving Fox alone with his last surviving child, Anna Maria. His final recorded correspondence with Sabine regarding dipping needles was in 1863, concerning the acquisition of Kew Observatory's Fox type, which had been supplied in 1861, itself as a replacement for the one despatched with Dr Bergama to Java.[58] This final order marked the end of Falmouth's supply of magnetic instruments and the Fox–Sabine network. In 1865 Sabine consulted Fox as to whether the RCPS at Falmouth might be appointed one of the Royal Society's six observation stations for a 'systematic scheme of meteorological observation' conducted in conjunction with the Board of Trade, but this was far from a magnetic venture.[59] When asked for recent magnetic results from West Cornwall in 1870, Fox confessed that he had none to offer, having no new data for the previous twenty years. He knew that the declination in his garden at Penjerrick had decreased since measuring it in 1837, but this was the extent of his knowledge.[60] Fox died in 1877, leaving Sabine and Lloyd as the only survivors of Britain's leading magnetic surveyors. It marked the end of an era in which terrestrial magnetism had been the foremost subject of scientific investigation.

---

[55] Ibid. 270–1.   [56] Ibid. 273.
[57] Edward Sabine, 'Contributions to Terrestrial Magnetism: No. XII, The Magnetic Survey of the British Islands, Reduced to the Epoch 1842.5', *Philosophical Transactions of the Royal Society of London*, 160 (1870), 265–75, at 266.
[58] RS, Robert Were Fox Papers, MS/710/94: Edward Sabine to Robert Were Fox (26 Nov. 1863).
[59] RS, Robert Were Fox Papers, MS/710/95: Edward Sabine to Robert Were Fox (24 June 1865).
[60] CUL, Add.9942/14: Robert Were Fox to 'Friend' (26 Aug. 1870).

7.4. Systems of Navigation

Undeniably, Fox's dipping needles had been prominent within the great scientific venture of the mid-nineteenth century. They had helped chart magnetic phenomena around the world and provided substantiating evidence for Sabine's claims over the relationship between terrestrial magnetism and the action of the Sun. What is harder to discern is their practical contribution to resolving the navigational problem of compass deviation in iron ships. In other words, where did all the magnet data obtained through dip-needle experiments fit within the system of correction for magnetic error developed during the second half of the nineteenth century? For all the surveying naval expeditions and Sabine's industrious production of maritime charts, the 1850s saw public fears surrounding iron-induced compass error escalate. Indeed, to Victorian audiences, it seemed that the loss of shipping due to the rising use of iron in shipbuilding was getting worse. Following the notorious beaching of Brunel's iron-hulled SS *Great Britain* in Dundrum Bay in 1846, shipping losses had grown, while ship architects were using more iron in their vessels, with the Peninsula and Oriental Steam Navigation Company (P&O) conducting its first experiments with the metal for steamships in 1843 with the launch of the *Pacha*. Over the next decade, P&O's oceanic rivals followed, including Cunard, the Pacific Steam Navigation Company, and the Royal Mail Steam Packet Company (RMSP), the last of which ordered its first iron-hulled steamship, the *Atrato*, in 1853. Throughout the 1850s the use of iron within British shipbuilding continued at an unprecedented rate. At 4,050 tons, the *Great Britain* had been the largest ship in the world in 1845, but by 1854 Brunel had begun work on an even bigger ship in collaboration with the naval architect John Scott Russell. SS *Great Eastern* boasted a displacement of 35,450 tons and constituted the greatest iron fabrication in existence. Although a complete experiment, this fitted within a wider adoption of iron among Britain's commercial shipping firms.

Despite this trend, the *Great Britain*'s wrecking had been one of several controversies surrounding magnetism to shake public confidence in oceanic navigation. This was not just a problem that affected iron-hulled ships; the rising use of steam engines and increasing amount of iron employed, even on wooden ships, appeared to risk compass deviation. In 1843, the loss of RMSP's *Solway* sparked a heated debate over these dangers. This was the third in a series of disasters to beset the company, established to convey mail from Britain to the West Indies and Spanish Main. In particular, the *Solway* caught the attention of London barrister Archibald Smith (1813–72), who wrote in *The Times* that the vessel's magnetic influence on its compass was to blame. A keen amateur sailor and Cambridge's Senior Wrangler for 1835, Smith asserted that on northerly or southerly courses in northern latitudes compasses were attracted to a ship's head, resulting in little deviation, but when taking an easterly or westerly bearing, error could amount to

5°–6°. For iron ships, he warned, this could escalate as much as 20°. On long voyages, with a ship's compass regularly checked to astronomical observations, this was not a problem. In such instances, Smith claimed that magnetic deviation was continually corrected, being indistinguishable from magnetic variation, which captains knew how to adjust their compasses for. The danger, however, was in shorter voyages or in instances when a ship changed course without making new observations for variation. Smith contended that, in the case of the *Solway*, the officers had not had chance to determine her course astronomically.[61]

Things came to a head at the 1854 BAAS meeting in Liverpool. Following the recent wrecking of the Australia-bound RMS *Tayleur*, blamed on the failure of her compasses and a huge navigational error in which the crew believed themselves to be heading south through the Irish Sea when they were in fact heading east into Ireland, Airy and Scoresby clashed over how best to correct for magnetic deviation. Airy's practice was to position magnets around a ship's compass as correctors, placed in reference to French mathematician Siméon Denis Poisson's (1781–1840) formula for calculating deviation in wooden ships. For this method to work, it was crucial that navigators refrained from interfering with the arranged magnets throughout their voyage. Scoresby, however, was critical of this, accusing the Astronomer Royal of overlooking the impact of waves on a ship's iron, known as 'heeling', as well as the effects of 'transient' magnetism, both of which changed a vessel's magnetic character over time and space. In short, he thought Airy naïve in treating a ship as a stable magnetic system and instead proposed the positioning of a compass high above the ship's iron, on a wooden mast, from which a navigator could take readings throughout a journey to compare to the vessel's main compass.[62] Assuming that the influence of a ship's iron on its compass changed during a voyage, the principle of Scoresby's solution was to ensure there was a constant on-board reference available to the crew. A year later, in 1855, the Scottish shipbuilder James Robert Napier (1821–79) concurred with Scoresby, informing the Glasgow Philosophical Society that although a ship's iron acquired its magnetic properties through hammering during construction, this character changed at sea from the effects of storms, roll, and the vibrations of its engines. Airy appeared to have ignored these practical environmental changes in his theoretical solution. As Napier put it, 'every system which makes the captain think he can rely on his compass without constant observation is liable to produce great evils'.[63]

---

[61] Crosbie Smith, *Coal, Steam and Ships: Engineering, Enterprise and Empire on the Nineteenth-Century Seas* (Cambridge: Cambridge University Press, 2018), 156–9.

[62] Alison Winter, '"Compasses All Awry": The Iron Ship and the Ambiguities of Cultural Authority in Victorian Britain', *Victorian Studies*, 38/1 (Autumn, 1994), 69–98, at 73–4; Crosbie Smith and M. Norton Wise, *Energy and Empire: A Biographical Study of Lord Kelvin* (Cambridge: Cambridge University Press, 1989), 766.

[63] Napier, quoted in Smith and Wise, *Energy and Empire*, 769.

Although the BMS had not resolved the crisis in iron shipping, with the data largely collected and Sabine publishing the investigation's findings, the focus of magnetic research and debates increasingly turned to the character of the iron ship themselves: now that the Earth's magnetic properties were known, it was possible to measure and compare the magnetic properties of individual vessels. To complete the rout of Airy's method of correcting compass error, Scoresby travelled to Australia on the maiden voyage of the *Royal Charter*, recording changes in the disturbing influence of the ship's iron on a magnetized needle. He took note of the percussive impact of the sea on the *Royal Charter*'s magnetic character as well as making comparisons between earlier measures of dip taken during Sabine's magnetic survey and those he observed resulting from the ship itself during its journey. For these experiments, Scoresby established eight points of observation on the *Royal Charter*, and employed a Fox dipping needle and Admiralty magnetic charts which Sabine and Belcher supplied him with prior to his departure. His results demonstrated that the vessel's magnetic character changed dramatically between Britain and Australia.[64] Here, then, was a very tangible example of how the BMS's results were contributing to new understandings of the magnetic influence of iron on ships' compasses: once dip measures were known for precise locations, it became possible for Scoresby to calculate the magnetic dip induced from a ship's iron.

Scoresby died before he had a chance to publish on account of his voyage, but he was not alone in emphasizing the problem iron presented for oceanic navigation, nor was he alone in drawing on Sabine's data to help resolve this challenge. Archibald Smith was also in Liverpool for the 1854 BAAS meeting, presenting the results of his own investigations conducted since the loss of the *Solway*. Having recorded the performances of compass needles on a host of naval vessels during the early 1850s, Smith used the fifth, sixth, and eighth of Sabine's 'Contributions' to differentiate between the Earth's magnetism and that of ships as they travelled to locations which had been the sites of BMS measurement. The iron steamer HMS *Torch*, for instance, stopped at Madeira in 1852, followed by Rio and Simon's Bay, Cape Town, all of which had been places of magnetic observations on the voyages of Ross, Belcher, and Stanley. Expeditions like those of *Terror*, *Erebus*, *Rattlesnake*, and *Samarang* had effectively established a global network of experimental locations with recorded, though changing, magnetic properties. Similar comparisons for magnetic intensity and dip were made on board the iron steamers *Vulcan*, *Hecla*, *Simoom*, and *Trident*, as well as several wooden vessels. Thanks to the wealth of magnetic data, measured largely with Fox's dipping needles, Smith was able to differentiate between the magnetic character of the Earth and that of

---

[64] William Scoresby, *Journal of a Voyage to Australia and Round the World for Magnetic Research*, ed. Archibald Smith (London: Longman, Green, Longman, & Roberts, 1859), 9 and 276–8; Winter, '"Compasses All Awry"', 89–90.

individual ships. He concluded that, for wooden ships in the northern hemisphere, the northern end of a compass's magnetic needle was drawn towards the vessel's head, with this influence diminishing towards the magnetic equator. For iron ships, however, the influence of the ferrous metal was less predictable, changing over time throughout a voyage.[65] In 1851, Smith published his mathematical solution to the problem of magnetic disruption. He reduced Poisson's fundamental equations, worked out in 1824, to a few simple and easily applied formulae, allowing navigators to calculate the error of their ships when swinging and then, throughout a voyage, to correct their readings in reference to tables previously prepared.[66] Unlike Airy, who treated each particle of iron in a ship to act as a magnet of which its direction was 'parallel to that of the dipping needle, and whose intensity is proportional to the intensity of terrestrial magnetism', Smith had originally envisaged what he believed to be a more practical approach, conceiving of an iron ship as an ellipsoid with three principal axes and three corresponding constants which determined its rotational behaviour.[67] To correct a compass, Smith reasoned, required knowing these three constants and placing soft iron bars, vertically, horizontally, and along a ship, to correct for the magnetic influence acting on its compass. By 1851, however, Smith had changed his approach to the problem, instead focusing on correcting readings, rather than the compass itself.

While it was clear that, due to changes in dip and intensity over time, knowledge of the Earth's magnetic field would not, in itself, solve the problem of compass error, the data from the survey was informing an increasingly complex understanding of the interaction between a ship's iron and its compass. During the 1860s and 1870s, this contributed to the fashioning of a new system of correction that would, to a great extent, resolve the challenge of magnetic deviation. Appointed in 1837 and comprised of Beaufort, James Clark Ross, Edward Johnson, Samuel Christie, and Sabine, the Admiralty Compass Committee oversaw the design and introduction of a standard Admiralty compass which remained in use until 1889. In 1842, this committee drew up rules for enabling naval officers to swing their ships and produce tables of deviation for each point on their compasses. This process of 'swinging' a ship consisted of first identifying true north in reference to some known observable point, such as a mountain or lighthouse, and then turning the vessel through 360° of the compass, steadying it on a succession of headings and ascertaining the deviation of the compass on each bearing. A year later the Admiralty established the Compass Department to check the navy's compasses and examine iron-induced deviation. Through this work,

---

[65] Archibald Smith, 'On the Deviations of the Compass in Wooden and Iron Ships', in *Report of the Twenty-Fourth Meeting of the British Association for the Advancement of Science, Held at Liverpool in September 1854* (London: J. Murray, 1855), 434–8, at 437–9.
[66] Smith and Wise, *Energy and Empire*, 760–1.
[67] Airy, quoted in Smith and Wise, *Energy and Empire*, 758.

the Admiralty provided much impetus to systematize navigational practices and reduce compass error.[68]

In contrast to Britain's mercantile companies, the Admiralty had remained cautious over the use of iron throughout the 1850s. While employing this material for some smaller vessels, fears that it would splinter when shot sustained a commitment to wood for the Royal Navy's bigger warships. However, following France's launching of *La Gloire* in 1859, a wooden ship with iron armour plating, the Admiralty took action, fearing this French innovation, and ordered its first ironclad, HMS *Warrior*. The launch of this battleship in 1860, which employed wrought-iron armour, was soon followed by the construction of her sister, HMS *Black Prince*. Nevertheless, anxieties over the new construction material endured well into the 1860s, with naval authorities fearing that ironclads might prove to be mere 'experiments', akin to the discredited *Great Eastern* which financially ruined all who invested in her.[69] But the commission of the Royal Navy's first iron warships stimulated fresh anxieties over the question of navigation and the Admiralty soon ordered the overhauling of its existing regime of compass correction. In 1862, the Compass Department's Frederick John Evans (1815–85) collaborated with Archibald Smith to publish the new *Admiralty Manual for Ascertaining and Applying the Deviations of the Compass*. Part I provided 'Practical Rules' for determining and correcting compass error by the swinging of a ship, as outlined in the Admiralty's 1842 manual, while Part II consisted of Smith's 1851 corrections for deviation, including calculations for heeling error, this being when a ship pressed to an extreme degree by sail, causing more iron to be on the ship's upward weather side and, therefore, unbalancing its magnetic influence on a compass.[70] To Smith's mathematical skill, Evans added his own experimental inquiries, conducted throughout the 1850s, which extended Smith's and Scoresby's earlier studies of the magnetic character of individual ships. In 1860, Evans had published the results of a series of voyages in which he had monitored the behaviour of compasses on forty-two warships, especially those with great masses of iron. From this study, Evans discovered that iron vessels built in Britain's dockyards, when heading south, found their needles drawn to the head, but when heading north, saw them attracted to the stern.[71] Producing almost 250 tables, divided into

---

[68] Smith and Wise, *Energy and Empire*, 759–60.

[69] Don Leggett, 'Neptune's New Clothes: Actors, Iron and the Identity of the Mid-Victorian Warship', in Don Leggett and Richard Dunn (eds.), *Re-Inventing the Ship: Science, Technology and the Maritime World, 1800–1918* (Farnham: Ashgate, 2012), 71–91, at 76 and 85.

[70] Frederick J. Evans and Archibald Smith, *Admiralty Manual for Ascertaining and Applying the Deviations of the Compass Caused by the Iron in a Ship* (London: The Hydrographic Office, Admiralty, 1862), 1–3 and 9; Crosbie Smith and M. Norton Wise, *Energy and Empire: A Biographical Study of Lord Kelvin* (Cambridge: Cambridge University Press, 1989), 761.

[71] Frederick J. Evans, 'Reduction and Discussion of the Deviations of the Compass Observed on Board of All the Iron-Built Ships, and a Selection of the Wood-Built Steam-Ships in Her Majesty's Navy, and the Iron Steam-Ship "Great Eastern"; Being a Report to the Hydrographer of the Admiralty', *Philosophical Transactions of the Royal Society of London*, 150 (1860), 337–78, at 341.

wooden-paddle, iron-hulled, and wooden-screw steamers, Evans delivered a systematic analysis of how the 'magnetic elements' of each vessel varied over time and space by recording the behaviour of an 'Admiralty Standard' compass in various positions on and above deck.[72] Using Sabine's 'Contributions' from 1849, these measures were compared to dip and intensity observations from the 1840s, including those obtained at Greenwich, Ascension Island, Hong Kong, Rio, Shanghai, and the West Indies, which were reprinted in Evans's tables to illustrate the difference between terrestrial magnetism measured during the BMS and that of individual naval warships as they traversed the world's oceans.[73] Added to this was data Evans had obtained through experiments on the *Great Eastern* during its trial voyage in the autumn of 1859. Leaving Deptford on 7 September 1859, Evans monitored the ship's compass throughout its passage down the Thames and westward through the English Channel, referencing these readings to accurate positions determined through astronomical observations of the Sun. On 10 September, the *Great Eastern*'s crew swung the monstrous vessel at Portland, repeating the process in reference to a distant Welsh mountain on the ship's arrival at Holyhead. Evans recorded a gradual diminution in the ship's magnetic force, losing a fifth of its initial value between Deptford and Portland. The tendency of the ship's magnetic force was to assume a fore-and-aft line along the hull, giving a north-and-south polarity which varied in reference to local dip measurements, and inducing an error of around 6° throughout its voyage (Figure 7.3).[74]

The 1862 *Admiralty Manual* combined mathematical theory with experimentally obtained data, but also included Sabine's BMS's findings, consisting of a chart for the magnetic variation of 1860, measured at 'every part of the navigable globe'.[75] From this, it could be seen how the world was divided into regions of westerly and easterly variation. As Smith and Evans put it, 'besides furnishing the navigator with the amount of the variation at any place', this chart would 'be useful as indicating changes most rapidly, as shown by the lines being crowded together, and lying in a direction at right angles to the usual tracks of navigation'.[76] They instructed navigators to use this in conjunction with astronomical observations to determine compass deviation, offering knowledge of both the ship's error and how the Earth's magnetic dip and intensity varied over space. These qualities fluctuated over time, but Sabine's chart helped to give mariners a sense of what to expect at different locations and how their compasses would deviate. Dip, for instance, changed more rapidly in lower latitudes, increasing by about thirty minutes per degree at the poles compared to 2° per degree of latitude at the magnetic equator.[77] Thanks to the BMS, changes in dip, intensity, and variation were known for 1840 to 1861, allowing general magnetic laws and navigational principles to be deduced. The survey had, for the northern hemisphere, revealed 'the systems

[72] Ibid. 359.   [73] Ibid. 343 and 359.   [74] Ibid. 350–2.
[75] Evans and Smith, *Admiralty Manual*, 49.   [76] Ibid. 50.   [77] Ibid. 51.

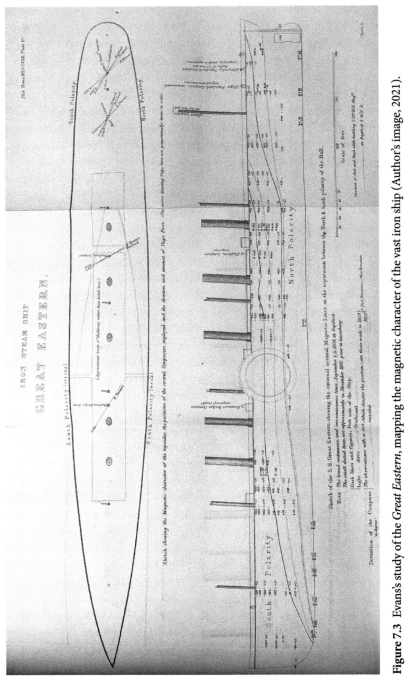

Figure 7.3 Evans's study of the *Great Eastern*, mapping the magnetic character of the vast iron ship (Author's image, 2021).

of lines' all sweeping from east to west, and vice versa in the southern hemisphere (Figure 7.4).[78] Sabine's survey and the data on which it relied had not made navigation safe, but they had contributed to a network of techniques for calculating deviation. As Evans and Smith warned, 'the mariner can have no absolutely safe guide, except in the system of actual and increasing observation which has been enjoined in the forgoing pages'.[79] Though not, in itself, a solution to the challenge of terrestrial magnetism, the BMS's data allowed navigators to anticipate variations that might be expected to occur with geographical change. The *Admiralty Manual* made practical use of the 'several thousand observations made by observers of all nations in all parts of the globe'.[80] Fox's instrument had contributed both to the formation of this organized account of the Earth's magnetic properties and to the navigational practices that resulted from it.

As valuable as the *Admiralty Manual* was for oceanic navigation, it was in itself only of limited use without a reliable maritime compass and it was not until the 1870s that the question of iron-induced magnetic error can be said to have been largely resolved. The completion of the Admiralty's regime of correction for deviation involved the refashioning of the navy's standard compass. Along with mathematical formulae for tabular constructions, the practices of swinging a ship, astronomical observations for fixing location, and knowledge of terrestrial magnetism, the design and implementation of accurate compasses were crucial to negotiating the effects of deviation. It was the new compass that William Thomson (1824–1907), later Baron Kelvin of Largs, patented in 1876 that marked the completion of this system of magnetic correction. Having collaborated with Archibald Smith on the question of compass error until the latter's death in 1872, Thomson continued to develop a compass that consisted of a lightweight card, fixed to a set of parallel magnetic needles arranged below the card to make it steady in a rough sea. Thomson's compass was a response to what he identified as the four key challenges of compass design: namely, that the card must be subject to little friction while sitting on its pivot and rest in a position of equilibrium when the ship was on a steady course; that the needle be short, so as to be corrected; that it have a small magnetic moment; and that the card have a long vibrational period to eradicate oscillations due to 'kinetic error', which arose from a ship's roll. This final element, that of the time in which the needle vibrated during a reading, was usually termed the 'power', or 'strength', of a compass and, with a longer duration, compensated for rolling error and heeling: a small magnetic moment and a long, nineteen-second-vibration period ensured that a precise magnetic reading was obtained without disruption from the sea.[81]

To meet these criteria, Thomson's compass consisted of eight small needles of thin wire fixed on two parallel silk threads surrounded by a light aluminium

---

[78] Ibid. 56.   [79] Ibid. 7.   [80] Ibid., p. iv.
[81] Smith and Wise, *Energy and Empire*, 770–2.

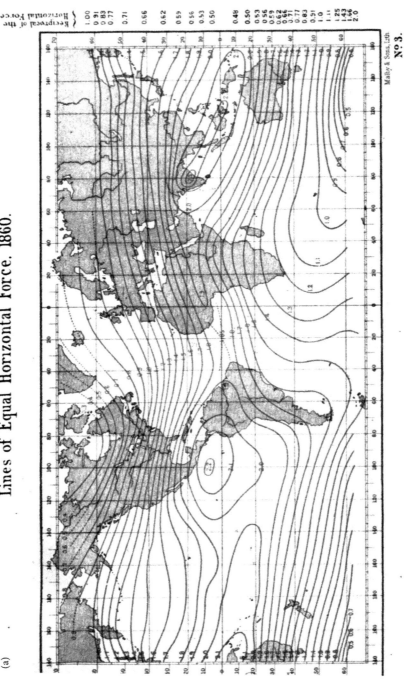

Figure 7.4 Sabine's global magnetic charts of 1860 for (*a*) dip, (*b*) horizontal force, and (*c*) variation, as represented in Smith and Evans's *Admiralty Manual* and revealing the magnetic system of the Earth (Author's images, 2021).

(b)

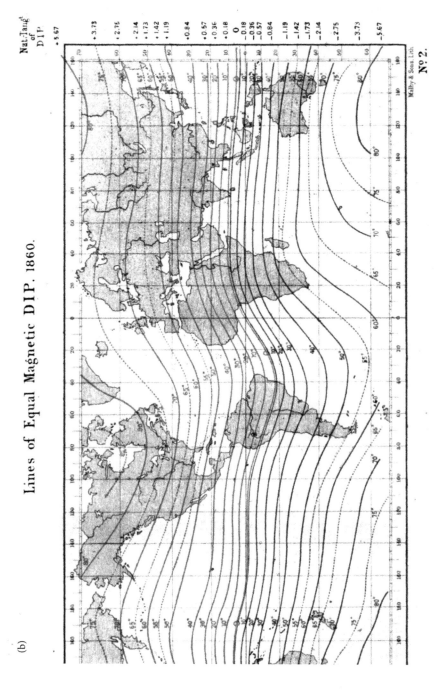

Figure 7.4 Continued

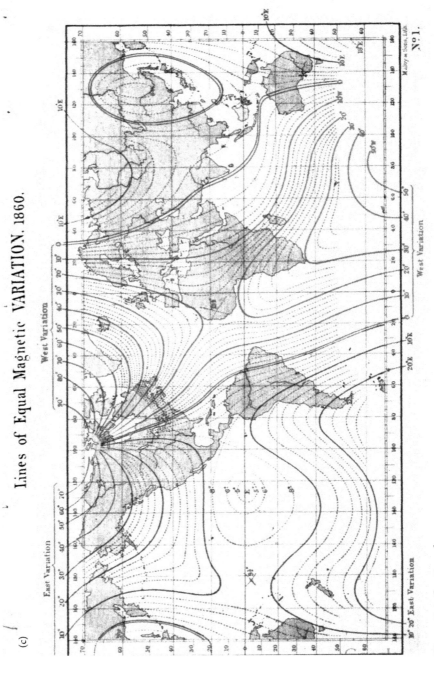

Figure 7.4 Continued

circular rim by four silk threads, connected by thirty-two silk spoke-like threads to form a wheel connected to a central disk. At 170 grains, it was very light, compared to the Admiralty Standard's 1,600 grains and the 2,900 grains of those in use by merchant steamers.[82] In particular, Thomson's compass resolved what he believed to be the two forms of magnetism that acted on a ship's compass. He distinguished between the 'permanent' magnetism of a hull, which resulted during a ship's construction and was determined by the position in which it laid when built, and 'transient' magnetism, which changed in reference to the heading of the ship when at sea. Two adjustable soft-iron spheres on either side of the compass neutralized the effects of a ship's 'transient' magnetism, while Thomson placed magnets beneath the compass, within the binnacle, to compensate for the ship's 'permanent' magnetism. The two adjustable iron globes corrected quadrangular error, while a steel magnet under the compass compensated for heeling to deliver a complete system of magnetic correction. This was, to some extent, Airy's solution, but Thomson recognized that navigators wanted to be able to adjust their apparatus as the magnetic character of their ships changed.[83]

Initially, the Admiralty was reluctant to adopt Thomson's new compass, but the device gradually secured patronage after 1876. The Oceanic Steam Navigation Company, better known as the White Star Line, was quick to adopt the device, issuing it to its *Britannic* and *Germanic* steamships between 1876 and 1877 and employing them until the 1900s, including for the *Olympic* in 1911 and *Titanic* in 1912. P&O issued the *Poonah* with a Thomson type in 1876 and, after an impressive performance, ordered a second for its new flagship, the *Kaiser-i-Hind* in 1878, with eight more orders following over the next twelve months. By the 1880s, Britain's leading oceanic navigation companies were all using Thomson's compasses, including P&O, White Star, Cunard, and the Castle Line.[84] It took longer for the Royal Navy to adopt the device, having acquired just two by 1877. However, after impressing during trials on the HMS *Minotaur*, which at 10,700 tons was the ultimate symbol of empire, naval support for the compass grew. The *Minotaur*'s captain, Walter Kerr (1839–1927), was especially enthusiastic following his experiences with the instrument. To Thomson's adjustable method of correction, Evans and the Admiralty favoured its existing system of assuming that compasses were always in error and constantly required correcting. Nevertheless, throughout the 1880s, the navy gradually adopted Thomson's compass in place of the Admiralty Standard, with five ordered in 1883 and eighteen in 1884. By the 1890s, almost all British warships carried Thomson's compasses, as did the world's foremost navies, including those of Russia, Germany, and Japan.[85] By the time of his death in 1907

---

[82] Ibid. 773.
[83] Ibid. 774; Ben Marsden and Crosbie Smith, *Engineering Empires: A Cultural History of Technology in Nineteenth-Century Britain* (Basingstoke: Palgrave Macmillan, 2005), 32–3.
[84] Smith and Wise, *Energy and Empire*, 778–85.   [85] Ibid. 787–96.

over 10,000 of Thomson's compasses had been manufactured, underpinning the world's shipping and solving the navigational challenge magnetic deviation.[86]

7.5. Conclusion

Traditionally, historians have ascribed a somewhat vague link between the BMS and the system of practices and apparatus for compass correction that emerged from the 1850s.[87] While Thomson's compass, Archibald Smith's mathematics, and Evans's Admiralty instructions feature prominently in narratives of nineteenth-century navigation, the magnetic data collected with Fox's instruments during the expeditions of the 1840s and 1850s appear to have taken on a somewhat nebulous role in resolving the challenges of compass deviation. However, on closer inspection, the worldwide surveying of terrestrial magnetism featured prominently within schemes to resolve the navigation challenge that iron shipbuilding presented. Scoresby, Evans, and Archibald Smith all employed Sabine's 'Contributions' as a point of reference for magnetic phenomena when analysing the behaviour of iron in individual ships. The BMS had established a global network of measured locations with known, if changing, magnetic properties, at which ships could be examined and their specific magnetic elements calculated. While dip and intensity readings changed over time, Sabine's data provided a sense of what navigators could expect and taught natural philosophers to think about how changes in one location related to variations in another, with the globe conceptualized as a colossal magnetic system. This, along with the BMS's philosophical contribution to contemporary understandings over the relationship between solar activity and terrestrial magnetism, represented the end results of Sabine and Fox's global system of magnetic investigation.

---

[86] Ibid. 754.
[87] As Alison Winter put it, magnetic work like Airy's 'was related, albeit ambivalently, to the so called "Magnetic Crusade"'. See Winter, '"Compasses All Awry"', 72; see also Marsden and Smith, *Engineering Empires*, 27–8.

Epilogue
Global Science in an Age of Empire

In September 1882, Falmouth was full of scientific bustle. Fifty years since its founding, the RCPS celebrated its golden jubilee with an extravagant exhibition. Along with fish casts from South Kensington's Natural History Department, meteorological instruments, and models of mining and agricultural contrivances, it was the wonders of electricity that provided the festival's greatest spectacle. Technicians set up telephone and telegraph connections between the RCPS building and Falmouth's Drill Hall, delivering instantaneous communication with the latest electrical apparatus from Messrs Siemens. Most celebrated of all, however, was the illumination of the RCPS Hall with Swan electric lights.[1] Dynamo machines supplied electricity from nearby New Street, as the hundred incandescent lamps within the hall transformed the polytechnic into a gleaming beacon of enlightenment. Along with these fantastical exhibits, a host of local MPs and Cornish scientific celebrities were in attendance, including John Couch Adams and Anna Maria Fox, the sole survivor of Robert Were Fox's three children (see Figure 8.1).[2] Yet behind the festivities, it was apparent to those who looked closely that this was not quite such a celebration of Cornish science as the earlier meetings of the 1830s and 1840s. Neither Carl Siemens (1823–83) nor Joseph Swan (1828–1914) had accompanied their electric apparatus to the Cornish port, and few from Britain's science elites participated. The 1882 jubilee was, in fact, a poignant moment, rich in nostalgia but implicitly revealing that Cornwall had fallen behind Britain's leading centres of science in London, Edinburgh, Oxford, Cambridge, and Glasgow.

Indeed, this decline had been evident at the RCPS since the 1840s. From November 1843, the society was running an annual deficit of about £15, with the committee requesting local mine owners subscribe to ensure its survival. However, the RCPS never mobilized the industrial support it had hoped to attain. After controversially opening up its building to public functions in 1846, the society's financial problems were somewhat resolved, but subscribers often failed to pay up and, after 1845, the RCPS's lecture programmes were abandoned.[3]

---

[1] Anon., *The Fiftieth and Jubilee Report of the Royal Cornwall Polytechnic Society* (Falmouth: Lake & Co., and R. C. Richards, 1882), 21.
[2] Ibid. 36.
[3] Alan Pearson, 'A Study of the Royal Cornwall Polytechnic Society', MA thesis, University of Exeter, 1973, 110–28.

258 AN EMPIRE OF MAGNETISM

Figure 8.1 By 1882, Anna Maria was Fox's sole surviving child but remained a central figure at the RCPS long after her father's death (Author's image, 2021).

It was as a civic venue, rather than a purely scientific institution, that the institution endured the financial hardships of the late nineteenth century. This decline was not limited to Falmouth but was endemic throughout the county's scientific societies, which found survival increasingly challenging during the second half of the nineteenth century. Some failed altogether, such as the Royal Horticultural Society of Cornwall in 1861, and the Penzance Natural History and Antiquarian Society, which was inactive by 1872. With scientific investigation increasingly conducted in, or coordinated from, Britain's centralized institutions and metropolitan laboratories, regional practitioners often struggled to contribute to wider national and international studies.[4]

Much of this decline in Cornish scientific activity accompanied the collapse of the region's global dominance of copper mining. Throughout the nineteenth century, Cornish output had grown from 59,640 tons in the 1800s to 114,326 tons

---

[4] A. Pearson, 'The Royal Horticultural Society of Cornwall', *Journal of the Royal Institution of Cornwall*, NS 7/2 (1974), 165–73, at 171; Simon Naylor, *Regionalizing Science: Placing Knowledge in Victorian England* (London: Pickering & Chatto, 2010), 180.

in the 1830s, out of a world production of 214,557 tons.[5] Although the county's copper produce peaked in 1856, this was largely due to the huge output of East Cornwall's mines, along the River Tamar. Between 1855 and 1856, Devon and Cornwall produced 205,999 tons of copper ore, this being a record, but of this, 28,602 tons came from a single mine, the Devon Great Consols. After then, between 1865 to 1870, Cornish copper output fell from 157,473 to 78,737 tons in the face of cheaper foreign exports, especially from Chile.[6] Excavated to astonishing depths and with their once rich veins exhausted, West Cornwall's mines were no longer profitable. By 1874, only three major Cornish copper mines had survived and when the Devon Great Consols closed in 1903, it left Levant as the region's last copper mine.[7] Only the mining of much less profitable tin and, latterly, arsenic, softened the economic depression of Cornwall's copper collapse. This eclipse of the region's mining wealth accompanied a decline in duty figures for the county's steam engines. By the 1850s, Lean's duty reports observed a flatlining in efficiency, with the limits of pumping engine economy apparently reached.[8]

8.1. Reputation and Endurance

That Cornwall's scientific and industrial fortunes were so closely linked is hardly surprising. Throughout the first half of the nineteenth century, the expansion of Cornish mining had accompanied the scientific innovation of which Fox's work had become so emblematic. It is not that the industrial extraction of ore had inevitably precipitated the expansion of the region's scientific culture, but rather that around the network of mines, steam engines, and engineering ventures a community of individuals were united by their shared interests in commerce and Nature. However, by the 1880s, few of these philosophically curious men and women were still alive. This loss was all too apparent at the RCPS's 1882 jubilee where the vast majority of the scientific instruments and electrical apparatus on display were not Cornish products but brought together from across the British Isles. Tellingly, the most iconic of Cornwall's philosophical contributions dated back to the 1840s and was all too familiar. Within the RCPS Hall, on a table alongside a collection of barometers and rain gauges, was 'Mr. R. W. Fox's own Dipping-needle Deflector, exhibited in 1834, and one made under his direction for the Austrian Government in 1845'.[9] Despite its age, Fox's circle was still

---

[5] John Rowe, *Cornwall in the Age of Industrial Revolution* (St Austell: Cornish Hillside Publications: 1993), 128.
[6] D. B. Barton, *A History of Copper Mining in Cornwall and Devon* (Truro: D. Bradford Barton Ltd, 1968), 75–9.
[7] Ibid. 90–3.
[8] D. B. Barton, *The Cornish Beam Engine* (Truro: D. Bradford Barton Ltd, 1965), 59–71.
[9] Anon., *The Fiftieth and Jubilee Report of the Royal Cornwall Polytechnic Society*, 28.

Falmouth's foremost scientific ware. In his address to the meeting, George Mathews Whipple (1842–93), a Devonian physicist and superintendent of Kew Observatory, declared that there was

> one instrument that I should like to refer to specially, as having been of great advantage to science, the dipping needle, invented by our late friend, Mr. Robert Were Fox. (Applause) That instrument was left by him in so perfect a state that no one has been able to improve upon it. (Applause) I have had occasion to see a number of them supplied to government expeditions, both British and Colonial, and to establishments where the most important observations are made...since Mr. Fox left it in so perfect a state that it appears incapable of further improvement.[10]

Thanks to this device, Whipple believed the benefits to practical navigation were beyond measure, having 'resulted in shipmasters being now able to navigate iron vessels over the world in every direction'.[11] As the secretary of the committee of Falmouth Observatory, Wilson Lloyd Fox (1847–1936) observed in his *Historical Synopsis of the Royal Cornwall Polytechnic Society*, published for the jubilee, Fox's dipping needles had been exported to the Russian, Austrian, and American governments, as well as to the Raja of Travancore.[12]

Indeed, Falmouth still owed much to the Fox type's reputation as was to become apparent with the Royal Society's selection of the port as the location for one of its new meteorological observatories in 1867. In 1866 the Board of Trade requested that the Royal Society appoint a committee to examine how the nation's collection of data relating to the weather might be enhanced and standardized. Reporting to Parliament, the committee was critical of the board's Meteorological Department and recommended the establishment of a network of meteorological observatories around the British Isles, which would be divided into three classifications. The 'first order' of these observation sites would collect data on an extensive scale, with hourly or continuous weather measurements recorded. Of the seven locations selected for these premier observatories, three would be in England, with the Royal Society inviting the RCPS to oversee the operation of such an establishment at Falmouth in 1867. The society eagerly embraced the suggestion, erecting a building before the end of the year, and accepted the annual grant of £250 for its maintenance. In 1882, Wilson Fox had no doubt that the port's selection was due to the enduring kudos of Fox's dipping needle.[13] Despite this celebration of Falmouth's continuing scientific prominence, just six months after

---

[10] Ibid. 34.   [11] Ibid.
[12] Wilson Lloyd Fox, *Historical Synopsis of the Royal Cornwall Polytechnic Society Prepared by Wilson Lloyd Fox, F.M.S., (Hon. Sec. of the Meteorological Committee) and Presented by Him to the Society on Its Year of Jubilee, 1882* (Falmouth: Lake & Co., 1882), 35.
[13] Naylor, *Regionalizing Science*, 150–8; Fox, *Historical Synopsis*, 35.

Wilson Fox's boasts, the Royal Society withdrew its funding of the observatory to focus on data analysis in London, although the RCPS's council managed to get this reinstated after an intense lobbying campaign throughout 1883.[14] For all this uncertainty over the port's scientific future, Fox's scientific standing remained high throughout the 1870s and 1880s. On learning that Falmouth required a superintendent for its meteorological observatory in 1873, for instance, Armagh astronomer Thomas Robinson (1792–1882) wrote to Fox, recommending a young assistant of his own for the post. Fox was, Robinson declared, 'the patriarch of science in the South West'.[15]

This enduring reputation of Fox's device owed much to its continuing use within expeditionary survey science, most famously during the voyage of HMS *Challenger* between 1872 and 1876. Under the command of George Strong Nares (1831–1915), the *Challenger* would circumnavigate the globe, performing experimental investigations into the physical conditions of the Earth's deepest oceans. The Admiralty equipped the expedition with a Fox dipping needle and gimbal, and ordered the crew to repeat many of the original magnetic observations made during the 1840s and 1850s.[16] In 1872, Frederick Evans, by now chief assistant to the Hydrographer of the Navy, wrote to Fox, recalling the experimental instruction the Cornish philosopher had provided in Rosehill garden to Captain Francis Blackwood and his assistant, Charles Shadwell (1814–86), before their magnetic survey of northern Australia on HMS *Fly* between 1842 and 1846.[17] Preparing the forthcoming *Challenger* expedition for determining the changes in the Earth's magnetic field that had occurred during the preceding thirty years, Evans was keen for advice from the aged magnetic authority. Fox replied to say that although he had moved permanently to Penjerrick, Evans was welcome to send the expedition's magnetic observers there for instrumental advice and trials. Nevertheless, it appeared to Fox that the instructions issued to the *Challenger*'s officers for the use of his circle had been 'admirably drawn up' and asserted that he knew of 'no instructions relative to my instrument so perfect as those issued by the Admiralty & General Sabine'.[18] As ever, he was confident that the Fox type would serve the *Challenger*'s magnetic observers perfectly. The *Challenger*'s officers were to remeasure the terrestrial magnetism of the Indian Ocean and Antarctic Circle:

---

[14] Naylor, *Regionalizing Science*, 164–6.

[15] RS, Robert Were Fox Papers, MS/710/87: Thomas Romney Robinson to Robert Were Fox (2 Oct. 1873).

[16] Trevor H. Levere, 'Magnetic Instruments in the Canadian Arctic Expeditions of Franklin, Lefroy, and Nares', *Annals of Science*, 43/1 (1986), 57–76, at 68–71.

[17] Having risen to the rank of vice-admiral, Shadwell continued his magnetic interests, becoming a fellow of the Royal Society and publishing an account of further magnetic observations made on HMS *Iron Duke*'s voyage to China and Japan. See Charles Frederick Alexander Shadwell, 'A Contribution to Terrestrial Magnetism; Being the Record of Observations of the Magnetic Inclination, or Dip, Made During the Voyage of H.M.S. "*Iron Duke*" to China and Japan, &c., 1871–1875', *Philosophical Transactions of the Royal Society*, 167 (1877), 137–47.

[18] CUL, Add.9942/16: Robert Were Fox to Capt. Evans (*c.*1872).

tracing the magnetic lines Ross had observed between 1839 and 1843 showed that they had altered considerably and that the south magnetic pole had migrated north. Around Cape Horn, measurements suggested a secular change in declination of over ten minutes of arc per year.[19] Evidently, the legacy of Cornwall's mid-century scientific culture endured well. Until his death, Fox enjoyed boasting of how 'he had corresponded with two of our great circumnavigators,—living widely apart in time—Sir Joseph Banks being the first, and Capt. Nares, of the Challenger and the late Arctic expedition, being the other'.[20]

Though lacking the enthusiasm of the 1830s and 1840s, the surveying of the Earth's magnetic qualities continued throughout the century, with fresh government lobbying efforts in the 1880s and 1890s leading to a new generation of Antarctic ventures during the 1900s, culminating in the *Discovery* (1901–4) and *Nimrod* (1907–9) expeditions.[21] Fox's dipping needles continued to have a role in this ongoing investigation well into the twentieth century. Following Nares's Arctic Expedition of 1875–6 aboard HMS *Alert* and *Discovery*, Captain Ettrick Creak (1835–1920) determined to improve the design of the Fox type, with John Dover (1824–81) of London building the first of these modified devices. This was still largely Fox's design but incorporated a glazed circular metal case in front of the circle's face, above which was mounted a deflection magnet. Known as a 'Lloyd-Creak' type, Captain Robert Scott (1868–1912) took such a device on his *Discovery* expedition of 1901 to 1904, this being the first official government exploration of the Antarctic since Ross's. Likewise, the German Antarctica expedition of 1901 to 1903, aboard the survey ship *Gauss*, employed a dipping needle to Lloyd-Creak's design. In 1909, the Carnegie Institution of Washington's Department of Terrestrial Magnetism launched a completely non-magnetized ship, the *Carnegie*, custom-built for magnetic investigation. Until she caught fire and sank off Western Samoa in November 1929, the *Carnegie* carried a modified Lloyd-Creak dipping circle. Britain planned a similar vessel for magnetic observations, the *Research*, but this was abandoned with outbreak of the Second World War in 1939.[22] Creak's modified Fox dipping needle would almost certainly have been equipped to the vessel had the project gone ahead.

8.2. Global Systems

Despite Cornwall's decline, throughout the second half of the nineteenth century, Fox's instruments and magnetic investigations were part of the network of data, devices, and practices on which the world's commerce and navigation depended.

---

[19] Edward J. Larson, 'Public Science for a Global Empire: The British Quest for the South Magnetic Pole', *Isis*, 102/1 (Mar. 2011), 34–59, at 45.
[20] Anon., 'Mr. Robert Were Fox', *The Athenaeum*, 2597 (4 Aug. 1877), 153–5.
[21] Larson, 'Public Science for a Global Empire', 45–59.
[22] Anita McConnell, 'Surveying Terrestrial Magnetism in Time and Space', *Archives of Natural History*, 32/2 (2005), 346–60, at 355–7.

Cornish natural philosophy and instrumentation had become fully incorporated within a body of imperial science which underpinned the trade and naval power sustaining Europe's colonial empires. There was no mention of Fox or his dipping needles in Evans and Smith's 1862 *Admiralty Manual*, Sabine's later 'Contributions', or Thomson's accounts of his compass, but the transformation from provincial to imperial knowledge was complete. It is unsurprising then, that at the 1882 RCPS jubilee, promoters of Cornish science were so eager to emphasize the enduring practical value of Fox's dipping needle. There was some truth to Whipple's boast that Fox's instrument had assisted shipmasters in being 'able to navigate iron vessels over the world in every direction'. For all that Fox's contribution to the BMS had merged seamlessly into a wider, international, accumulation of magnetic knowledge, in Cornwall it was still celebrated for its local character.

The magnetic data collection that Fox had helped orchestrate constituted a component part within an increasingly well-established regime for trustworthy navigation. Yet his philosophical contribution is best understood as a critical element within the BMS's systematic scrutinizing of the Earth's magnetic properties. From Falmouth, Fox mobilized local resources, including experimental locations like mines, artisan instrument makers, the experience of miners and ship captains, and his own philosophical work, within Sabine's survey. It was one thing to maintain a dipping needle for experiments in a Cornish mine, but it was quite a different challenge to extend this examination of natural phenomena around the world. Fox fashioned a centre of magnetic science at Falmouth from which to direct magnetic experiments over vast geographical distances and that could be performed by a body of practitioners of varying experimental skill and experience. Fox and Sabine had to eradicate error, standardize measurements, and bring order to the capriciousness of expeditionary science. To realize this, Fox established a robust system of instrument construction and maintenance, experimental training and instruction for naval officers, and techniques for ensuring precise measurements, including mathematical formulae for correcting instrumental errors and temperature variations, and the use of his subtropical gardens as places of instrument trial and testing. In the magnetic isolation of Rosehill and Penjerrick, Fox oversaw the construction and examination of new dipping circles and met with passing naval officers, exchanging philosophical knowledge and experimental guidance. Here he drafted written instructions for magnetic observers, with strict regimes of ordered serialized measurements and uniformed instrumental arrangements, and published the results of his latest inquiries. But this Cornish scientific centre was also prominent within the network of ongoing magnetic survey investigations. Broken needles and devices returned to Falmouth for repair, while old devices arrived in the port for checking against instruments of known magnetic qualities. And data and correspondence from Sabine and naval officers went back and forth between Falmouth, London, and the worldwide network of ships and observatories.

This book has shown how, in the nineteenth century, the British state's naval, financial, and colonial resources transformed Fox's provincial knowledge, practices, and instruments into a powerful organ of global investigation. In throwing these resources behind Fox's work, the government enhanced the authority of his philosophical programme and increased its claims to being universal: in other words, the material apparatus of the state presented the chance for Fox's dipping needle and experimental practices to be trialled all over the world and, if successful, demonstrate that what was true of natural phenomena in Cornwall was also true anywhere on Earth. In a sense, it was this genius of government, its exceptional creative power, that turned Fox's localized speculations into an elite science of empire. This amounted to what we might think of as the governmentalization of Cornish science. *An Empire of Magnetism*'s account of the government's role within the investigation of terrestrial magnetism is at variance with traditional accounts of the Victorian British state. Usually, the nineteenth century is heralded as the great age of low-state intervention and liberal, laissez-faire-style governance. What is clear from the story of the Fox dipping needle is how active the state was in the promotion of scientific endeavour and the cultivation of philosophical knowledge. The 1839–43 Antarctic voyage, for instance, was very much a state-orchestrated expedition, but contemporaries celebrated this for James Clark Ross's heroic endeavours. Likewise, it was the government that financed the commission of new Fox types, but these were acclaimed as a triumph of Fox's creative ingenuity. For all that the government, through Parliament, the army, and the Admiralty, financed and sustained the BMS, this was heralded as a project of entrepreneurial achievement. This promotion of the individual within the framework of state patronage is, nevertheless, revealing of what the nineteenth-century British state actually was. Often, historical studies treat the 'state' as a disembodied institution. In this book, however, the state has been unpacked as a broad collection of human agents, with interconnected scientific and political interests. It operated through social networks, with individuals like Belcher, Ross, Franklin, and FitzRoy eager to employ Fox's dipping needle and emphasize its worth with influential authorities in the Admiralty, like Beaufort, and trusted government advisers, notably Sabine and Herschel. Similarly, MPs such as Lemon and Gilbert exerted influence on Fox's behalf in the elite centres of political and scientific patronage at Parliament and the Royal Society. It was these human networks that constituted the 'British state' and it was their contributions that newspapers and scientific journals celebrated, rather than the role of the state itself within the BMS. In this way, while a government venture, Fox's work fitted neatly within a scientific programme that was highly compatible with Britain's liberal political culture.

Fox's instruments and practices had, by the 1840s, become widespread and generally accepted. With endorsements from the Admiralty, the Royal Society, and the BAAS, and with powerful promoters including Sabine and a host of local MPs, Fox's dipping circles became the trusted magnetic devices of the Royal Navy.

As the standard issue for expeditions, these instruments were adopted within the broader expansion of government science witnessed during the nineteenth century: the apparatus of the British state, including the army and navy, recast Fox's work from provincial philosophical knowledge into an internationally acclaimed science. It took on new authority and credibility from this transformation. It was promoted as cutting-edge science, typical of a progressive, industrial, imperial nation: it was knowledge for a modern government. Yet behind all the state patronage, plaudits of scientific elites, and increasing systematization within the trials of expeditionary science, Fox's instrument continued to carry the local values and character of its place of production. This was science built on underground experiments in mines, Fox's experiences of magnetic observations in Cornwall and beyond, and the practical working know-how of miners. To transform this Cornish science beyond its regional connotations required a demonstration that the same practices and instruments worked anywhere in the world and in the hands of those beyond Fox and his close associates. This transition from the regional to the universal was critical to securing Cornish science authority throughout Britain and the world.

After living and working with one of Fox's surviving dipping needles for three months and over thousands of nautical miles, I feel that I have a particular appreciation for the effectiveness and importance of these instruments for the systematic production of magnetic knowledge. The business of measuring the planet's magnetism with Fox's instruments—be it in the 1840s or the 2020s—was, and is, full of uncertainty: it is a constant concatenation of disturbing elements, be it rain, wind, and salt spray, rough seas or bad light preventing the precise reading of the vernier, or human interferences (Figure 8.2). At Lisbon, port authorities came close to stopping our onshore experiments, suspicious as to what we were doing; they did eventually allow the observations to be completed, but only after careful negotiation. At Cape Town, fears over the security of the instrument prevented us taking it ashore, while bad light and a strong swell made our on-board measurements difficult (Figure 8.3). Even before the outbreak of COVID-19 curtailed our voyage, the spread of the pandemic meant that many of our intended ports of call were closed to us on arrival. While vibration at sea tended to diminish our magnetic needle's strength, a far greater disturbance was encountered during overland travel. At sea, the needle's tapping on the circle's axle coincided with the ocean's swell, but this was a gentle motion for all but the roughest of conditions. On land, however, transporting the instrument by car caused a far harsher impact between the needle and the axle. On the Bay of Biscay, just before returning home, we inspected the instrument in the hope of a final set of experiments. However, the previous months of travel had, through the needle's constant tapping on the circle's axle, demagnetized the device. When passing the Scilly Isles the following day, rough weather caused the needle to tap so violently on this axle as to render it completely inert, delivering a lesson on how contingent the instrument's fitness

Figure 8.2 Performing a magnetic measurement with a Fox-type dipping needle on a calm Indian Ocean, off the east coast of Madagascar in February 2020 (Author's image, 2020).

Figure 8.3 The capricious nature of magnetic survey work. Having travelled about 5,000 miles from Bristol, rough sea prevented a disembarkation at St Helena, confining our measurements to our ship (Author's image, 2020).

was on good weather and calm seas. Performing magnetic survey work in 2020 was very different to conducting similar experimental inquiries in the 1840s, but our encounters echoed the experiences of nineteenth-century users.[23] The experience of reworking experiments from the BMS, and use of the Fox needle in an array of different circumstances, has emphasized not only the importance of Fox and Sabine's system of magnetic data collection in fulfilling the enterprise's grand ambitions, but also the fragility of this system when instigated on a global scale.

[23] Edward J. Gillin, 'The Instruments of Expeditionary Science and the Reworking of Nineteenth-Century Magnetic Experiment', *Notes and Records of the Royal Society*, 76/3 (Sep., 2022), 565–592.

Select Bibliography

Place of publication for primary and secondary material is London unless otherwise stated.

1. Primary Material

1.1. Manuscripts
Cambridge, Cambridgeshire
Robert Were Fox Box, Cambridge University Library (CUL)
    Add.9942/14: Robert Were Fox to 'Friend' (26 Aug. 1870).
    Add.9942/16: Robert Were Fox to Capt. Evans (c.1872).
    Add.9942/18(i): Robert Were Fox, 'Notice of an Instrument for Ascertaining Various Properties of Terrestrial Magnetism, and Affording a Permanent Standard Measure of Its Intensity in Every Latitude' (c.1833).
    Add.9942/18(ii): Robert Were Fox, 'Notice of an Instrument for Ascertaining Various Properties of Terrestrial Magnetism, and Affording a Permanent Standard Measure of Its Intensity in Every Latitude' (c.1833).
    Add.9942/19: 'Variation of Magnetic Needle at Falmouth, March 22nd 1838'.
    Add.9942/21: 'Extracts from Cap Ross' Report: Magnetic Dip & Intensity Observations On Shore & On Board the Erebus with Needle' (1839–40).
    Add.9942/23: 'Results of Experiments Made by R. W. Fox in the Spring of 1839 on the Magnetic Intensity and Dip in Italy & Some Other Parts of Europe; His Perran Apparatus Was Employed as in His Previous Experiments in the British Islands the Needle Only Having Been Since Altered' (1839).
    Add.9942/26: 'Letter to Major Sabine from Capt Jms Clerk-Ross, HMS Erebus off C Verd' (10 Nov. 1839).
    Add.9942/27: untitled note dated '27th Jan., 1844' and '30th Jan., 1844'.
    Add.9942/34: 'Instructions for Using Mr. Fox's Instrument for Determining the Magnetic Inclination and Intensity' (c.1842–3), proof with Fox's annotations.
Christ College Library
    STAN1/32: '32. Safe Home'.
Scott Polar Research Institute Library (Scott)
    Ms248/291: 'Journal of Sir John Franklin, Tasmania, 8 April to 18 May 1841'.
    Ms248/295/3: J. Franklin to Beaufort (23 May 1843).
    Ms248/316/6: J. Franklin to J. C. Ross (4 Aug. 1841).
    Ms248/316/7: J. Franklin to J. C. Ross (16 Sept. 1841).
    Ms248/316/10: J. Franklin to J. C. Ross (21 July 1843).
    Ms248/316/17: John Franklin to James Clark Ross (9 Jan. 1845).
    Ms248/364/26: Francis Crozier to James Ross (9 July 1845).
    Ms248/469/2: Edward Sabine to Lady Franklin (Sept. 1859).
    Ms248/469/4: E. Becker to Edward Sabine (26 Oct. 1859).
    Ms438/18/7: John Franklin to Edward Parry (10 July 1845).
    Ms862: Edward Sabine, 'Journal 30 April to 2 Nov 1818, and Letters, 1807–1836'.

Ms1226/27/1: Edward Sabine to James Clark Ross (24 July 1858).
Ms1524: James Clark Ross to John Franklin (25 Mar. 1834).

Falmouth, Cornwall
Royal Polytechnic Society of Cornwall Library
   Fox family archives: Robert Were Fox, 1789–1877.

London
The National Archives (TNA), Edward Sabine Papers
   BJ3/17: Miscellaneous Letters to Edward Sabine.
   BJ3/19: Robert Were Fox to Edward Sabine Correspondence.
Royal Society Library (RS)
   AP/15/20: Robert Were Fox, 'On Certain Irregularities in the Magnetic Needle Produced by Partial Warmth, and the Relations Which Appear to Subsist between Terrestrial Magnetism and the Geological Structure, and Thermo:electric Currents of the Earth' (1832).
   MS/258: Sabine Correspondence, Vol. 2, D–J.
   MS/710: Robert Were Fox Papers.
   RR/1/73: Samuel Christie, 'Report on Mr Fox's Paper, 27th June 1832' (1832).

Ottawa, Canada
Library and Archives Canada (CAN)
   R3888-0-7-E, J. Franklin Correspondence, 1834–6.

Oxford, Oxfordshire
Bodleian Library (BOD)
   Ms Dep. Papers of the British Association for the Advancement of Science (BAAS) 5, Miscellaneous Papers, 1831–1869, Folio 39, 'First Resolution of the York Scientific Meeting' (1831).
Oxford University Museum Library (OUM), Papers of John Phillips, 1800–1874
   Box 101, Item 12: 'Questions Relative to Mineral Veins, Submitted to Practical Miners, by Robert Were Fox' (c.1836).
   Phillips 1836/7: Robert Were Fox to John Phillips (11 Feb. 1836).
   Phillips 1836/16: Robert Were Fox to John Phillips (9 Apr. 1836).
   Phillips 1845/36: Robert Were Fox to John Phillips (29 Nov. 1845).
   Phillips 1857/43: Robert Were Fox to John Phillips (13 Nov. 1857).
   Phillips 1857/45: Robert Were Fox to John Phillips (19 Nov. 1857).

Redruth, Cornwall
Kresen Kernow (Cornwall Centre) (KKW)
   X114/1: 'Patent to Robert Were Fox and Joel Lean, Improvements on Steam Engine' (1812).
   FOX/B/2/60: R. W. Fox senior to R. W. Fox junior (1 Apr. 1814).
   FOX/B/2/63: 'Part of A. Fox's Diary' (1816).

1.2. Parliamentary Papers

Papers Relative to the Expedition to the River Niger, No. 472 (1843).

1.3. Periodicals

Abstracts of the Papers Communicated to the Royal Society of London
Annales de chimie et de physique (Paris)

Annals of Electricity, Magnetism, & Chemistry; and Guardian of Experimental Science (Sherwood, Gilbert and Piper)
Annals of Philosophy (Baldwin, Cradock and Joy)
The Athenaeum
Calcutta Journal of Natural History: And Miscellany of the Arts and Sciences in India, 7 (Calcutta: Bishop's College Press)
Friend of Africa
Friends' Review: A Religious, Literary and Miscellaneous Journal (1847–94)
Illustrated London News
Journal of the Royal Institution of Great Britain
Leisure Hour (Jan. 1877–Oct. 1903)
London, Edinburgh, and Dublin Philosophical Magazine and Journal of Science
Nature
Philosophical Magazine
Philosophical Transactions of the Royal Society of London
Proceedings of the Royal Society of London
Punch
The Times
Transactions of the Royal Geological Society of Cornwall (Penzance: T. Vigurs)

1.4. Printed

Airy, George Biddell, 'On the Computation of the Effect of the Attraction of Mountain-Masses, as Disturbing the Apparent Astronomical Latitude of Stations in Geodetic Surveys', *Philosophical Transactions of the Royal Society of London*, 145 (1855), 101–4.

Airy, Wilfrid (ed.), *Autobiography of Sir George Biddell Airy, K.C.B., M.A., LL.D., D.C.L., F.R.S., F.R.A.S., Honorary Fellow of Trinity College, Cambridge, Astronomer Royal from 1836 to 1881* (Cambridge: Cambridge University Press, 1896).

Allen, William, and Thomson, T. R. H., *A Narrative of the Expedition Sent by Her Majesty's Government to the River Niger, in 1841; Under the Command of Captain H. D. Trotter, R.N.*, 2 vols. (Richard Bentley, 1848), vol. i.

Anon., 'Review: *On the Electro-Magnetic Properties of Metalliferous Veins in the Mines of Cornwall*, By Robert Were Fox (Read June 10, 1830)', *Journal of the Royal Institution of Great Britain*, 1 (Oct. 1830–May 1831), 345–6.

Anon., 'A General Outline of the Various Theories Which Have Been Advanced for the Explanation of Terrestrial Magnetism', *Annals of Electricity, Magnetism, & Chemistry; and Guardian of Experimental Science*, 1 (Oct. 1836–Oct. 1837), 117–23.

Anon., 'Niger Expedition', *Friend of Africa*, 1/4 (25 Feb. 1841), 57–8.

Anon., 'The Niger Expedition', *The Times*, 17887 (22 Jan. 1842), 5.

Anon., 'Fifteenth Meeting of the British Association for the Advancement of Science: Cambridge, 18th June 1845', *Calcutta Journal of Natural History: And Miscellany of the Arts and Sciences in India*, 7 (1847), 81–142.

Anon., 'Memorial of Maria Fox, Wife of Robert Were Fox, of Falmouth, England', *Friends' Review: A Religious, Literary and Miscellaneous Journal*, 12/22 (Feb. 1859), 337.

Anon., 'Mr. Robert Were Fox', *The Athenaeum*, 2597 (4 Aug. 1877), 153–5.

Anon., 'Robert Were Fox, F. R. S.', *Leisure Hour*, 1370 (30 Mar. 1878), 197–9.

Anon., *The Fiftieth and Jubilee Report of the Royal Cornwall Polytechnic Society* (Falmouth: Lake & Co., and R. C. Richards, 1882).

'A Surgeon', 'Niger Expedition.—Prevention of Fever', *The Times*, 17885 (20 Jan. 1842), 5.

Belcher, Edward, *Narrative of a Voyage Round the World, Performed in Her Majesty's Ship Sulphur, During the Years 1836–1842, Including Details of the Naval Operations in China, from Dec. 1840, to Nov. 1841*, 2 vols. (Henry Colburn, 1843).

Belcher, Edward, *Narrative of the Voyage of H.M.S. Samarang, During the Years 1843–1846; Employed Surveying the Islands of the Eastern Archipelago*, 2 vols. (Reeve, Benham, and Reeve, 1848).

Bettany, G. T., 'The Late W. J. Henwood, F.R.S.', *Nature*, 12 (1875), 293.

Collins, J. H., *A Catalogue of the Works of Robert Were Fox, F.R.S, &c. Chronologically Arranged, with Notes and Extracts, and a Sketch of His Life* (Truro: Lake & Lake, 1878).

Conrad, Joseph, *Heart of Darkness* (Macmillan Collector's Library, 2018).

Daniell, John, 'On the Waters of the African Coast', *Friend of Africa*, 1/2 (15 Jan. 1841), 18–23.

Daniell, John, 'A Probable Cause of Miasma', *Friend of Africa*, 1/3 (1 Feb. 1841), 40.

Dayman, Joseph, 'Observations of the Mean Magnetic Inclination Made on Shore in the Voyage of H.M.S. Rattlesnake, by Lieut. Joseph Dayman, R.N.; Instruments Employed: Robinson's 6-inch Inclinometer; Fox's Dipping Apparatus', in John MacGillivray (ed.), *Narrative of the Voyage of H.M.S. Rattlesnake, Commanded by the Late Captain Owen Stanley, R.N., F.R.S. &c. During the Years 1846–1850; Including Discoveries and Surveys in New Guinea, the Louisiade Archipelago, etc. To Which Is Added the Account of Mr. E. B. Kennedy's Expedition of the Cape York Peninsula*, 2 vols. (T. & W. Boone, 1852), i. 337–42.

Davy, Humphry, 'Electro-Chemical Researches, on the Decomposition of the Earths; With Observations in the Metals Obtained from the Alkaline Earths, and on the Amalgam Procured from Ammonia', *Philosophical Transactions of the Royal Society*, 98 (1808), 333–70.

De la Beche, Henry T., *Report on the Geology of Cornwall, Devon, and West Somerset* (Longman, Orme, Brown, Green, and Longmans, 1839).

Evans, Frederick J., 'Reduction and Discussion of the Deviations of the Compass Observed on Board of All the Iron-Built Ships, and a Selection of the Wood-Built Steam-Ships in Her Majesty's Navy, and the Iron Steam-Ship "Great Eastern"; Being a Report to the Hydrographer of the Admiralty', *Philosophical Transactions of the Royal Society of London*, 150 (1860), 337–78.

Evans, Frederick J., and Smith, Archibald, *Admiralty Manual for Ascertaining and Applying the Deviations of the Compass Caused by the Iron in a Ship* (The Hydrographic Office, Admiralty, 1862).

Faraday, Michael, *Experimental Researches in Electricity* (Richard and John Edward Taylor, 1839).

Forbes, John, 'On the Temperature of Mines', *Transactions of the Royal Geological Society of Cornwall*, 2 (1822), 159–215.

Fourier, Joseph, 'Remarques générales sur les températures du globe terrestre et des espaces planétaires', *Annales de chimie et de physique*, 27 (1824), 136–67.

Fox, Robert Were, 'Observations on the Temperature of Mines in Cornwall', *Annals of Philosophy*, NS (Jan.–June 1822), 381–3.

Fox, Robert Were, 'On the Temperature of Mines', *Transactions of the Royal Geological Society of Cornwall*, 2 (1822a), 14–18.

Fox, Robert Were, 'On the Temperature of Mines', *Transactions of the Royal Geological Society of Cornwall*, 2 (1822b), 19–28.

Fox, Robert Were, *Further Observations on the Temperature of Mines; Communicated to the Royal Geological Society of Cornwall and Published in Their 3rd Volume of Transactions* (Penzance: T. Vigurs, 1827).

Fox, Robert Were, 'On the Electro-Magnetic Properties of Metalliferous Veins in the Mines of Cornwall', *Philosophical Transactions of the Royal Society*, 120 (1830), 399–414.

Fox, Robert Were, *Observations on Metalliferous Veins and Their Electro-Magnetic Properties* (Penzance: T. Vigurs, 1831).

Fox, Robert Were, 'On the Discharge of a Jet of Water Under Water', *Journal of the Royal Institution of Great Britain*, 1 (Oct. 1830–May 1831), 368.
Fox, Robert Were, 'On the Discharge of a Jet of Water Under Water', *Journal of the Royal Institution of Great Britain*, 1 (Oct. 1830–May 1831), 599–600.
Fox, Robert Were, 'On the Variable Intensity of Terrestrial Magnetism, and the Influence of the Aurora Borealis upon It', *Philosophical Transactions of the Royal Society*, 121 (1831), 199–207.
Fox, Robert Were, 'On Certain Irregularities in the Magnetic Needle, Produced by Partial Warmth, and the Relations Which Appear to Subsist between Terrestrial Magnetism and the Geological Structure and Thermo-Electric Currents of the Earth', *Proceedings of the Royal Society*, 10 (1831–2), 123–5.
Fox, Robert Were, 'On Certain Irregularities in the Vibrations of the Magnetic Needle Produced by Partial Warmth; And Some Remarks on the Electro-Magnetism of the Earth', *Philosophical Magazine*, 1 (1832), 310–14.
Fox, Robert Were, 'Some Facts Which Appear to Be at Variance with the Igneous Hypothesis of Geologists', *Philosophical Magazine*, 1 (1832), 338–40.
Fox, Robert Were, 'Notice of an Instrument for Ascertaining Various Properties of Terrestrial Magnetism, and Affording a Permanent Standard Measure of Its Intensity in Every Latitude', *Philosophical Magazine*, 3rd ser., 4/20 (Feb. 1834), 81–8.
Fox, Robert Were, 'On Magnetic Attraction and Repulsion and on Electrical Action', *Philosophical Magazine*, 3rd ser., 5/25 (July 1834), 1–11.
Fox, Robert Were, 'Account of Some Experiments on the Electricity of the Copper Vein in Huel Jewel Mine', in *Report of the Fourth Meeting of the British Association for the Advancement of Science; Held at Edinburgh in 1834* (John Murray, 1835), 572–4.
Fox, Robert Were, *Description of R. W. Fox's Dipping Needle Deflector* (Falmouth: Jane Trathan, c.1835).
Fox, Robert Were, *Observations on Mineral Veins* (Falmouth: J. Trathan, 1837).
Fox, Robert Were, 'Report of Some Experiments on the Electricity of Metallic Veins, and the Temperature of Mines', in *Report of the Seventh Meeting of the British Association for the Advancement of Science; Held at Liverpool in September 1837* (John Murray, 1838), 133–7.
Fox, Robert Were, 'Report on Some Observations on Subterranean Temperature', *Report of the Tenth Meeting of the British Association for the Advancement of Science; Held at Glasgow in August* 1840 (John Murray, 1841), 309–19.
Fox, Robert Were, *Report on Some Observations on Subterranean Temperature* (Richard and John E. Taylor, 1841).
Fox, Robert Were, *Report on the Temperature of Some Deep Mines in Cornwall* (Taylor and Francis, 1858).
Fox, Wilson Lloyd, *Historical Synopsis of the Royal Cornwall Polytechnic Society Prepared by Wilson Lloyd Fox, F.M.S. (Hon. Sec. of the Meteorological Committee) and Presented by Him to the Society on Its Year of Jubilee, 1882* (Falmouth: Lake & Co., 1882).
Hawkins, John, 'On Some Advantages Which Cornwall Possesses for the Study of Geology, and on the Use Which May Be Made by Them', *Transactions of the Royal Geological Society of Cornwall*, 2 (1822), 1–13.
Herschel, John, 'Light', *Encyclopaedia Metropolitana*, Vol. IV, (London: Baldwin and Cradock, 1830), 341–586.
Herschel, John, *A Preliminary Discourse on the Study of Natural Philosophy* (Longman, Rees, Orme, Brown, and Green, 1830).
Herschel, John, 'The President's Address', *The Athenaeum*, 921 (21 June 1845), 612–18.

274 SELECT BIBLIOGRAPHY

Herschel, John F. W. (ed.), *Manual of Scientific Enquiry; Prepared for the Use of Her Majesty's Navy: And Adapted for Travellers in General* (John Murray, 1849).
Jordan, T. B., 'Description and Use of a Dipping Needle Deflector Invented by Robert Were Fox, Esq. by Mr. T. B. Jordan, Philosophical Instrument Maker, Falmouth', *Annals of Electricity*, 3 (July 1838–Apr. 1839), 288–97.
Lean, Thomas, 'On the Temperature of the Mines in Cornwall', *Philosophical Magazine: A Journal of Theoretical, Experimental and Applied Physics*, 52 (1818), 204–6.
Lean, Thomas, & Brother, *Historical Statement of the Improvements Made in the Duty Performed by the Steam Engines in Cornwall, from the Commencement of the Publication of the Monthly Reports* (Simpkin, Marshall, and Co., 1839).
Leifchild, J. R., *Cornwall: Its Mines and Miners; With Sketches of Scenery; Designed as a Popular Introduction to Metallic Mines* (Longman, Brown, Green Longmans, Roberts, 1857).
MacGillivray, John, *Narrative of the Voyage of H.M.S. Rattlesnake, Commanded by the Late Captain Owen Stanley, R.N., F.R.S. &c. During the Years 1846–1850; Including Discoveries and Surveys in New Guinea, the Louisiade Archipelago, etc.; To Which Is Added the Account of Mr. E. B. Kennedy's Expedition of the Cape York Peninsula*, 2 vols. (T. & W. Boone, 1852), vol. i.
Melville, Herman, *Moby-Dick, or The Whale* (1851; Macmillan, 2004).
Moyle, M. P., 'Observations on the Temperature of Mines in Cornwall', *Annals of Philosophy*, NS (Jan.–June 1822), 308–10.
Moyle, M. P., 'On the Temperature of Mines in Cornwall', *Annals of Philosophy*, NS (Jan.–June 1822), 415–16.
Moyle, M. P., 'On the Temperature of the Cornish Mines', *Transactions of the Royal Geological Society of Cornwall*, 2 (1822), 404–15.
Pole, William, *A Treatise on the Cornish Pumping Engine; in Two Parts* (John Weale, 1844).
Pym, Horace N. (ed.), *Memoirs of Old Friends: Being Extracts from the Journals and Letters of Caroline Fox of Penjerrick, Cornwall from 1835 to 1871* (Smith, Elder, & Co., 1882).
Riddell, C. J. B., *Magnetic Instructions for the Use of Portable Instruments Adapted for Magnetical Surveys and Portable Observatories, and for the Use of a Set of Small Instruments Adapted for a Fixed Magnetic Observatory; With Forms for the Registry of Magnetical and Meteorological Observations* (Her Majesty's Stationery Office, 1844).
Ross, James Clark, *A Voyage of Discovery and Research in the Southern and Antarctic Regions, During the Years 1839–1843*, 2 vols. (John Murray, 1847), vol. i.
Sabine, Edward, 'On Irregularities Observed in the Direction of the Compass Needles of H.M.S. *Isabella* and *Alexander*, in Their Late Voyage of Discovery, and Caused by the Attraction of the Iron Contained in the Ships', *Philosophical Transactions of the Royal Society of London*, 109 (1819), 112–22.
Sabine, Edward, 'Observations on the Dip and Variation of the Magnetic Needle, and on the Intensity of the Magnetic Force; Made during the Late Voyage in Search of a North West Passage', *Philosophical Transactions of the Royal Society of London*, 109 (1819), 132–44.
Sabine, Edward, 'The Bakerian Lectures: An Account of Experiments to Determine the Amount of the Dip of the Magnetic Needle in London, in August 1821; With Remarks on the Instruments Which Are Usually Employed in Such Determinations', *Philosophical Transactions of the Royal Society of London*, 112 (1822), 1–21.
Sabine, Edward, 'On the Temperature at Considerable Depths of the Caribbean Sea', *Philosophical Transactions of the Royal Society of London*, 113 (1823), 206–10.
Sabine, Edward, 'Experiments to Ascertain the Ratio of the Magnetic Forces Acting on a Needle Suspended Horizontally, in Paris and in London', *Philosophical Transactions of the Royal Society of London*, 118 (1828), 1–14.

SELECT BIBLIOGRAPHY

Sabine, Edward, 'On the Dip of the Magnetic Needle in London, in August 1828', *Philosophical Transactions of the Royal Society of London*, 119 (1829), 47–53.

Sabine, Edward, *Report on the Magnetic Isoclinal and Isodynamic Lines in the British Islands: From Observations by Professors Humphrey Lloyd, and John Phillips; Robert Were Fox, Esq; Capt. James Clark Ross; and Major Edward Sabine* (Richard and John E. Taylor, 1839).

Sabine, Edward, 'Contributions to Terrestrial Magnetism', *Philosophical Transactions of the Royal Society of London*, 130 (1840), 129–55.

Sabine, Edward, 'Contributions to Terrestrial Magnetism: No. II', *Philosophical Transactions of the Royal Society of London*, 131 (1841), 11–35.

Sabine, Edward, 'Instructions for Magnetic Observations in Africa', *Friend of Africa*, 1/4 (25 Feb. 1841), 55–7.

Sabine, Edward, 'Contributions to Terrestrial Magnetism: No. III', *Philosophical Transactions of the Royal Society of London*, 132 (1842), 9–41.

Sabine, Edward, 'Contributions to Terrestrial Magnetism: No. IV', *Philosophical Transactions of the Royal Society of London*, 133 (1843), 113–43.

Sabine, Edward, 'Contributions to Terrestrial Magnetism: No. V', *Philosophical Transactions of the Royal Society of London*, 133 (1843), 145–231.

Sabine, Edward, 'Remarks on M. de La Rive's Theory for the Physical Explanation of the Causes Which Produce the Diurnal Variation of the Magnetic Declination', *Abstracts of the Papers Communicated to the Royal Society of London*, 5 (1843–50), 821–5.

Sabine, Edward, 'Contributions to Terrestrial Magnetism: No. VI', *Philosophical Transactions of the Royal Society of London*, 134 (1844), 87–118, 115*–118*, 119–224.

Sabine, Edward, 'Contributions to Terrestrial Magnetism: No. VII', *Philosophical Transactions of the Royal Society of London*, 136 (1846), 237–336.

Sabine, Edward, 'Contributions to Terrestrial Magnetism: No. VIII', *Philosophical Transactions of the Royal Society of London*, 136 (1846), 337–432.

Sabine, Edward, *Observations Made at the Magnetic and Meteorological Observatory at St Helena*, i: *1840, 1841, 1842, and 1843, with Abstracts of the 1847 Observations from 1840 to 1845 Inclusive* (Longman, Brown, Green, & Longmans, 1847).

Sabine, Edward, 'On the Diurnal Variation of the Magnetic Declination at St Helena', *Philosophical Transactions of the Royal Society of London*, 137 (1847), 51–7.

Sabine, Edward, 'Contributions to Terrestrial Magnetism: No. IX', *Philosophical Transactions of the Royal Society of London*, 139 (1849), 173–234.

Sabine, Edward, 'Terrestrial Magnetism', in John F. W. Herschel (ed.), *Manual of Scientific Enquiry; Prepared for the Use of Her Majesty's Navy: And Adapted for Travellers in General* (John Murray, 1849), 14–53.

Sabine, Edward, 'On Periodical Laws Discoverable in the Mean Effects of the Larger Magnetic Disturbances', *Philosophical Transactions of the Royal Society of London*, 141 (1851), 123–39.

Sabine, Edward, 'On the Annual Variation of the Magnetic Declination, at Different Periods of the Day', *Philosophical Transactions of the Royal Society of London*, 141 (1851), 635–41.

Sabine, Edward, 'On Periodical Laws Discoverable in the Mean Effects of the Larger Magnetic Disturbances: No. II', *Philosophical Transactions of the Royal Society of London*, 142 (1852), 103–24.

Sabine, Edward, 'Report of the Joint Committee of the Royal Society and the British Association, for Procuring a Continuance of Magnetic and Meteorological Observatories', *Proceedings of the Royal Society of London*, 9 (1857–9), 457–74.

Sabine, Edward, 'Report on the Repetition of the Magnetic Survey of England, Made at the Request of the General Committee of the British Association', in *Report of the Thirty-First Meeting of the British Association for the Advancement of Science* (John Murray, printed by Taylor & Francis, 1862), 250–79.

Sabine, Edward, 'Results of Hourly Observations of the Magnetic Declination Made by Sir Francis Leopold McClintock, and the Officers of the Yacht "Fox," at Port Kennedy, in the Arctic Sea, in the Winter of 1858–9; and a Comparison of These Results with Those Obtained by Captain Rochfort Maguire, and the Officers of Her Majesty's Ship "Plover," in 1852, 1853, and 1854, at Point Barrow', *Philosophical Transactions of the Royal Society*, 153 (1863), 649–63.

Sabine, Edward, 'Contributions to Terrestrial Magnetism: No. X', *Philosophical Transactions of the Royal Society of London*, 156 (1866), 453–543.

Sabine, Edward, 'Contributions to Terrestrial Magnetism: No. XI', *Philosophical Transactions of the Royal Society of London*, 158 (1868), 371–416.

Sabine, Edward, 'Contributions to Terrestrial Magnetism: No. XII, The Magnetic Survey of the British Islands, Reduced to the Epoch 1842.5', *Philosophical Transactions of the Royal Society of London*, 160 (1870), 265–75.

Sabine, Edward, 'Contributions to Terrestrial Magnetism: No. XIII', *Philosophical Transactions of the Royal Society of London*, 162 (1872), 353–433.

Sabine, Edward, 'Contributions to Terrestrial Magnetism: No. XV', *Philosophical Transactions of the Royal Society of London*, 167 (1877), 461–508+v.

Scoresby, William, *Journal of a Voyage to Australia and Round the World for Magnetic Research*, ed. Archibald Smith (Longman, Green, Longman, & Roberts, 1859).

Shadwell, Charles Frederick Alexander, 'A Contribution to Terrestrial Magnetism; Being the Record of Observations of the Magnetic Inclination, or Dip, Made During the Voyage of H.M.S. "*Iron Duke*" to China and Japan, &c., 1871–1875', *Philosophical Transactions of the Royal Society*, 167 (1877), 137–47.

Smith, Archibald, 'On the Deviations of the Compass in Wooden and Iron Ships', in *Report of the Twenty-Fourth Meeting of the British Association for the Advancement of Science, Held at Liverpool in September 1854* (J. Murray, 1855), 434–38.

Somerville, Mary, *On the Connexion of the Physical Sciences* (John Murray, 1834).

Trevithick, Francis, *Life of Richard Trevithick, with an Account of His Inventions*, 2 vols. (E. & F. N. Spon, 1872).

Whewell, William, 'Report on the Recent Progress and Present Condition of the Mathematical Theories of Electricity, Magnetism, and Heat', in *Report of the Fifth Meeting of the British Association for the Advancement of Science; Held at Dublin in 1835* (John Murray, 1836), 1–34.

Whewell, William, *History of the Inductive Sciences, from the Earliest to the Present Time*, 3 vols. (3rd edn., John W. Parker and Son, 1857), vol. iii.

1.5. Private Papers
Letters in author's possession
Robert Were Fox to Davies Gilbert, 27 July 1827.
Robert Were Fox to Charles Lemon, 30 Mar. 1830.
Robert Were Fox to Charles Lemon, 9 Apr. 1834.
Robert Were Fox to Gilbert Davies, 28 Jan. 1837.
Robert Were Fox to [Gilbert Davies], 20 Nov. 1837.
Letter in private ownership from James Clark Ross to Alfred Fox (*c.*1839), sold 12 June 2019 in New York by Invaluable. Lot 126: 'James C. Ross orders a needle deflection compass for research & expeditions'. Printed copy in author's possession.

2. Secondary Material

2.1. Published

Alter, Peter, *The Reluctant Patron: Science and the State in Britain, 1850–1920* (Oxford: Berg, 1987).
Anon., *People from Falmouth* (Books LLC, 2010).
Anthony, Patrick, 'Mining as the Working World of Alexander von Humboldt's Plant Geography and Vertical Cartography', *Isis*, 109/1 (Mar. 2018), 28–55.
Aubin, David, Brigg, Charlotte, and Sibum, H. Otto (eds.), *The Heavens on Earth: Observatories and Astronomy in Nineteenth-Century Science and Culture* (Durham, NC: Duke University Press, 2010).
Baber, Zaheer, *The Science of Empire: Scientific Knowledge, Civilization, and Colonial Rule in India* (Albany: State University of New York, 1996).
Baigent, Elizabeth, 'Arctic Council (act. 1851)', *Oxford Dictionary of National Biography*, https://www.oxforddnb.com/view/10.1093/ref:odnb/9780198614128.001.0001/odnb-9780198614128-e-95281, 28 Sept. 2006; accessed 24 Mar. 2020.
Baigent, Elizabeth, 'Crozier, Francis Rawdon Moira (1796–1848)', *Oxford Dictionary of National Biography*, https://www.oxforddnb.com/view/10.1093/ref:odnb/9780198614128.001.0001/odnb-9780198614128-e-6840, 23 Sept. 2004; accessed 24 Mar. 2020.
Baigent, Elizabeth, 'Ross, Sir James Clark (1800–1862)', *Oxford Dictionary of National Biography*, https://www.oxforddnb.com/display/10.1093/ref:odnb/9780198614128.001.0001/odnb-9780198614128-e-24123, 23 Sept. 2004; accessed 24 Mar. 2020.
Barton, D. B., *The Cornish Beam Engine* (Truro: D. Bradford Barton Ltd, 1965).
Barton, D. B., *A History of Copper Mining in Cornwall and Devon* (Truro: D. Bradford Barton Ltd, 1968).
Bate, David G., 'Sir Henry Thomas De la Beche and the Founding of the British Geological Survey', *Mercian Geologist*, 17/3 (2010), 149–65.
Bennett, Brett M., and Hodge, Joseph M. (eds.), *Science and Empire: Knowledge and Networks of Science across the British Empire, 1800–1970* (Basingstoke: Palgrave Macmillan, 2011).
Bijker, Wiebe E., Hughes, Thomas P., and Pinch, Trevor (eds.), *The Social Construction of Technological Systems: New Directions in the Sociology and History of Technology* (Cambridge, MA: MIT, 2012).
Bradley, Margaret, and Perrin, Fernand, 'Charles Dupin's Study Visits to the British Isles, 1816–1824', *Technology and Culture*, 32/1 (Jan. 1991), 47–68.
Brett, R. L. (ed.), *Barclay Fox's Journal* (Bell & Hyman, 1979).
Brewer, John, *The Sinews of Power: War, Money and the English State, 1688–1783* (Cambridge, MA: Harvard University Press, 1989).
Bühler, W. K., *Gauss: A Biographical Study* (Berlin: Springer-Verlag, 1981).
Bulstrode, Jenny, 'Cetacean Citations and the Covenant of Iron', *Notes and Records*, 73/2 (2019), 167–85.
Burchell, Graham, Gordon, Colin, and Miller, Peter (eds.), *The Foucault Effect: Studies in Governmentality with Two Lectures by and an Interview with Michel Foucault* (Chicago: University of Chicago Press, 1991).
Burton, Anthony, *Richard Trevithick: Giant of Steam* (Aurum Press, 2000).
Cantor, Geoffrey, Gooding, David, and James, Frank A. J. L., *Michael Faraday* (New York: Humanity Books, 1991).
Cardwell, Donald S. L., *From Watt to Clausius: The Rise of Thermodynamics in the Early Industrial Age* (Heinemann, 1971).
Carlyle, E., rev. Payton, P., 'Woolf, Arthur (bap. 1766, d. 1837)', *Oxford Dictionary of National Biography*, https://www.oxforddnb.com/view/10.1093/ref:odnb/9780198614128.001.0001/odnb-9780198614128-e-29953, 23 Sept. 2004; accessed 30 Nov. 2019.

Carter, Christopher, *Magnetic Fever: Global Imperialism and Empiricism in the Nineteenth Century* (Philadelphia: American Philosophical Society, 2009).
Cawood, John, 'Terrestrial Magnetism and the Development of International Collaboration in the Early Nineteenth Century', *Annals of Science*, 34/6 (1977), 551–87.
Cawood, John, 'The Magnetic Crusade: Science and Politics in Early Victorian Britain', *Isis*, 70/4 (Dec. 1979), 493–518.
Chancellor, V. E., 'Fox, Caroline Fox (1819–1871)', *Oxford Dictionary of National Biography*, https://www.oxforddnb.com/view/10.1093/ref:odnb/9780198614128.001.0001/odnb-9780198614128-e-10019, 23 Sept. 2004, accessed 5 Dec. 2019.
Chakrabarti, Pratik, *Medicine and Empire, 1600–1960* (Basingstoke: Palgrave Macmillan, 2014).
Chakrabarti, Pratik, *Inscriptions of Nature: Geology and the Naturalization of Antiquity* (Baltimore: Johns Hopkins University Press, 2020).
Chapman, Allan, 'George Biddell Airy, F. R. S. (1801–1892): A Centenary Commemoration', *Notes and Records of the Royal Society of London*, 46/1 (Jan. 1992), 103–10.
Chapman, Allan, 'Airy, Sir George Biddell (1801–1892)', *Oxford Dictionary of National Biography*, https://www.oxforddnb.com/view/10.1093/ref:odnb/9780198614128.001.0001/odnb-9780198614128-e-251, 23 Sept. 2004; accessed 11 Jan. 2018.
Coen, Deborah R., *Climate in Motion: Science, Empire, and the Problem of Scale* (Chicago: University of Chicago Press, 2018).
Cooper, Alix, 'Homes and Households', in Katherine Park and Lorraine Daston (eds.), *Cambridge History of Science*, iii: *Early Modern Science* (Cambridge: Cambridge University Press, 2006), 224–37.
Craciun, Adriana, *Writing Arctic Disaster: Authorship and Exploration* (Cambridge: Cambridge University Press, 2016).
Craik, D. D., 'Victorian "Applied Mathematics"', in Raymond Flood, Adrian Rice, and Robin Wilson (eds.), *Mathematics in Victorian Britain* (Oxford: Oxford University Press, 2011), 177–98.
Crook, Denise, 'Fox, Robert Were (1789–1877)', *Oxford Dictionary of National Biography*, https://www.oxforddnb.com/view/10.1093/ref:odnb/9780198614128.001.0001/odnb-9780198614128-e-10042, 23 Sept. 2004; accessed 5 Dec. 2019.
Crook, Denise, 'Henwood (William) Jory (1805–1875)', *Oxford Dictionary of National Biography*, https://www.oxforddnb.com/view/10.1093/ref:odnb/9780198614128.001.0001/odnb-9780198614128-e-12997, 23 Sept. 2004; accessed 24 Mar. 2020.
Crosland, Maurice, *Science Under Control: The French Academy of Science, 1795–1914* (Cambridge: Cambridge University Press, 1992).
Crowe, Michael J., 'Herschel, Sir John Frederick William, first baronet (1792–1871)', *Oxford Dictionary of National Biography*, https://www.oxforddnb.com/view/10.1093/ref:odnb/9780198614128.001.0001/odnb-9780198614128-e-13101, 23 Sept. 2004; accessed 17 July 2019.
Daunton, Martin, *State and Market in Victorian Britain* (Woodbridge: Boydell Press, 2008).
Delbourgo, James, 'The Knowing World: A New Global History of Science', *History of Science*, 57/3 (2019), 373–99.
Dunn, Richard, and Higgitt, Rebekah (eds.), *Navigational Enterprises in Europe and Its Empires, 1730–1850* (Basingstoke: Palgrave Macmillan, 2016).
Enebakk, Vidar, 'Hansteen's Magnetometer and the Origin of the Magnetic Crusade', *British Journal for the History of Science*, 47/4 (Dec. 2014), 587–608.
Fan, Fa-Ti, 'The Global Turn in the History of Science', *East Asian Science, Technology and Society: An International Journal*, 6 (2012), 249–58.

Flood, Raymond, Rice, Adrian, and Wilson, Robin (eds.), *Mathematics in Victorian Britain* (Oxford: Oxford University Press, 2011).
Forgan, Sophie, 'Faraday—From Servant to Savant: The Institutional Context', in David Gooding and Frank A. J. L. James (eds.), *Faraday Rediscovered: Essays on the Life and Work of Michael Faraday, 1791–1867* (Basingstoke: Macmillan Press Ltd, 1985), 51–67.
Foucault, Michel, 'Governmentality', in Graham Burchell, Colin Gordon, and Peter Miller (eds.), *The Foucault Effect: Studies in Governmentality with Two Lectures by and an Interview with Michel Foucault* (Chicago: University of Chicago Press, 1991), 87–104.
Fox, Charles, *On the Brink: The Story of G. C. Fox and Company; Quaker Business in Cornwall through Eight Generations* (Zuleika, 2019).
Galison, Peter, 'The Many Faces of Big Science', in Peter Galison and Bruce Hevly (eds.), *Big Science: The Growth of Large-Scale Research* (Stanford, CA: Stanford University Press, 1992), 1–17.
Galison, Peter, and Hevly, Bruce (eds.), *Big Science: The Growth of Large-Scale Research* (Stanford, CA: Stanford University Press, 1992).
Gay, Susan E., *Old Falmouth: The Story of the Town from the Days of Killigrews to the Earliest Part of the 19th Century* (Headley Brothers, 1903).
Gillin, Edward J., 'Science on the Niger: Ventilation and Tropical Disease during the 1841 Niger Expedition', *Social History of Medicine*, 31/3 (Aug. 2018), 605–26.
Gillin, Edward J., 'Cornish Science, Mine Experiments, and Robert Were Fox's Penjerrick Letters', *Notes and Records of the Royal Society*, 76/1 (Mar. 2022), 49–65.
Gillin, Edward J., 'The Instruments of Expeditionary Science and the Reworking of Nineteenth-Century Magnetic Experiment', *Notes and Records of the Royal Society*, 76/3 (Sep. 2022), 565–592.
Good, Gregory A., 'Sabine, Sir Edward (1788–1883)', *Oxford Dictionary of National Biography*, https://www.oxforddnb.com/view/10.1093/ref:odnb/9780198614128.001.0001/odnb-9780198614128-e-24436, 23 Sept. 2004; accessed 24 Mar. 2020.
Gooday, Graeme, 'Precision Measurement and the Genesis of Physics Teaching Laboratories in Victorian Britain', *British Journal for the History of Science*, 23/1 (Mar. 1990), 25–51.
Gooday, Graeme, 'Placing or Replacing the Laboratory in the History of Science', *Isis*, 99/4 (Dec. 2008), 783–95.
Gooding, David, '"In Nature's Schools": Faraday as an Experimentalist', in David Gooding and Frank A. J. L. James (eds.), *Faraday Rediscovered: Essays on the life and Work of Michael Faraday, 1791–1867* (Basingstoke: Macmillan Press Ltd, 1985), 105–35.
Gooding, David, and James, Frank A. J. L. (eds.), *Faraday Rediscovered: Essays on the Life and Work of Michael Faraday, 1791–1867* (Basingstoke: Macmillan Press Ltd, 1985).
Goodman, Matthew, 'Follow the Data: Administrating Science at Edward Sabine's Magnetic Department, Woolwich, 1841–1857', *Notes and Records*, 73/2 (2019), 187–202.
Grattan-Guinness, I., 'Mathematics and Mathematical Physics from Cambridge, 1815–1840: A Survey of the Achievements and of the French Influences', in P. M. Harman (ed.), *Wranglers and Physicists: Studies on Cambridge Physics in the Nineteenth Century* (Manchester: Manchester University Press, 1985), 84–111.
Griffiths, Tom, *Slicing the Silence: Voyaging to Antarctica* (Cambridge, MA: Harvard University Press, 2007).
Guerrini, Anita, 'The Ghastly Kitchen', *History of Science*, 54/1 (2016), 71–97.
Gunn, Simon, and Vernon, James, 'Introduction: What Was Liberal Modernity and Why Was It Peculiar in Imperial Britain?', in Simon Gunn and James Vernon (eds.), *The Peculiarities of Liberal Modernity in Imperial Britain* (Berkeley and Los Angeles: University of California Press, 2011), 1–18.

280 SELECT BIBLIOGRAPHY

Gunn, Simon, and Vernon, James (eds.), *The Peculiarities of Liberal Modernity in Imperial Britain* (Berkeley and Los Angeles: University of California Press, 2011).

Hardenberg, Wilko Graf von, and Mahony, Martin, 'Introduction—Up, Down, Round and Round: Verticalities in the History of Science', *Centaurus*, 62 (2020), 595–611.

Harman, P. M. (ed.) *Wranglers and Physicists: Studies on Cambridge Physics in the Nineteenth Century* (Manchester: Manchester University Press, 1985).

Headrick, Daniel R., *The Tools of Empire: Technology and European Imperialism in the Nineteenth Century* (Oxford: Oxford University Press, 1981).

Hecht, Gabrielle, *Being Nuclear: Africans and the Global Uranium Trade* (Cambridge, MA: MIT Press, 2012).

Hills, Richard L., *Power from Steam: A History of the Stationary Steam Engine* (Cambridge: Cambridge University Press, 1994).

Hills, R., 'Hornblower, Jonathan (1717–1780)', *Oxford Dictionary of National Biography*, https://www.oxforddnb.com/view/10.1093/ref:odnb/9780198614128.001.0001/odnb-9780198614128-e-13783, 23 Sept. 2004; accessed 30 Nov. 2019.

Howard, Bridget, *Mr. Lean and the Engine Reporters* (Penryn: Trevithick Society, 2002).

Hughes, Thomas P., *Human-Built World: How to Think about Technology and Culture* (Chicago: University of Chicago Press, 2004).

Hughes, Thomas P., 'The Evolution of Large Technological Systems', in Wiebe E. Bijker, Thomas P. Hughes, and Trevor Pinch (eds.), *The Social Construction of Technological Systems: New Directions in the Sociology and History of Technology* (Cambridge, MA: MIT, 2012), 45–76.

Hunt, Bruce J., 'Doing Science in a Global Empire: Cable Telegraphy and Electrical Physics in Victorian Britain', in Bernard Lightman (ed.), *Victorian Science in Context* (Chicago: University of Chicago Press, 1997), 312–33.

Inkster, Ian, and Morrell, Jack (eds.), *Metropolis and Province: Science in British Culture, 1780–1850* (Hutchinson, 1983).

Irving, Sarah, *Natural Science and the Origins of the British Empire* (Routledge, 2016).

James, Frank A. J. L. (ed.), *The Correspondence of Michael Faraday*, ii: *1832–December 1840: Letters 525–1333* (Institution of Electrical Engineers, 1993).

James, Frank A. J. L. (ed.), *The Correspondence of Michael Faraday*, iv: *Jan. 1849–Oct. 1855: Letters 2146–3032* (Institution of Electrical Engineers, 1996).

Jenkin, A. K. Hamilton, *The Cornish Miner* (George Allen & Unwin, 1962).

Josefowicz, Diane Greco, 'Experience, Pedagogy, and the Study of Terrestrial Magnetism', *Perspectives on Science*, 13/4 (Winter, 2005), 452–94.

Joyce, Patrick, *The State of Freedom: A Social History of the British State since 1800* (Cambridge: Cambridge University Press, 2013).

Knight, D., 'Davy, Sir Humphry, baronet (1778–1829)', *Oxford Dictionary of National Biography*, https://www.oxforddnb.com/view/10.1093/ref:odnb/9780198614128.001.0001/odnb-9780198614128-e-7314, 23 Sept. 2004; accessed 1 Dec. 2019.

Lambert, Andrew, *Franklin: Tragic Hero of Polar Navigation* (Faber and Faber, 2009).

Lambert, Erin, *Singing the Resurrection: Body, Community, and Belief in Reformation Europe* (Oxford: Oxford University Press, 2017).

Larson, Edward J., 'Public Science for a Global Empire: The British Quest for the South Magnetic Pole', *Isis*, 102/1 (Mar. 2011), 34–59.

Laughton, J. K., rev. Lambert, Andrew, 'Belcher, Sir Edward (1799–1877)', *Oxford Dictionary of National Biography*, https://www.oxforddnb.com/view/10.1093/ref:odnb/9780198614128.001.0001/odnb-9780198614128-e-1979, 23 Sept. 2004; accessed 24 Mar. 2020.

Leggett, Don, 'Neptune's New Clothes: Actors, Iron and the Identity of the Mid-Victorian Warship', in Don Leggett and Richard Dunn (eds.), *Re-Inventing the Ship: Science, Technology and the Maritime World, 1800–1918* (Farnham: Ashgate, 2012), 71–91.

Leggett, Don, and Dunn, Richard (eds.), *Re-Inventing the Ship: Science, Technology and the Maritime World, 1800–1918* (Farnham: Ashgate, 2012).

Levere, Trevor H., 'Magnetic Instruments in the Canadian Arctic Expeditions of Franklin, Lefroy, and Nares', *Annals of Science*, 43/1 (1986), 57–76.

Levere, Trevor H., *Science and the Canadian Arctic: A Century of Exploration, 1818–1918* (Cambridge: Cambridge University Press, 1993).

Lightman, Bernard (ed.), *Victorian Science in Context* (Chicago: University of Chicago Press, 1997).

Livingstone, David N., and Withers, Charles W. J. (eds.), *Geographies of Nineteenth-Century Science* (Chicago: University of Chicago Press, 2011).

Locher, Fabien, 'The Observatory, the Land-Based Ship and the Crusade: Earth Sciences in European Context, 1830–1850', *British Journal for the History of Science*, 40/4 (Dec. 2007), 491–504.

McConnell, Anita, 'Jordan, Thomas Brown (1807–1890)', *Oxford Dictionary of National Biography*, https://www.oxforddnb.com/view/10.1093/ref:odnb/9780198614128.001.0001/odnb-9780198614128-e-15123, 23 Sept. 2004; accessed 24 Mar. 2020.

McConnell, Anita, 'Surveying Terrestrial Magnetism in Time and Space', *Archives of Natural History*, 32/2 (2005), 346–60.

McCook, Stuart, 'Introduction', *Isis*, 104/4 (Dec. 2013), 773–6.

Macdonald, Lee T., 'Making Kew Observatory: The Royal Society, the British Association and the Politics of Early Victorian Science', *British Journal for the History of Science*, 48/3 (Sept. 2015), 409–33.

Macdonald, Lee T., '"Solar Spot Mania": The Origins and Early Years of Solar Research at Kew Observatory, 1852–1860', *Journal for the History of Astronomy*, 46/4 (2015), 469–90.

Macdonald, Lee T., *Kew Observatory and the Evolution of Victorian Science* (Pittsburgh: University of Pittsburgh Press, 2018).

Macdonald, Lee T., 'The Origins and Early Years of the Magnetic and Meteorological Department at Greenwich Observatory, 1834–1848', *Annals of Science*, 75/3 (2018), 201–33.

Mandler, Peter, 'Introduction: State and Society in Victorian Britain', in Peter Mandler (ed.), *Liberty and Authority in Victorian Britain* (Oxford: Oxford University Press, 2006), 1–21.

Mandler, Peter (ed.), *Liberty and Authority in Victorian Britain* (Oxford: Oxford University Press, 2006).

Marsden, Ben, *Watt's Perfect Engine: Steam and the Age of Invention* (Cambridge: Icon Books, 2002).

Marsden, Ben, and Smith, Crosbie, *Engineering Empires: A Cultural History of Technology in Nineteenth-Century Britain* (Basingstoke: Palgrave Macmillan, 2005).

Mawer, Granville Allen, *South by Northwest: The Magnetic Crusade and the Contest for Antarctica* (Edinburgh: Birlinn Ltd, 2006).

Miller, D., 'Gilbert [formerly Giddy], Davies (1767–1839)', *Oxford Dictionary of National Biography*, https://www.oxforddnb.com/view/10.1093/ref:odnb/9780198614128.001.0001/odnb-9780198614128-e-10686, 23 Sept. 2004; accessed 1 Dec. 2019.

Morrell, Jack, 'Economic and Ornamental Geology: The Geological and Polytechnic Society of the West Riding of Yorkshire, 1837–1853', in Ian Inkster and Jack Morrell (eds.), *Metropolis and Province: Science in British Culture, 1780–1850* (Hutchinson, 1983), 230–56.

Morrell, Jack, 'Phillips, John (1800–1874)', *Oxford Dictionary of National Biography*, https://www.oxforddnb.com/view/10.1093/ref:odnb/9780198614128.001.0001/odnb-9780198614128-e-22163, 23 Sept. 2004; accessed 24 Mar. 2020.

Morrell, Jack, *John Phillips and the Business of Victorian Science* (Ashgate: Aldershot, 2005).

Morrell, Jack, and Thackray, Arnold, *Gentlemen of Science: Early Years of the British Association for the Advancement of Science* (Oxford: Clarendon Press, 1981).

Morris, Richard (ed.), *A Journal of Sir Henry De la Beche: Pioneer Geologist (1796–1855)* (Swansea: Royal Institution of South Wales, 2013), 29–31.

Naylor, Simon, *Regionalizing Science: Placing Knowledge in Victorian England* (Pickering & Chatto, 2010).

Naylor, Simon, 'Geological Mapping and the Geographies of Proprietorship in Nineteenth-Century Cornwall', in David N. Livingstone and Charles W. J. Withers (eds.), *Geographies of Nineteenth-Century Science* (Chicago: University of Chicago Press, 2011), 345–70.

Naylor, Simon, and Schaffer, Simon, 'Nineteenth-Century Survey Sciences: Enterprises, Expeditions and Exhibitions: Introduction', *Notes and Records*, 73/2 (2019), 135–47.

Nuvolari, Alessandro, and Verspagen, Bart, '*Lean's Engine Reporter* and the Development of the Cornish Engine: A Reappraisal', *Transactions of the Newcomen Society*, 77 (2007), 167–89.

Ogle, Vanessa, *The Global Transformation of Time, 1870–1950* (Cambridge, MA: Harvard University Press, 2015).

O'Hara, James Gabriel, 'Gauss and the Royal Society: The Reception of His Ideas on Magnetism in Britain (1832–1842)', *Notes and Records*, 38/1 (Aug. 1983), 17–78.

O'Hara, James G., 'Lloyd, Humphrey (1800–1881)', *Oxford Dictionary of National Biography*, https://www.oxforddnb.com/view/10.1093/ref:odnb/9780198614128.001.0001/odnb-9780198614128-e-16840, 23 Sept. 2004; accessed 24 Mar. 2020.

Park, Katherine, and Daston, Lorraine (eds.), *Cambridge History of Science*, iii: *Early Modern Science* (Cambridge: Cambridge University Press, 2006).

Parry, J. P., 'Liberalism and Liberty', in Peter Mandler (ed.), *Liberty and Authority in Victorian Britain* (Oxford: Oxford University Press, 2006), 71–100.

Payton, P., 'Fox, Robert Were (1754–1818)', *Oxford Dictionary of National Biography*, https://www.oxforddnb.com/view/10.1093/ref:odnb/9780198614128.001.0001/odnb-9780198614128-e-42083, 23 Sept. 2004; accessed 5 Dec. 2019.

Pearson, A., 'The Royal Horticultural Society of Cornwall', *Journal of the Royal Institution of Cornwall*, NS 7/2 (1974), 165–73.

Porter, Dale H., *The Life and Times of Sir Goldsworthy Gurney: Gentleman Scientist and Inventor, 1793–1875* (Bethlehem: Lehigh University Press, 1998).

Poskett, James, 'Phrenology, Correspondence, and the Global Politics of Reform, 1815–1848', *Historical Journal*, 60/2 (2017), 409–42.

Poskett, James, *Materials of the Mind: Phrenology, Race, and the Global History of Science, 1815–1920* (Chicago: University of Chicago Press, 2019).

Poskett, James, *Horizons: A Global History of Science* (Viking, 2022).

Ratcliff, Jessica, *The Transit of Venus Enterprise in Victorian Britain* (Pickering & Chatto, 2008).

Ratcliff, Jessica, 'Travancore's Magnetic Crusade: Geomagnetism and the Geography of Scientific Production in a Princely State', *British Journal for the History of Science*, 49/3 (Sept. 2016), 325–52.

Riffenburgh, B. A., 'Franklin, Sir John (1786–1847)', *Oxford Dictionary of National Biography*, https://www.oxforddnb.com/view/10.1093/ref:odnb/9780198614128.001.0001/odnb-9780198614128-e-10090, 23 Sept. 2004; accessed 24 Mar. 2020.

Roberts, Lissa, 'Situating Science in Global History: Local Exchanges and Networks of Circulation', *Itinerario*, 33/1 (2009), 9–30.

Rolt, L. T. C., *The Cornish Giant: The Story of Richard Trevithick, Father of the Steam Locomotive* (Lutterworth Press, 1960).

Rowe, John, *Cornwall in the Age of Industrial Revolution* (St Austell: Cornish Hillside Publications, 1993).

Rudwick, Martin J. S., *The Great Devonian Controversy: The Shaping of Scientific Knowledge among Gentlemanly Specialities* (Chicago: University of Chicago Press, 1985).

Rudwick, Martin J. S., *Worlds Before Adam: The Reconstruction of Geohistory in the Age of Reform* (Chicago: University of Chicago Press, 2008).

Savours, Ann, and McConnell, Anita, 'The History of the Rossbank Observatory, Tasmania', *Annals of Science*, 39 (1982), 527–64.

Schaffer, Simon, 'Physics Laboratories and the Victorian Country House', in Crosbie Smith and Jon Agar (eds.), *Making Space for Science: Territorial Themes in the Shaping of Knowledge* (Basingstoke: Macmillan Press Ltd, 1998), 149–80.

Secord, James A., *Visions of Science: Books and Readers at the Dawn of the Victorian Age* (Oxford: Oxford University Press, 2014).

Seth, Suman, 'Colonial History and Postcolonial Science Studies', *Radical History Review*, 127 (Jan. 2017), 63–85.

Shapin, Steven, 'The House of Experiment in Seventeenth-Century England,' *Isis*, 79/3, special issue on Artifact and Experiment (Sept. 1988), 373–404.

Shapin, Steven, *The Scientific Life: A Moral History of a Late Modern Vocation* (Chicago: University of Chicago Press, 2008).

Sharpe, T., and McCartney, P. J. (eds.), *The Papers of H. T. De la Beche (1796–1855) in the National Museum of Wales* (Cardiff: National Museums & Galleries of Wales, 1998).

Siegfried, Robert, and Dott, Jr., Robert H. (eds.), *Humphry Davy on Geology: The 1805 Lectures for the General Audience* (Madison: University of Wisconsin Press, 1980).

Sivasundaram, Sujit, 'Science and the Global: On Methods, Questions, and Theory', *Isis*, 101/1 (Mar. 2010), 146–58.

Smith, Crosbie, *Coal, Steam and Ships: Engineering, Enterprise and Empire on the Nineteenth-Century Seas* (Cambridge: Cambridge University Press, 2018).

Smith, Crosbie, and Agar, Jon (eds.), *Making Space for Science: Territorial Themes in the Shaping of Knowledge* (Basingstoke: Macmillan Press Ltd, 1998).

Smith, Crosbie, and Wise, M. Norton, *Energy and Empire: A Biographical Study of Lord Kelvin* (Cambridge: Cambridge University Press, 1989).

Temperley, Howard, *White Dreams, Black Africa: The Antislavery Expedition to the River Niger, 1841–1842* (New Haven: Yale University Press, 1991).

Todd, A. C., *Beyond the Blaze: A Biography of Davies Gilbert* (Truro: D. Bradford Barton Ltd, 1967).

Tresch, John, 'Even the Tools Will Be Free: Humboldt's Romantic Technologies', in David Aubin, Charlotte Brigg, and H. Otto Sibum (eds.), *The Heavens on Earth: Observatories and Astronomy in Nineteenth-Century Science and Culture* (Durham, NC: Duke University Press, 2010), 253–84.

Trounson, J. H., *Cornish Engines and the Men Who Handled Them* (St Ives: J. H. Trounson & the Trevithick Society, 1985).

Vernon, James, *Politics and the People: A Study in English Political Culture, c.1815–1867* (Cambridge: Cambridge University Press, 1993).

Warwick, Andrew, *Masters of Theory: Cambridge and the Rise of Mathematical Physics* (Chicago: University of Chicago Press, 2003).

Weinberg, Alvin M., 'Impact of Large-Scale Science in the United States', *Science*, NS 134/3473 (21 July 1961), 161–4.

Weindling, Paul, 'The British Mineralogical Society: A Case Study in Science and Social Improvement', in Ian Inkster and Jack Morrell (eds.), *Metropolis and Province: Science in British Culture, 1780–1850* (Hutchinson, 1983), 120–50.

Werrett, Simon, *Thrifty Science: Making the Most of Materials in the History of Experiment* (Chicago: University of Chicago Press, 2019).

Williamson, Fiona, 'Weathering the Empire: Meteorological Research in the Early British Straits Settlements', *British Journal for the History of Science*, 48/3 (Sept. 2015), 475–92.

Wilton, Robert, *The Wiltons of Cornwall* (Chichester: Phillimore, 1989).

Winter, Alison, '"Compasses All Awry": The Iron Ship and the Ambiguities of Cultural Authority in Victorian Britain', *Victorian Studies*, 38/1 (Autumn, 1994), 69–98.

Yeo, Richard, 'Whewell, William (1794–1866)', *Oxford Dictionary of National Biography*, https://www.oxforddnb.com/view/10.1093/ref:odnb/9780198614128.001.0001/odnb-9780198614128-e-29200, 21 May 2009; accessed 17 July 2018.

2.2. Unpublished

Bulstrode, Jenny, 'Men, Mines, and Machines: Robert Were Fox, the Dip-Circle and the Cornish System', Part III dissertation, History and Philosophy of Science Department, University of Cambridge, 2013.

Goodman, Matthew, 'From "Magnetic Fever" to "Magnetical Insanity": Historical Geographies of British Terrestrial Magnetic Research, 1833–1857', PhD thesis, University of Glasgow, 2018.

Maxted, Julie, 'Scientific and Cultural Networks in Nineteenth-Century Falmouth: The Fox Family and the Royal Cornwall Polytechnic Society', BA dissertation, University of Exeter, n.d.

Pearson, Alan, 'A Study of the Royal Cornwall Polytechnic Society', MA thesis, University of Exeter, 1973.

Qidwai, Sarah Ahmed, 'Sir Syed (1817–1898) and Science: Popularization in Nineteenth-Century India', PhD thesis, University of Toronto, 2021.

Index

For the benefit of digital users, indexed terms that span two pages (e.g., 52–53) may, on occasion, appear on only one of those pages.

Aaron Manby 96–7
Abyssinia 230
Académie des sciences 137–8
Adams, John Couch 257
Aden 144–5
Admiralty 12–13, 16, 101–2, 105–6, 110, 134, 136–7, 140, 145, 148, 153–4, 168–9, 181–2, 215–18, 222–4, 261–2, 264–5
  Compass Committee 247–8
  Compass Department 247–51
  compasses 247–56
Admiralty Jetty 154
Africa 8–9, 133, 168–71, 173–6, 228–9, 231, 235, 246–51, 255–6
African Civilization Society (ACS) 169–71, 175–6
  inaugural meeting 169–70, 230
Ahab 57, 95
  makes a compass 95
Airy, George Biddell 97, 129–32, 165–6, 235, 245–7, 251–5
Aix 137–8
Albemarle Street 18–19, 40
Albert 170–1, 192–3
Albert, Prince 169–70
Alcala, Captain-General 197
Alert, HMS 262
Alexander 102, 105–6
Algiers 8–9, 144–5, 171
Allen, Bird 170–6
Allen, William 170–8
All Souls College, Oxford 243
Altona Observatory 106–7
America 107–8
  North 134, 140, 168–9, 171, 181–2, 186–8, 191–3
  South 191–2
Ampère, André-Marie 132
anatomy 40
Anglesey 29–31
Anglo-Chinese Opium War 192–3, 208–9, 212
Anglo-Japanese Friendship Treaty 202
Annales de chimie et de physique 49–50

Annals of Electricity 115–18
Annals of Philosophy 49–50
Antarctica 8–9, 184–5, 190–1, 204–5, 217, 226, 261–2
Antarctic Expedition of 1839–43 16, 21–2, 129–30, 133–5, 142–66, 177–80, 184–6, 192–3, 196–7, 205, 216–17, 227–8, 232–3
Arago, François 99–101, 107–8, 125, 163–4
Arctic xx–xxi, 8–10, 20–1, 24–5, 103, 105–6, 110, 119–20, 123–5, 128–9, 143–5, 184–5, 187–8, 190–1, 221–2, 261–2
  1818 expedition to xxi, 102–3, 119
  1836 expedition to 167, 173, 177
  Franklin Expedition to 190, 212–24
  Victory expedition 108, 148
Arctic Council 217–19
Armagh 260–1
Arnhem 137–8
Arrow, HMS 139–40
Ascension Island 192–3, 248–9
Ashmolean Museum 238–9
Asia 8–9, 68
Astrolabe 99–100
Athenaeum Club 119–20
Atlantic Ocean 5–9, 22, 27, 68, 98–9, 102, 105–7, 133–4, 139–42, 146–7, 149, 151, 165–6, 168–9, 185–6, 192–3, 204–5, 211–12, 226
Aurora Borealis 62–6
Australia 8–9, 119, 147–8, 157, 167–9, 177–9, 187–8, 208–11, 246
Austro-Hungarian Empire 6–7
Azores 210–11

Back, George 102–3, 137, 167, 177, 219
Baden 137–8
Baffin Bay 102–3, 216
Bahia 140–1
Bakerian Lecture 7–8, 57, 63–6, 106
Balambangan Island 200
Banks, Joseph 41–2, 261–2
Barclay, Arthur 175
Barclay, Maria 43–5, 238–9
Baring, Francis 16–17

Barnard Island 210-11
Barnett, Edward 137, 177, 180-1, 186-7
Barlow, Peter 108
Barrow, John 102-3, 105-6, 110, 119, 212, 217-19, 230
Barrow, John junior 217-19
Batavia 206
Bay of Islands 210-11
Beaufort, Francis 16, 121-2, 137, 142-4, 148-52, 157, 191-2, 217-18, 247-8, 264
Beagle, HMS xxi, 135-6, 139
Beechey, William Frederick 191-2, 219
Belcher, Edward xxi, 10-12, 191-205, 209, 218-20, 246-7, 264
 fashions temporary magnetic observatories 197-200, 203-4
 magnetic experiments at Datoo Danielle's home 200-1
 magnetic surveying of Japan 201-3
 reports on Fox's dipping needle 202-4
 tea at Rosehill 193-5
Bellerophon, HMS 119
Bengal 205-8
Bere Alston Mine 51
Bergama, Dr 240, 243
Berkeley Sound 210-11
Berlin 5-6
Bermuda 189-90
Big Science 13-14
Biot, Jean-Baptiste 100-1
Bird, Edward 219
Birmingham 31-3, 126
Bisaynn 197-200
Biscay 265-7
Black Prince 248-9
Blackwood, Francis Price 177, 182-4, 186-7, 261-2
Bligh, William 101-2
Board of Longitude 12-13
Board of Trade 260-1
 Meteorological Department 260-1
Bombay
 Observatory 144-5
Bonaparte, Napoleon 2-3, 102
 Egyptian Expedition 47-8
 exile on St Helena 151
Borneo 8-10, 190-1, 193, 197-208, 226
Botallack Mine 27, 48-9, 75-8
botany 39-40, 222-4
Boulton, Matthew 31-4
Bramble, HMS 177, 182-4
Brazil 78-9, 110-11, 134, 186-7
Breslau 8-9, 144-5
Brewer, John 15-16

Bridget, Howard 37
Bristol 40-2, 78-9, 84-8, 115-19, 266
 BAAS meeting 80-1
Britannic 255-6
British America 12-13
British Army 133, 264
British Association for the Advancement of Science (BAAS) 10-12, 80-8, 115-19, 121-2, 128-33, 142-3, 156, 158, 189-90, 205-8, 228, 233-5, 238-9, 245-7
 Fox needle at 121-2
 Kew Observatory 235
 magnetic lobby of Parliament 134, 142-6
 Magnetic Congress of 1845 189-90
 Magnetic Survey of the British Isles 121-32, 136-9, 141-3, 237-9, 243
 Magnetic Survey of the British Isles of 1857-1861 228, 237-43
British Empire xx-xxi, 8-9, 169-70, 227-8, 235, 262-3
 and magnetic science 8-9
British Geological Survey 12-13, 89
British Isles 7-8, 21, 24-5, 97, 121-3, 260-1
British Isles magnetic survey 121-32, 136-9, 141-3, 237-9, 243
British Isles magnetic survey of 1857-1861 228, 237-43
British Magnetic Scheme (BMS) 2-5, 7-8, 13-15, 18, 21-2, 24, 142-4, 164-6, 262-7
 Antarctic Expedition 133-5, 142-63, 177-80, 184-6, 227-9, 231-3, 248-9, 256
 as a system 164-6, 189-91, 204-12, 222-6, 256, 262-7
 and the state 15-18
 Canada survey 184-7, 190
 challenges faced by 8
 controversy surrounding 233-5
 Elliot's Singapore and Borneo data 204-8
 finance 12-13, 134, 142-3
 Franklin Expedition 190, 212-24
 global ambitions of 8-12, 237-40, 256
 government extends funding of 232-5
 HMS *Thunder* expedition 177, 180-3
 Kew Observatory 235, 240
 lobby for 142-6
 Niger Expedition 171-7, 184-5, 187-8, 190
 observatories 144-5
 Pagoda Expedition 161-3, 165-6
 problem of iron shipping 246-51, 256
 published outputs 160-1, 163-6, 191-3, 198, 207, 211-12, 232-7, 241-3, 246, 252, 256
 Rattlesnake Expedition 190, 208-12
 Samarang Expedition 190-1, 193-205
Solar spot findings 235-7

Sulphur results 191–3
systematic data collection 21, 24–5, 167–9, 177, 180–1, 184, 187–8, 237–8, 243, 252, 256, 262–7
Trivandrum data 206–8
British Magnetic Survey of the British Isles 121–32
British shipping 96–7, 101–2
British state 14–18, 21, 24, 264–5
and the BMS 15–18
liberal politics 14–15, 17–18, 264
relationship with science 14–15, 189–90
British Queen 118–19
Brown, Allen 206–8
Brown, Henry 103
Brumi Island 211
Brunel, Isambard Kingdom 96–7, 118–19, 231–2, 244
Brussels 8–9, 144–5
Observatory 126–8
Buckland, William 84–8
Bulstrode, Jenny 186–7
Buxton, Thomas 169–71, 175–6

Cadiz 8–9, 144–5, 171
Cagayan Sulu 203–4
Cairo 144–5, 171
Calcutta 206–8
Caldecott, John 206–8
Camborne 33–4, 38
School of Mines 54–5
Cambridge 5–6, 47–8, 93–4, 97, 242, 244–5, 257
BAAS meeting 80–1, 122
Cambridge, Massachusetts 8–9, 144–5
Canada xx–xxi, 21–2, 133–4, 138–9, 144–5, 184–6, 190
Canton River 191–2
Cape Coast Castle 175–6
Cape Crozier 156–7
Cape Felix 220–1
Cape of Good Hope 141–3, 154–5, 159–60, 180–1, 196–7, 210–12, 235–8
Observatory 144–5, 154, 161–5, 189–90, 196–7
Cape Horn 261–2
Cape Palmas 175–6
Cape Town 1–2, 7–10, 133, 145, 150–1, 156–8, 161, 164–5, 171, 187–8, 196–7, 204–5, 208–11, 227–8, 235–6, 246–7, 265–7
for the observatory, see *Cape of Good Hope Observatory*
Cape Verde 150–3, 156, 173–4, 192–3, 196–7
Cape York 210–11
Carclew House 84–8, 167–8
Cardon Island 191–2

Caribbean 8–9, 97
Carter, Christopher 8–9
Castle Line 255–6
Cawood, John 2–3, 134
Ceylon 189–90, 192–3
Chacewater 32–3
Mine 31–2
Chakrabarti, Pratik 6–7
Challenger, HMS 261–2
Expedition 243, 261–2
Charing Cross 112
Chartist 233–5
Chatham 184–5
Cheltenham
BAAS meeting 238–9
chemistry 38–40
Chesney, Francis 120–1
Chile 61
copper output 258–9
China 191–5, 201, 203–4, 235–6
Sea 192–3
Chiswick 107–8
Christiania 115–18
University of 110
Christie, Samuel Hunter 69–73, 93–4, 110, 128–9, 247–8
Christmas Day 200
Christmas Harbour 154
Clausen, Henrik 110–11
Clerk, H 161, 196–7
Cliffdown Mine 75
Clyde, River 30–1
coal 3–4, 27–9, 42–3, 50, 53–4
economy 30–2, 35–7
mines 32
Coalbrookdale 33–4
Coalbrookdale Iron Company 118–19
Coco Islands 206–8
Cocos Islands 191–2
Coen, Deborah 6–7
Cologne 137–8
Colombo 1–2, 7–8
colonialism 190, 201, 226
colonial observatories 12–13
Colonial Office 16, 107–8
colonies 197
and magnetic survey work 133, 189–91, 201
compass 95, 98–9, 105–7, 118–19, 195–6
error 2–3, 95–7, 101–2, 105–6, 118–19
solving magnetic-induced error 244–56
compound engine 32–8, 42–3, 45–6, 49, 52–3, 55
Conrad, Joseph 1
Consolidated Mines 51–2, 75–8

Consols Mine 35–7
Cook, James 12–13, 101–2
Cook's Kitchen Mine 36–7, 46–7
Coppermine River Expedition 119–20, 216–17
copper trade 29, 258–9
Coquille 99–100
Coral Haven 210–11
Coral Sea 210–11
Cordier, Pierre Louis Antoine 47–8, 79
Cornish 4–6, 18–21, 24–5, 78–9, 161, 165–7, 177–8, 185–6
  copper 29, 55, 258–9
  engineering 27–8, 33–9, 42–3, 45–6, 52–5
  mining 3–4, 19–20, 26–8, 36–7, 91–3, 132, 258–60, 262–3
  industry 3–4
  science 3–8, 21, 28, 38–45, 59, 63–6, 88–9, 118–19, 124, 130–1, 134–5, 141, 146–7, 149–50, 167–8, 190, 228, 232, 257–64
  scientific networks 40–2, 59, 263
Cornish boiler 34–5
Corn Laws 169–70
Cornwall 3–8, 18–21, 24, 27–35, 46–8, 53–4, 71–2, 94, 97–8, 131–3, 177–8, 240, 243, 264
  climate 38–9
  geothermic gradient 54–5
  industrial science 43–5
  mines 21, 26–8, 30–2, 35–7, 42–51, 57–61, 79–80, 93–4, 111–12, 258–60
  steam power 55
Cornwall Geological Society, see *Royal Geological Society of Cornwall*
Cornwall Philosophical Society 39–40
Cosmos 236–7
Coulman, Ann 217–18, 238–9, 243
County Down 145
Cove, HMS 145
COVID-19 1–2, 265–7
Creak, Ettrick 262
credibility 131–2
  scientific 57, 71, 80–1, 93–4
Crocker Mountains 102–3, 108
Cromer 175–6
Crowan 36–7
  Mine 36–7
Crozier, Francis xxi, 16, 140–1, 173–4, 192–3, 209, 228–9
  biography 145
  on Antarctic Expedition, 1839–1843 145, 147–54, 156–8, 160
  on Franklin Expedition 212–17, 220–1
Cumberland House 119–20
Cunard 244, 255–6

d'Abbadie, Thomson 229–31
Darwin, Charles xxi, 135–6, 218–20, 222–4
Datoo Danielle 200–1
Davey, John 36–7, 43–5
Davey, William 43–5
Davis Strait 102–3, 220
Davy, Humphry 40–2, 59, 106–7
Dayman, Joseph 208–12
decline of Cornish science 257–9
deep time 6–7
Delbourgo, James 6–7
De la Beche, Henry 82–3, 88–91, 131–2, 177–8
  on mineral formation 91–3
  Robert Were Fox relations 89–91
Deptford 248–9
Denmark 12–13
Derbyshire 32–3
Devon 7–8, 43–5, 78–9, 88–9, 91–3
  copper output 258–9
Devon Great Consols Mine 258–9
Devonport 173–4
Devonshire Street 220–1
Dickens, Charles 217–18
Ding Dong Mine 33–4, 48–9
dip circle, see *dipping needle*
dipping needle xix–xx, xxi, 3–5, 7–8, 21–2, 24–5, 58, 97–101, 107–8, 110–15, 130–1, 241–2, 246–7
  failings 9–10, 106, 123, 130
  Hansteen's 110–11, 123, 193–5, 205–6
  Robinson's 172–3, 184–6, 191–2, 208–11, 214, 216, 241
Discovery, HMS
  Expedition 262
diurnal variation 163–4
Dolcoath Mine 33–7, 42–3, 45–7, 49–51, 69–70, 75
Dolland 106–8, 110–11, 123
Dover, John 262
Drummondville 138–9
Duchâteau Islands 210–11
Duchy of Cornwall 78–9
Dublin 103, 122, 124–5, 173–4, 233–5
  BAAS meeting 80–1, 121–2, 126, 129–30
  Trinity College 122–5, 238–9
Dudley Castle 29
Dundrum Bay 244
Dunlop, James 141
Dunn, Sarah 115–18
Dunstan, Richard 75
Dutch East Indies 206, 240
duty
  as a measure of steam engine efficiency 32, 34–7, 258–9
d'Urville, Dumont 154–5

INDEX

Eastbourne 130–1, 240
East India Trading Company 142–5, 204–5
East Liscombe Mine 51
École des mines 137–8
Edinburgh 5–6, 69–70, 93–4, 257
   BAAS meeting 81–2, 125
   University of 16–17, 73
Eggarah 173–4
Egypt 47–8, 230
   Pasha of 8–9
Eisenhower, Dwight D. 13–14
electromagnetic induction 18–19, 56–7
electromagnetic rotation 18–19, 56–7
Elliot, Charles 204–8
   survey work 206–8
empire 4–5, 22–5
   as a term 9–12
   and science 10–12
Endeavour, HMS 12–13
England 191–3, 210, 212, 217–18, 220
   magnetic character 240, 242–3
English Channel 133, 248–9
Enys, John 88–9, 131–2, 141
Erebus, HMS xxi, 208–12, 228–9, 246–7
   Antarctic Expedition, 1839–1843 133, 145, 147–54, 156–8, 160–6, 178–80, 184–5, 187–8
   Franklin Expedition 212–20
Estcourt, James Bucknell 121, 138–42
Ethiopian Expedition 230
Eugenie 231
Euphrates, River 120–1
   Expedition 121, 138–9, 212, 214
Europe 21, 68, 94, 107–8, 110–11, 141, 150–1, 165–6, 169–71, 235–6
Evans, Frederick John 242–3, 248–51, 255–6, 261–3
Exeter
   University of 54–5
Exeter Hall 169–71
experiment 18–21, 24–5
   capricious nature 24–5
   challenges of 9–10
   electrical 56–70
   in mines 57–60, 62–3, 91, 93–4
   laboratory 56–8, 61–2, 93–4
   on subterranean heat 45–7, 51, 54–5
   synchronization of 9–10
   systematic 10–12
   visual regulation of 19–20
Experimental Researches in Electricity 56–7, 63–6

Falkland Islands 134, 139–41, 158, 161, 208–11, 237–8
Falmouth 3–4, 7–8, 21–2, 24, 31–2, 38–9, 42–5, 57, 62–4, 68, 71–2, 79, 91, 97–8, 114–22, 130–41, 145–6, 149–52, 156, 168–9, 172, 176–84, 186–8, 193–7, 203, 227–32, 236–41, 243, 257–60, 263
   Meteorological Observatory 260–1
Faraday, Michael 7–8, 18–19, 56–66, 71–3, 93–4, 132
   Fox relations 71–2
   on Fox's dipping needle 114–15
Far East 8–9, 24–5, 191, 196–7
Felix Harbour 108
Fernando Po 176
Figueroa, Colonel 197
Fiscal-military 2–3
   state 15–16
Fishbourne, Commander 192–3
Fitzjames, James 212, 214–15, 217–18, 220–1, 228–9
   reports on Fox dipping needles 215–17
FitzRoy, Robert xxi, 16, 97–8, 130, 135–6, 139, 185–6, 264
Flinders, Matthew 101–2, 105–6, 119
Fly, HMS 177–9, 184, 261–2
Fountainbleu 137–8
Forbes, James 26–7, 128–9, 235
Forbes, John 48–50
Forster, Westgarth 75–8
Fort Chipewyan 119–20, 185–7
Foucault, Michel
   governmentality 15–16
Fourier, Joseph 47–50, 55
   theory on the Earth's heat 47–8
Fox 218–20
Fox, Alfred 43–5, 52–3, 147–8
Fox, Anna Maria 43–5, 72–3, 84–8, 169–70, 238–9, 243, 257–8
Fox, Barclay 43–5, 53–4, 72–3, 80–1, 84–8, 112–13, 115–22, 135–6, 169–70, 175–6, 193–5, 238–9, 243
Fox, Caroline 43–5, 72–3, 84–8, 137–8, 158, 167–70, 175–6, 238–9, 243
Fox, Charles 43–5, 78–9
Fox Dipping Needle 3–5, 8–9, 18–19, 21–2, 24–5, 169, 189–90, 194–5, 225, 227–31, 233, 236–7, 244, 246–7, 256
   advantages of on-board measurements 151, 184–7, 189–90, 208–9
   attack on experiments with 197–200
   cost of 178–9
   damage to 161–3, 180–2, 185–6
   determines the south magnetic pole 156–7
   development of 97–8, 111–15
   Ettrick Creak redevelops 262
   experiments with 114–15

Fox Dipping Needle (*cont.*)
  fragility of 167–9, 184–7
  global use of 263–7
  illustration of experiment with 209
  instruction 193–6, 214
  instrument-maker controversies 177–84, 193, 212–13
  intensity experiments 123–4
  magnetic tour of Europe 137–9, 141
  make-shift repair 202–3
  on Antarctic Expedition of 1839–1843 133–5, 138–42, 145–61, 164–6, 177–80, 184–6
  on Arctic Expedition of 1836 167–9, 177
  on British Isles magnetic survey of 1857–1861 240–3
  on Lefroy's Canadian survey 168–9, 184–7
  on expedition 167–9
  on Franklin Expedition 190, 212–22
  on HMS *Challenger* Expedition 261–2
  on HMS *Fly* 177–9
  on HMS *Pagoda* 161–3
  on HMS *Rattlesnake* 190, 208–12
  on HMS *Samarang* 190–1, 193–205
  on HMS *Sulphur* 191–3
  on HMS *Thunder* 177, 180–3
  on Niger Expedition 168–9, 172–9, 184–5, 187–8, 190
  quality control 180–1
  RCPS discussion of 141
  reputation of 135–8, 145–6, 160–1, 165–6, 168–9, 180–1, 231–2, 259–62
  temperature variations 140–1, 147, 157–8, 188
  tested at the Admiralty 137
  twentieth-century use 262
  use by Charles Elliot 204–8
  using 222–6
  value to navigation 246–7, 249–51, 256, 260, 262–3
  voyage in 2020 265–7
  written instructions 213–14, 222–6
Fox, Elizabeth 167
Fox family 43–5, 83–4, 111–12, 158, 167, 169–70, 175–6, 228–9, 238–9, 257
Fox, G. C. 118–19
Fox, Robert Were 1–2, 6–7, 10–12, 17–22, 26–8, 42, 44–7, 52–4, 62, 64, 69, 72–80, 93–4, 128–9, 142–5, 169–70, 172–4, 227–32, 236–7, 244, 249–51, 257, 259–60, 262–7
  BAAS activity 83–8
  Belcher communications 196–7, 202–4
  biography 43–5
  British Isles magnetic survey 121–8
  British Isles magnetic survey of 1857–1861 238–43

Brunel communications over *Great Eastern's* iron 231–2
control over magnetic experiments 187–8
credibility and reputation 80–1, 231, 264–5
De la Beche relations 88–93
dinner at Carclew House 167–9
dipping needle 3–5, 18–19, 21–2, 24–5, 124, 129–35, 138–42, 145–61, 164–70, 172–87, 240–3, 259–63
development of dipping needle 97–8, 111–15
electrical experiments 57–60, 62–3, 89
experiments with dipping needles 114–15
Faraday relations 71–2
FitzRoy visits 135–6
Franklin supports 119–22
industrial and commercial interests 43–5
instruction on using dipping needle 184, 188, 193–6
instrument makers 193, 212–13, 228–9, 232, 240–1
later life 261–2
magnetic experiments 59–60, 62–8, 80–1, 94
magnetic needles 62–3, 66–71, 94, 112–13
magnetic tour of Europe 137–9, 141
mine experiments 57–9, 73, 79–83, 91–4, 111–12, 118–19
on mineral veins 59–62, 70–1, 75, 78–9, 89, 92–4, 111–12, 130
on subterranean heat 45–55, 58–9, 74, 97–8
on terrestrial magnetism 61–2, 69–71, 79, 97–8
Oxford BAAS performance 115–18
Plymouth BAAS 158
Rattlesnake Expedition 208–9
Royal Society dealings 66, 69–71, 93–4
Sabine collaborations 140, 145–8, 177–84, 186–7, 193, 212–13, 228–9, 235, 240
social networks 88–9, 91
steam engine patent 42–3, 45–6, 52–4
system of magnetic experiments 9–10, 22, 24–5, 211–12, 224–6, 256, 262–7
tropical gardens 43–5
Fox, Wilson Lloyd 260–1
France 48–9, 72–3, 99–100, 124, 134, 171, 231, 248–9
Franklin, John xxi, 16, 97–8, 102–3, 107–8, 119–20, 123, 126–8, 135–7, 156–7, 228–9, 264
  biography 119–20
  expedition to the Arctic of 1845 212, 214, 216–22
  Governor of Van Diemen's Land 154–7
  patronage of Fox dipping needle 119–22

Franklin, Lady Jane, see *Jane Griffen*
Franklin Expedition xxi, 17–18, 24–5, 190, 212–24, 228, 231
  disappearance of 217–18
  poor performance of Fox dipping needles on 215–17
  relics of 220–3
  search for 217–20
French state 12–13
Frieberg Mining Academy 18
Friend of Africa 169–70, 172, 175
Frodsham, Charles 130
Funchal 150–1, 209
Fury, HMS 102–3, 145
Fury Beach 108

Galileo 235–6
Gambey, Henri-Prudence 4–5, 110, 130, 135–6, 180–1, 185–7, 229–30
Garden Island 157, 210–11
Gauss 262
Gauss, Carl 8, 99–102, 107–8, 110, 122, 126–9, 143–5, 161–3, 186–7, 189–90, 228–9, 232
Genesis, Book of 28
Geneva 137–8
geodesy 2–3
Geographical Society 136
Geological and Polytechnic Society of the West Riding of Yorkshire 125
Geological Society of London 38–9
geology 38–9, 58, 62, 68, 82–3, 88–9, 222–4
George III, King 142–3, 235
George IV, King 16–17
George, William 172, 175, 177–87, 193–4, 212–13, 225
geothermal energy 54–5
Germanic 255–6
German Antarctic Expedition 262
German Lands, see *Germany*
Germany 12–13, 99–101, 134, 255–6
Giant's Causeway 124–5
Gibraltar 7–8
Giddy, Davies, see *Gilbert, Davies*
Gilbert, Davies 16–17, 33–4, 38, 41–3, 51–2, 60, 62, 66, 68, 79, 121–2, 131–2, 141, 264
Gilbert, Mary Ann 41–2
Gilbert, William 98–9
Gilolo 197–200
Glasgow 124–5, 257
  Philosophical Society 245
  University of 30–1, 152–4
Glenelg, Lord 142–3
Glendurgan 39–40, 43–5

global 3–6, 8–9, 21
  as an actor category 10–12
  histories of science 5–8
  history 18
  knowledge production 14
  magnetic enterprise 9–10, 15–16, 129–31, 135, 141–2, 167–9, 176–7, 187–91, 211–12, 224–6, 228, 237–40, 243, 256
  network of observatories 154, 164–6
  science 10–12, 24–5, 118–19, 262–7
  trade 38–9, 96–7
global turn 6–7
Gondor 230
Goodman, Matthew 2–3, 123, 164–5, 185–6
Gold Coast 175–6
Göttingen 5–6, 99–101, 132, 172
  Magnetisher Verein 8–9, 14, 100–1, 107–8, 161–3
  Observatory 144–5
  Time 161–3, 213–14
government 4–5, 13–14, 21–2, 24–5, 53–4
  and terrestrial magnetism 97–8, 189–90, 222–4, 232–5
  science 10–18
governmentality 15–16
Graham, George 98–9
Gray, George 147–8
Great Cumbrae 123
Great Britain, SS 96–7, 118–19, 244, 256
Great Eastern, SS 231–2, 244, 248–50
Great Exhibition 232–3
Great Ice Barrier 156–7
Great St George Mine 60–1
Great Western, SS 115–18
Great Western Steamship Company 118–19
Greenhithe 215
Greenock 30–1
Greenwich 33–4, 106–7, 192–3, 248–9
  Royal Observatory 165–6, 235
Grenfell, Nicholas 75–8
Grenoble 137–8
Griffen, Jane 119–22, 154–5, 217–18, 221–2, 228–9
GroveHill 84–8
Guatemala 59
Gunang Taboor, Sultan of 200
Gurney, Goldsworthy 38, 53–4
Gwennap 36–7, 51, 83–4
Gwinear 36–7

Habsburgs
  geology 6–7
Halifax 138–9
Halley, Edmond 98–101, 110, 128–9
Hamilton, William 219

Hammond, Robert 136
Hanover 100–1
Hansteen, Christopher 4–5, 100–1, 107–8, 110, 128–9, 145–6, 173, 189–90, 215
  dipping needles 110–11, 123, 193–5, 205–6
Harris, William Snow 131–2
Harvey & Co 35–6
Hayle 35–6
Hawkins, John 46–7
heat of the Earth, see *subterranean heat*
Heart of Darkness 1
Hecla 102–3, 145, 246–7
Helmreichen, Virgil von 186–7
Helston 49–50
Henwood, William Jory 53–4, 78–9, 112–13, 132
Herschel, John 14, 16, 72–3, 107–8, 119–20, 128–9, 144–5, 154, 156–7, 222–6, 264
  lobbying for the BMS 142–4, 232–3
  on the BMS 10–12, 189–90, 227–8, 233–6
Herschel, William 235–6
high-pressure steam, see *compound engines*
Hindustan, HMS 118–19
History of the Inductive Sciences 10–12
Hobart 133, 154–7, 209, 227–8, 235–6
  for the Observatory, see *Van Diemen's Land Observatory*
Hodge, Captain 43–5
Hong Kong 9–10, 197, 202–5, 248–9
Hope, Thomas Charles 73
Hornblower family 32–3, 37
Hornblower, Jonathan (1717–1780) 32–3
Hornblower, Jonathan junior (1753–1815) 32–3, 41–2, 52–3, 55
  high-pressure compound steam engine 32–5
Hornblower, Joseph 32–3
Horticultural Society 107–8
House of Commons 120, 142–3
Hudson's Bay 12–13, 185–6
Hughes, Thomas
  systems analysis 22–3
Humboldt, Alexander von 12–13, 18, 99, 101–2, 107–8, 129–30, 163–4, 189–90, 236–7
  meets Robert Were Fox 48–9
  study of terrestrial magnetism 20–1
Hutchinson, Lieutenant 167
Hutton, James 59

Iddah, Attah of 173–4
Illustrated London News 221–3
imperial 8–9, 24–5
  as a term 9–12
  order 9–10, 12–13
  science 8–12, 262–3
imperialism
  nineteenth-century 4–5

India 120–1, 204–8
  Geology 6–7
  Great Trigonometrical Survey of 12–13
Indian Ocean 7–9, 22, 68, 133, 154, 191, 197–8, 226, 261–2
industrial capitalistic science 14
industrialization 3–4, 26–7
Industrial Revolution 26–7, 37
instrumentation 2–5, 7–8, 20–1, 23, 67–8, 94
Institute de France 106–7
Inuit 108, 119–20, 220–1
Investigator, HMS 119
Iphigenia 106–7
Ireland 145
Irish Sea 245
iron 98–9, 118–19, 143–4, 161–3, 191–2
  influence of 24–5
  influence on a ship 196–7
  influence on HMS *Terror* and *Erebus* 150–1
  influence on HMS *Samarang* 193–5
  on the *Great Eastern* 231–2
  shipbuilding 97, 105
  solving influence on compasses 244–56
  weapons interfering with dipping needles 197–200
Isabella 102–3, 105–6, 108
Isle of Dogs 231–2
Italy 134, 230
ivory xx

Jamalul-Kiram I, Sultan 197–200
Jamestown 151, 160
Japan 8–9, 24–5, 133, 191–2, 201–5, 226, 255–6
  Emperor of 201–2
Java 205–8, 243
Jermyn Street 177–8
Johnson, Edward 247–8
Jordan, Thomas Brown 115–18, 130–2, 135, 137–40, 145–50, 152–4, 156, 193, 228–9, 232
  end of collaboration with Robert Were Fox 177–81
Joyce, Patrick 15–16

Kafu 210
Kaiser-i-Hind 255–6
Kater, Henry 103
Kay, Joseph Henry 154–7, 161–5
Kelvin, Baron of Largs 251–6, 262–3
  mariner's compass 251–6
Kent 242
Kerala 206–8
Kerguelen Island 154–5, 158–60
Kerr, Walter 255–6

INDEX

Kew Observatory 235, 238–43, 259–60
King William Island 220–1
Kororareka Bay 210–11
Korea 8–9, 24–5, 190–1, 201–5
Kuo, Shen 98–9
Kynance Cove 84–8

La Gloire 248–9
laissez-faire 264
  government 14–15
Lalande gold medal 106–7
Lambeth 34–5
Lancaster Sound 102–3, 216
Landseer, Edwin 218
Land's End 3–4, 135–6, 240
Laplace, Cyrille 154
Laplace, Pierre-Simon 72–3
La Rive, Auguste 164
Lavoisier, Antoine 40
Lean family 47–8
Lean, Joel (1749–1812) 36–7
Lean, Joel the younger (1779–1856) 42–3, 45–8, 52–5
Lean, John 36–7
Lean Thomas 36–7, 45–7, 55
Lean's Engine Reporter 36–7, 42–3, 258–9
Leeves, Elizabeth Juliana 236–7
Lefroy, Henry 151, 161–5, 168–9, 177, 184–7, 211–12
Leifchild, J. R. 75
Lemon, Charles 16–17, 39–42, 59–60, 62, 71, 93–4, 141, 145–6, 167, 264
Levant Mine 75–8, 83–4, 258–9
Leviathan 231–2
Lewis, Colonel 161–3
Liberia 175–6
liberalism 14–15, 264
Lisboa, João de 98–9
Lisbon 7–8, 169, 265–7
Liverpool 83–4, 124–5, 173–4
  BAAS meeting 245–7
Lizard, the 38–9, 133
Lloyd, Humphrey 121–2, 130, 144–5, 157–8, 165–6, 173–4, 186–7, 204–6, 212–13, 216, 229–30, 233–5, 243
  British Isles Magnetic Survey 121–6, 128–30
  British Isles Magnetic Survey of 1857–1861 238–42
Loo-Choo 201–2
London xix, 19–20, 24–5, 34–5, 38–42, 51–3, 63–6, 93–4, 98–9, 103, 106–8, 110–12, 115–19, 123, 130, 132, 137–40, 144–5, 148, 151–2, 157, 160, 177–8, 180–1, 193, 196–7, 206–8, 212–13, 215–16, 229–30, 235, 257, 261–3

Lubbock, John 128–30
Lucknow 8–9, 144–5
Luther, Martin 54–5

Macao 197
MacGillvray, John 210–12
Madagascar 12–13, 192–3
Madeira 133, 150–1, 173–4, 187–8, 209–11, 246–7
Madras 206–8
  Observatory 144–5, 203–4
magnetic 24–5
  charts 14, 24–5, 137–8, 177, 224, 233–5, 252
  declination 99, 213–14, 224
  deviation 4–5, 97–8, 101–2, 118–19, 160, 244–8, 255–6
  dip xix, 4–5, 98–9, 103, 105–8, 110, 112–20, 122, 137–42, 149, 154, 172–3, 175–6, 181–3, 189–90, 195–6, 203–5, 210–11, 213–14, 224, 228–9, 235–6, 241–2, 247–51, 256
  error 95–7
  equatorial 203–8
  experiment 9–12, 108–15, 190, 193–201, 203, 206
  intensity xix, 4–5, 99–100, 105–6, 110, 112–18, 137–42, 154, 159–60, 172–3, 178–9, 183–4, 189–92, 195–7, 203, 210–11, 213–14, 224, 228–9, 235–6, 242–3, 256
  observatories 143–5, 197–200, 203–6, 235–8
  science 10–12, 161, 165–6, 176, 189–90
  science as Big Science 14
  storms 99–100, 164–5, 171, 176–7, 236–7
  survey 7–10, 21–2, 99, 110–11, 133–5, 137–66, 189–92, 227–8, 237–43, 256
Magnetic Congress 189–90, 227–8, 233–5
Magnetic Crusade 2–3, 134, 165–6
Magnetic Island 191–2
Magnetisher Verein 8–9, 14, 100–1, 107–8, 126–8, 144–5, 161–3, 235
  collaboration with the BMS 8
magnetometer 99, 110–11, 186–7, 189–90, 201
Malay 197–201
Mandler, Peter 14–15
Manhattan Project 13–14
Manila 10–12, 190, 197, 202–3, 226
Man Proposes, God Disposes 218
Manual of Scientific Enquiry 222–6
Marsden, Ben 22–3
Margate 242
Mathesius, Johannes 54–5
Mauritius 209
Mazatlán 191
McClintock, Leopold 218–22

McCook, Stuart 18
Mediterranean 7–8
Melbourne, Lord 16–17, 142–3, 170–1
Melvill, James 142–3
Melville, Herman 95–7
metalliferous veins 18–21, 57, 60–1, 63–6, 69, 71, 73–5, 78–83, 93–4, 111–12
 and electricity 7–8, 61–2
 formation 59–61, 70–1, 73, 88–9, 91–3
meteorological observatories 260–1
meteorology 39–40
Mexican Sea 106–7
Mexico 61
Meyer, Tobias 106
Middlemore, George 151
Milan 8–9, 144–5
Mindoro Sea 203–4
mine 3–4, 18–21, 26–7, 29, 32–3, 35–40, 47–50, 62, 74–5, 189–90, 264–5
 epistemology of the 19–21
 experiments in 45–7, 51, 57–60, 97–8, 111–12
 heat 59, 74
 in Devon 43–5, 51
 knowledge of 58, 61–2, 75–8
Mineral veins, see *metalliferous veins*
miners 26–8, 75, 79–80, 91–3
 on subterranean heat 49–51
 knowledge 19–21, 45–6, 73–4, 79–80, 91–4
 Saxon 48–9
Mining Chronicle 89–92
Minto, Earl of 142–4
Minotaur, HMS 255–6
Moby-Dick 95–7
Molesworth, William 41–2
Monthly Duty Paper, see *Lean's Engine Reporter*
Montreal 138–9, 185–6
monumentalism 13–14
Moore, Lieutenant 161, 211–12
Mt Erebus 156–7
Mt Terror 156–7
Moyle, Matthew 49–51
Mumbai 1–2, 7–8
Munich 8–9, 144–5
Murdock, William 33
Museum of Economic Geology 177–8

Nagasaki 201–3
Nairne & Blunt 103
Napier, George 154
Napier, James Robert 245
Napoleon Bonaparte, see *Bonaparte*
Napoleonic Wars 20–1
Nares, George Strong 261–2

Nature 3–6, 17–20, 27–8, 39–45, 47–8, 54–5, 58, 71–2, 93–4, 99–100, 132, 165–6, 221–2, 224, 227–8, 236–7, 259–60
navigation 96–8, 101–2, 105–6, 118–20, 238
 charts 98–9, 244
 problem of iron 239–40
 solution to compass error 244–56
 value of Fox dipping needle for 161, 246–7, 249–51, 256, 260
Naylor, Simon 12–13, 38–9
Neath Abbey 43–5
Netherlands 12–13, 134
New Brunswick 189–90
Newcastle on Tyne 32, 34–5, 121
 BAAS meeting 142–3
Newcomen, Thomas 29, 37
 engines 29–34
Newfoundland 189–90, 237–8
New Guinea 206–11
New Holland 105–6, 148
New Orleans 119
New South Wales 141
Newton, Jane 34–5
New Zealand 8–9, 156–7, 167, 210–11
Niagara Falls 138–9
Nichol, John 152–4
Niger Expedition 168–9, 174–7, 192–3, 222–4
 conception of 169–71
 magnetic survey on 171–9, 184, 187–8
 failure of 173–6
Niger, River 9–10, 169–71, 173–4, 176–7, 187–8
Nimrod
 Expedition 262
Norman, Robert 98–9
Northampton, Marquess of 16–17, 143–4, 171
north magnetic pole 20–1, 98–9, 128–9
 Ross reaches in 1831 108
North-West Passage xx–xxi, 97–8, 102, 105, 110, 119–20, 136
Notre Dame de Paris 13–14

Oatfield Mine 36–7
Oceanic Steam Navigation Company 255–6
Old Trevenen Mine 49–50
Olinda 230
Olympic 255–6
Opossum 139
Ordnance Survey 12–13, 131–2, 140
Ørsted, Christian 18–19, 56–7, 61–2, 93–4, 99–100, 132
Oslo 4–5
Ottawa, River 185–6
Oudh 8–9
Oxbridge 15–16

Oxford 93–4, 238–9, 257
  BAAS meeting 80–1, 115–18, 128–9, 156
  University Museum 238–9
  University of 16–17, 84–8, 119–20

Pacha 244
Pacific Ocean xx–xxi, 8–10, 22, 24–5, 68, 102, 133, 145–6, 191–2, 198, 208–11, 226
Pacific Steam Navigation Company (PSNC) 244
Pagoda, HMS 161–3, 165–6
Palmerston, Viscount 119–20
Panama 191–2
Pantai River 200
Paramour 98–9
Paris 4–6, 19–20, 47–50, 55, 99–101, 106–7, 110, 136, 144–5, 230
  Observatory 99, 106–7
Parliament 2–3, 15–17, 31–2, 38, 120, 222–4, 260–1, 264
  and the Niger Expedition 170–1
  Cornish Members of 41–2, 257
  magnetic lobby of 133–4, 142–6, 201
  system of government 14–15
Parramatta Observatory 141
Parry, William Edward 102–3, 105–6, 110, 145, 217–19
Parry, Jonathan 14–15
Parys Mountain 29
Peacock, George 128–9
Pearce, Stephen 217–19
Peel, Robert 16–17, 119–20, 169–70, 232–3
Peel Sound 108
Peking 144–5
Pellew, Edward 39–40
Pembroke College, Oxford 16
Pencalenick 39–40
Pendennis Castle 124–5
Peninsular and Oriental Steam Navigation Company (P&O) 244, 255–6
Penjerrick 43–5, 84–8, 141–2, 179, 186–7, 228–9, 238–43, 261–3
Pekin 237–8
Penryn 39–40, 59
Penwith Agricultural Society 39–40
Penzance 33–4, 38–40, 53–4, 78–9
  Grammar School 41–2
Penzance Natural History and Antiquarian Society 39–40, 257–8
Pequod 95–6
Perran 75, 112–13
  Iron Foundry 43–5
  Wharf 78–9
Perranarworthal 43–5, 112–13
Persian Gulf 120–1

Petherick, William 75
Persian Gulf 7–8
Pheasant, HMS 106–7
Philadelphia 8–9, 144–5
Philippines 8–12, 191, 193, 197, 201–2, 204–5
Phillips, John 82–3, 124–5, 203–4, 238–9, 243
Philosophical Magazine 46–7, 70–1, 113–15
phrenology 10–12
Physical Observatories 142–4
  construction of 161–3
Pico des Bodes 209
Pig Island 210–11
Playfair, Lyon 16–17
Plymouth 173–4
  BAAS meeting 118–19, 158
Point Victory 220–1
Poisson, Siméon Denis 245–7
polar 22
  exploration 16
Poldice Mine 51–2
Polynesia 190
Pondicherry 12–13
Pool 33–5
Poonah 255–6
Port Jackson 210–11
Portland 248–9
Portland Place 220–1
Porto Grand 173–4
Porto Praya 150–1, 196–7
Port Stanley 210–11
Portugal 96–7
  navigators 98–9
Poskett, James 10–12
Possession Island 156–7
Prague 8–9, 144–5
Price, Tregelles 43–5
Prince Regent 185–6
Protestant Reformation 54–5
Pulo Panjang 200

Quaker 42–5, 115–19, 169–71
Quebec 185–6
Quetelet, Aolphe 126

railways 22–3
Ratcliffe, Jessica 9–10, 13–14
  on state science 12–13
Rat Island 209
Rattlesnake, HMS xxi, 24–5, 190, 208–12, 229, 246–7
Redruth 33–4, 43–5
Regent's Park 53
  magnetic testing site 103, 107–8, 123, 130–1
Research 262

Resurrection 54-5
Richardson, John 219
Riddell, Charles 184-5, 213-14
Rio 140, 145-6, 158, 186-7, 209, 246-9
Rio Grande 106-7
Robinson, Thomas Charles 4-5, 118-19, 123, 130, 180-1, 260-1
  dipping needles 172-3, 184-6, 191-2, 208-11, 214, 216, 241
  Franklin Expedition needle 220-3
Robinson, Thomas Romney 260-1
Rodriguez 12-13
Rome 41-2
Rosehill 43-5, 84-9, 112-13, 115-18, 131-2, 135-7, 139-42, 178-9, 181-3, 186-7, 193-5, 197-200, 261-3
Roskar Mine 75
Rossbank Observatory 154-7, 161-5, 189-90, 209
Ross Cairn 217-18
Ross, James Clark xxi, 16, 21-2, 97-8, 102-3, 107-11, 119-25, 129-30, 145-6, 148, 173-4, 177-81, 184-6, 192-3, 196-7, 208-9, 211-12, 216-20, 261-2, 264
  Antarctic Expedition of 1839-1843 133-5, 142-66, 168-9, 177-80, 184-6, 192-3, 196-7, 205, 216-17, 227-8, 232-3
  British Isles magnetic survey of 1857-1861 238-40, 243, 246-8
  attacked by penguins 156-7
  at the Antarctic 156-7
  location of the north magnetic pole 108, 113
  magnetic lobbying of Parliament 142-4
  trails a Fox type 137
Ross, John xxi, 16, 102-5, 108, 110, 217-18
  1829 Arctic Expedition 108, 119
Ross Sea 156-7
Rotterdam 137-8
Rowse, John 75-8
Royal Artillery 103, 120-1, 161-3, 196-7, 213-14, 232-3
Royal Astronomical Society 145
Royal Charter 231, 246
Royal Cornwall Polytechnic Society (RCPS) 84-8, 93-4, 115-18, 124-5, 141, 243
  decline of 257
  dipping needle in possession of 7-8, 265-7
  exhibits Fox dipping needle 114-15
  golden jubilee exhibition 257, 259-60, 262-3
  meteorological observatory 260-1
Royal Engineers 161-3
Royal Geological Society of Cornwall (RGSC) 38-9
  Transactions of 46-52, 55, 78-9

Royal Horticultural Society of Cornwall 39-40, 257-8
Royal Institution 18-19, 40, 56-7, 59, 71-2, 114-15
Royal Institution of Cornwall 39-40
Royal Mail Steam Packet Company (RMSP) 244-5
Royal Navy xx-xxi, 2-3, 15-16, 102-3, 118-21, 133-6, 139, 142-4, 148, 170-1, 189-90, 211-14, 222-4, 242-3, 248-9, 255-6, 261-2
  compasses 242-3, 247-56
  science 102-3, 106-7
Royal Society 7-8, 12-13, 16-17, 42-3, 51-2, 56-7, 59-60, 62-3, 68-70, 78-9, 81-2, 93-4, 98-9, 103, 106-8, 112-13, 119-20, 129-30, 141, 145, 158, 161, 171-2, 184-5, 189-90, 204-8, 213-14, 243, 264-5
  Copley Medal 40, 105-6
  Cornish influence at 41-2
  magnetic lobby of Parliament 142-3, 145-6
  meteorological observatories 260-1
  Philosophical Transactions of 40, 51-2, 57, 62-6, 69-71, 88-9, 105, 141-2, 160, 164, 191-3, 232-3, 235, 242-3
Rule, John junior 45-7
Russell, John 16-17, 170-1, 175-6
Russell, John Scott 118-19, 231-2, 244
Russia 12-13, 237-8, 255-6
Ryukyu Islands 201-3

Sabine, Edward 4-5, 10-12, 16, 24-5, 97-8, 102-3, 110, 119, 121-2, 128-9, 131-2, 134-5, 137-41, 145-6, 149, 151-2, 156-8, 161-6, 170-1, 184, 190, 219, 228-9, 231-3, 244, 246-8, 256, 262, 264
  biography 103
  British Isles magnetic survey 121-32, 137-8
  British Isles magnetic survey of 1857-1861 237-43
  Canadian magnetic survey 184-7
  Contributions to Terrestrial Magnetism 141-2, 152, 160-1, 165-6, 191-3, 198, 207, 211-12, 232-6, 242-3, 246-52, 256, 262-3
  domineering control over the BMS 233-5
  exchanges with Fox 140, 145-8, 153-4, 158, 177-84, 186-7, 212-13, 228-9, 235, 240
  Franklin Expedition magnetic work 212-14, 218-22
  Hansteen relations 110-11
  Kew Observatory 235-6
  Magnetic Department 232-5
  Magnetic lobbying of Parliament 142-4
  Magnetic Report of 1839 131-2

magnetic research 103–8
Niger Expedition magnetic data
  collection 171–3, 175–7
on dipping needles 106
publication of magnetic observatory
  results 163–4, 191–3, 198, 207, 211–12
Rattlesnake Expedition magnetic
  work 208–12
Samarang Expedition magnetic
  work 193, 196–7
Singapore station data 204–8
solar spots 235–7, 240, 244, 256
system of magnetic data collection 168–9,
  187–8, 190–1, 237–8, 243, 252, 256, 262–7
testing a Fox type 137
written instructions on magnetic
  experiments 222–6
St Agnes Mine 51
St Austell 43–5
  The White Hart 89
St Erth 41–2
St Helena 1–2, 7–8, 145–6, 150–4, 158–60,
  164–5, 169, 171, 176, 187–8, 191–3,
  205, 266
  diurnal variation at 163–4
  Longwood House 151, 154, 161–3, 192–3
  Observatory 144–5, 151, 156–7, 161–5,
    184–5, 235–6
  Plantation House 151, 192–3
  rough weather off coast of 9–10
  volcanic character 151–3, 161–3
St Ives 75, 79–80
St Jago 196–7
St Just 75–9
St Katharine Docks 53
Salem 32–3
Samarang, HMS viii–ix, xxi, 24–5, 190–1,
  193–205, 212, 246–7
  at Datoo Danielle's home 200–1
  at Japan 201–3
  at Manilla 197
  magnetic preparation of 193–6
  makeshift magnetic observatories 197–200
Samboango 197
Samoa 262
Sandwich Islands 191–2
San Francisco 191–2
Sarawack River 199, 206
Saskatchewan 119–20
Savery, Thomas 29
Schaffer, Simon 12–13
Schwabe, Heinrich 236–7
science
  and empire 189–90, 262–3

and the state 4–5, 12–18, 134, 142–3,
  189–90, 264–5
European 5–7
global 262–7
in Cornwall 38–45
modern 6–7
progressiveness 14–15
spatial contingency of 5–6
scientist
  the term 10–12
Scilly Isles 124–5
Scoresby, William 96–7, 102, 122, 231,
  245–9, 256
Scotland 30–1, 123
Scott, Robert 262
Second World War 262
  and Big Science 13–14
Sedgwick, Adam 79
separate condenser 30–1, 35–6, 38
Seychelles 192–3
Shackleton, Ernest 158
Shadwell, Charles 184, 261–2
Shanghai 248–9
Shapin, Steven 19–20
Shrewsbury 135–6
Shropshire 32–3
Siberia 12–13, 110, 128–9
Siemens, Carl 257
Sierra Leone 173–6
Simla 206–8
  Observatory 144–5
Simon's Bay 154, 192–3, 196–7, 203–4, 246–7
Simoom, HMS 246–7
Singapore 9–10, 133, 144–5, 163, 191–2, 204–8
  magnetic observations at 204–6, 227–8
slave trade 169–71, 173–4
Smith, Archibald 244–51, 256, 262–3
Smith, Crosbie 1–2, 9–10, 169
  on systems builders 22–3
Smyth, James 161–3
Smyth, William 192–3
Soho 31–2
solar spots 227–8, 235–7, 256
Solway 244–7
Somerville, Mary 72–3
Soudan 169–72, 176–7
South Africa 133
South Kensington Natural History
  Department 257
south magnetic pole 17–18, 97–8, 133, 142–5
  located with a Fox dipping needle 156–7
  Shackleton reaches 158
South Wheal Towan 51
South, William 82–3

INDEX

Spain 12–13
spatial analysis
  in the history of science 5–7, 18
Spithead 193–5
Spitsbergen 106–7, 110–11
Spring-Rice, Thomas 16–17, 142–4
Staffordshire 29
Stag, HMS 139
standardisation 23–5
Stanley, Owen xxi, 137, 167, 184–5, 229
  Rattlesnake Expedition 208–11
  recollects Fox type's performance on expedition 167–9
Starbuck 95–6
state
  British 4–5, 14–18, 264–5
  definition of 14–16
  science 4–5, 12–15, 134, 142–3, 189–90
steam engine 3–4, 22–3, 27–9, 31–4, 42–3, 46–7, 49, 52–5, 258–60
  analogy to earth's central heat 49, 55
  economy 3–4, 27–8, 32–9
  high-pressure 32–7, 42–3, 45–6, 52–3
  in Cornish mines 45–55
  locomotion 38, 53–4
steamships 22–3
Sterling, John 169–70
Stewart, Balfour 239–40
Strait of Patientia 197–200
Stray Park Mine 33–4
Stromness 215
Sturgeon, William 132
Subterranean heat 3–4, 18, 20–1, 26–8, 47–8, 51–2, 55, 58, 60, 62, 68, 71, 82–3, 90, 97–8
Suez 7–8
Sulivan, Bartholomew James 139–41, 145–9, 184, 211–12, 222–4
Sulphur, HMS xxi, 191–3, 198
Sulu 197–200
  Sea 203–4
  Sultan of 201
Sumatra 205
Surrey Institution 38
survey sciences 12–13
Sussex, Duke of 16–17, 142–4
Swan, Joseph 257
Sweden 12–13
  government of 231
Sydney 9–10, 157
synchronization
  of magnetic observations 8, 100–1, 169
system 23–5, 99–100, 227–8
  for making knowledge 22–3

  of magnetic investigation 10–12, 14, 21–2, 24–5, 164–9, 177, 180–1, 184, 187–90, 204–5, 237–8, 243, 252, 256, 262–7
  of navigation 244–5, 247–56
  ship as a magnetic 250
systems analysis 22–3

Tahiti 12–13, 119
Taiwan 193
Tamar, River 258–9
Tasmania, see *Van Diemen's Land*
Tawi-Tawi Island 197–200
Tayleur 97, 245
Taylor, John 28, 75–8
Teague, William 43–7
Tenerife 150–1
telegraphy 22–3, 257
Term Days 8–10, 14, 144–5, 154–7, 173, 176, 203–4, 213–14
terrestrial magnetism xix–xxi, 2–4, 6–7, 20–2, 66, 68, 91–4, 97–8, 105–6, 110, 112–13, 118–20, 122, 126, 128–32, 190, 193, 200, 222–4, 238, 243, 246–51, 264
  action on a ship 97, 101–2, 105–6, 118–19
  cause of 98–9, 171, 235–8
  chart 98–100, 224, 233–5, 252
  deviation 97–8, 101–2, 118–19, 160
  dip 98–9, 103, 105–8, 110, 112–20, 122, 137–42, 149, 154, 172–3, 175–6, 181–3, 189–90, 195–6, 203–5, 210–11, 213–14, 224
  diurnal variation of 163–4
  experiments on 59–60, 62–6, 68–71, 103, 107–8, 190, 193–201, 203, 206
  global examination of 7–8
  history of 98–100
  in Africa 171–7, 184–5
  in Canada 184–7
  instruments for measurement of 99, 110–20, 123
  intensity measurements 137–42, 154, 159–60, 189–92, 195–7, 203, 210–11, 213–14, 224
  in 2020 265–7
  isoclinic maps of 99, 224
  isodynamic maps of 99, 224, 242–3
  isogonic maps of 99
  science of 98–103, 108, 120, 165–6, 189–90, 212, 217–18, 227–8
  survey of 99, 110–11, 133–5, 137–66, 189–92, 227–8, 237–8, 243, 256, 263
  systematic investigation of 2–3, 24, 180–1, 187–8, 237–8, 243, 252, 256, 262–7
  theory of 100–1, 110, 143–4, 163–4

Terror, HMS xxi, 136–7, 167, 184–5, 187–8, 246–7
   Antarctic Expedition, 1839–1843 133, 145, 147–54, 156–7, 160–6
   Franklin Expedition 210–20
Texel, Isle of 96–7
The Times 175–7, 244–5
Thétis 99–100
Thomas, Martin 75–8
Thomson, Thomas 176
Thomson, William 251–6, 262–3
   mariner's compass 251–6
Thunder, HMS 177, 180–2
Tierra del Fuego 136
Tincroft Mine 33–4, 46–7
Ting Tang Mine 31–2, 51
Tipton 31–2
Titanic 255–6
Tonnelier's Island 209
Torch, HMS 246–7
Toronto 165–6, 185–6, 228–9, 237–8
   Observatory 144–5, 161–5, 184–5, 213–14, 235–6
Torres Straits 177
Tory (party) 16–17
Trafalgar, Battle of 119
Traill, Thomas Stewart 122
Trasavean Mine 59, 61–2
Travancore 8–9
   Raja of 206–8, 260
Treasury 12–13, 129–30
Trebah 39–40
Tregaskis, Richard 75–8
Tregrew 139
Trelissick 39–40
Trent 119
Treskerby 46–7
Trevithick, Richard 33–7, 41–3, 49
   biography 33–4
   steam locomotion 38
Trident, HMS 246–7
Trinidad 150–1
Trinity College, Cambridge 46–7
Trivandrum 8–9
   Observatory 144–5, 206–8
Trotter, Henry Dundas 170–1, 173–4
Truro 88–9
   Grammar School 38
Tunbridge Wells 107–8

United Mines 46–7, 51
United States 14
   Big Science 13–14
   and magnetic science 8–9

Unsang 197–200
Usher, Henry 99–100

Vancouver 191–2
   Island 237–8
Van Diemen's Land 133, 145, 151–7, 159–61, 212
   Observatory 144–5, 154–7
Varma, Swathi Thirunal Rama 206–8
Venus
   Transit of 9–10, 12–13
Vertical turn 18
Victoria Land 156–7
Victoria, Queen 16, 84–8, 142–3, 221–2
Victory 108
Vienna 12–13
Vivian, John 39–40
Vivian, Nicholas 75
Vivian, Richard Hussey 233–5
Vulcan, HMS 246–7
Vyvyan, Richard 41–2

Wales 30–3, 240
Warrior, HMS 248–9
Washington
   Carnegie Institution's Department of Terrestrial Magnetism 262
Waterloo, Battle of 102
Watkin & Hill 112, 114–18, 121, 156, 172, 229–32
Watt, James 30–3, 38
   engines 31–7, 53–4
Weber, Wilhelm 8, 99–101, 107–8, 126–8, 144–5, 161–3, 172–3, 189–90, 233–5
Weinberg, Alvin 13–14
Wesleyan 36–7
Welsh, John 239–40
West Africa 22
Westbourne Green
   magnetic testing site 123, 130–1, 150–1
West Indies 177, 187–8, 244–5, 248–9
Westminster 41–2, 142–3
Whalefish Islands 216–18, 230
whaling 96–7
Wheal Abraham 34–7, 45–9
Wheal Alfred 35–7, 43–5
Wheal Busy 35–6
Wheal Fanny 36–7
Wheal Fortune 29, 34–5
Wheal Friendship 51
Wheal Jewel 59–61, 80–1
Wheal Mary 75
Wheal Sparnon 35–6
Wheal Towan 75

Wheal Trenowith 49–50
Wheal Trumpet 49–50
Wheal Unity 49–52
Wheal Vyvyan 75–8
Wheal Vor 36–7, 43–5, 48–9
Wheatstone, Charles 143–4, 240–1
Whewell, William 82–3, 128–9, 134, 156–7, 222–4
  on the BMS 10–12
Whig (party) 16–17, 39–40, 59, 71, 170–1
Whipple, George Matthews 259–60
Whitehall 15–16, 142–3
White Star Line 255–6
Wightwick, George 84–8
Wilberforce 170–4, 176–7
Wilberforce, Samuel 169–70
Wildman, Robert 109
William III, King 98–9, 103
William IV, King 16–17
Wilmot, Frederick 161–3
Wilton, William 228–9, 232–3, 241
Wilton, William junior 241

Windsor Castle 221–2
Winter, Alison 97
Woolf, Arthur (1766–1837) 43–5, 49
  biography 34–5
  steam engines 35–7, 42–3
Woolf, Arthur senior 34–5
Woolwich 167–8, 176–7, 182–3, 187–8, 192–3, 213–14, 220, 227–8, 232–3, 238–9
  Arsenal 4–5
  Magnetic Department 4–5, 232–5
  Royal Military Academy 103
Wyndham 201

Younghusband, Charles 161–3
York 125–6
  BAAS meeting 80–1
Yorkshire 161–3
  Museum 82–3

Zanzibar 1–2
Zoology 39–40